Lecture Notes in Computer Science 6478

Commenced Publication in 1973
Founding and Former Series Editors:
Gerhard Goos, Juris Hartmanis, and Jan van Leeuwen

Venkatesh Raman   Saket Saurabh (Eds.)

Parameterized and Exact Computation

5th International Symposium, IPEC 2010
Chennai, India, December 13-15, 2010
Proceedings

 Springer

Volume Editors

Venkatesh Raman
The Institute of Mathematical Sciences, Chennai, 600113, India
E-mail: vraman@imsc.res.in

Saket Saurabh
The Institute of Mathematical Sciences, Chennai 600113, India
E-mail: saket@imsc.res.in

Library of Congress Control Number: 2010939810

CR Subject Classification (1998): F.2, G.1-2, I.3.5, F.1, G.4, E.1, I.2.8

LNCS Sublibrary: SL 1 – Theoretical Computer Science and General Issues

ISSN   0302-9743
ISBN-10   3-642-17492-2 Springer Berlin Heidelberg New York
ISBN-13   978-3-642-17492-6 Springer Berlin Heidelberg New York

springer.com

© Springer-Verlag Berlin Heidelberg 2010
Printed in Germany

Typesetting: Camera-ready by author, data conversion by Scientific Publishing Services, Chennai, India
Printed on acid-free paper   06/3180

Preface

The International Symposium on Parameterized and Exact Computation (IPEC, formerly IWPEC) is an international symposium series that covers research in all aspects of parameterized and exact algorithms and complexity. Started in 2004 as a biennial workshop, it became an annual event from 2008.

The four previous meetings of the IPEC/IWPEC series were held in Bergen, Norway (2004), Zürich, Switzerland (2006), Victoria, Canada (2008) and Copenhagen, Denmark (2009). On recommendations of the Steering Committee, from this year, the word 'symposium' replaces the word 'workshop' in the name, and it gets a new abbreviation IPEC (rhyming with the old one IWPEC).

IPEC 2010 was the fifth symposium in the series, held in Chennai, India, during December 13-15, 2010. The symposium was co-located with the 30th Foundations of Software Technology and Theoretical Computer Science conference (FSTTCS 2010), a premier theory conference in India.

At IPEC 2010, we had three plenary speakers: Anuj Dawar (University of Cambridge, UK), Fedor V. Fomin (University of Bergen, Norway) and Toby Walsh (NICTA and University of New South Wales, Australia). The abstracts accompanying their talks are included in these proceedings. We thank the speakers for accepting our invitation and for their abstracts.

In response to the call for papers, 32 papers were submitted. Each submission was reviewed by at least four reviewers. The reviewers were either Program Committee members or invited external reviewers. The Program Committee held electronic meetings using the EasyChair system, went through extensive discussions, and selected 19 of the submissions for presentation at the symposium and inclusion in this LNCS volume.

From this year IPEC also started the tradition of awarding excellent student paper awards. This year the award was shared by two papers, "Small Vertex Cover Makes Petri Net Coverability and Boundedness Easier" by M. Praveen and "Inclusion/Exclusion Branching for Partial Dominating Set and Set Splitting" by Jesper Nederlof and Johan M. M. van Rooij. We thank Frances Rosamond for sponsoring the award.

We are very grateful to the Program Committee, and the external reviewers they called on, for the hard work and expertise which they brought to the difficult selection process. We also wish to thank all the authors who submitted their work for our consideration. Special thanks go to Meena Mahajan and her team for the local organization of IPEC 2010. We thank, in particular, Neeldhara Misra and Geevarghese Philip for help in maintaining the website of IPEC 2010.

Finally, we thank the Institute of Mathematical Sciences for the financial support, its staff for the organizational support, and the editors at Springer for their cooperation throughout the preparation of these proceedings.

December 2010   Venkatesh Raman
                                                            Saket Saurabh

Conference Organization

Programe Committee

Rod Downey	Victoria University of Wellington, New Zealand
Mike Fellows	University of Newcastle, Australia
Martin Fürer	The Pennsylvania State University, USA
Serge Gaspers	CMM Universidad de Chile, Chile
Jiong Guo	Max Planck Institute, Germany
Gregory Gutin	Royal Holloway University of London, UK
Petr Hliněný	Masaryk University Brno, Czech Republic
Falk Hüffner	Humboldt University of Berlin, Germany
Petteri Kaski	Helsinki Institute for Information Technology, Finland
Dieter Kratsch	University of Metz, France
Daniel Lokshtanov	University of Bergen, Norway
Dániel Marx	Tel Aviv University, Israel
Ramamohan Paturi	University of California San Diego, USA
Christophe Paul	LIRMM, University of Montpellier, France
Venkatesh Raman	Institute of Mathematical Sciences, India (Co-chair)
Peter Rossmanith	RWTH Aachen University, Germany
Saket Saurabh	Institute of Mathematical Sciences, India (Co-chair)
C. R. Subramanian	Institute of Mathematical Sciences, India
Dimitrios M. Thilikos	National and Kapodistrian University of Athens, Greece
Ryan Williams	IBM Almaden Research Center, USA

Steering Committee

Hans Bodlaender	Utrecht University, The Netherlands (Chair)
Jianer Chen	Texas A&M University, USA
Mike Fellows	University of Newcastle, Australia
Fedor V. Fomin	University of Bergen, Norway
Martin Grohe	Humboldt University, Germany
Michael Langston	University of Tennessee, USA
Dániel Marx	Humboldt University, Germany
Venkatesh Raman	Institute of Mathematical Sciences, India
Saket Saurabh	Institute of Mathematical Sciences, India

IPEC Publicity Chair

Frances Rosamond	University of Newcastle, Australia

Local Organization

K. Narayan Kumar	Chennai Mathematical Institute, India
Meena Mahajan	Institute of Mathematical Sciences, India (Chair)
R. Ramanujam	Institute of Mathematical Sciences, India
Saket Saurabh	Institute of Mathematical Sciences, India
Vikram Sharama	Institute of Mathematical Sciences, India

External Reviewers

Anders Yeo
Anthony Perez
Archontia Giannopoulou
Catherine Mccartin
Christian Komusiewicz
David Bryant
Dimitris Zoros
Gabor Erdelyi
Geevarghese Philip
Geoff Whittle
George B. Mertzios
Holger Dell
Hubie Chen
Ignasi Sau
Ildikó Schlotter
Isolde Adler
Iyad Kanj
Jan Arne Telle
Jeremiah Blocki
Jianer Chen
Klaus Reinhardt
Magnus Wahlström
Marcin Kaminski
Marek Cygan

Mark Jones
Mathieu Liedloff
Matthias Mnich
Matthias Killat
Meena Mahajan
Michael Lampis
N.R. Aravind
N.S. Narayanaswamy
Pascal Schweitzer
Petr Golovach
Piotr Faliszewski
René van Bevern
Robert Bredereck
Robert Ganian
Roland Meyer
Sharon Bruckner
Somnath Sikdar
Stefan Kratsch
Thore Husfeldt
Valia Mitsou
Virginia Vassilevska Williams
William Gasarch
Yixin Cao
Yngve Villanger

Table of Contents

The Complexity of Satisfaction on Sparse Graphs

Anuj Dawar

University of Cambridge Computer Laboratory, Cambridge CB3 0FD, UK
anuj.dawar@cl.cam.ac.uk

We consider the complexity of deciding, given a graph G and a formula φ of first-order logic in the language of graphs, whether or not $G \models \varphi$. In other words, we are interested in the complexity of the satisfaction relation for first-order logic on graphs. More particularly, we look at the complexity of this problem parameterized by the length of the formula φ. This problem (which is known to be $\mathsf{AW}[\star]$-complete) includes as special cases many important graph-theoretic problems, including INDEPENDENT SET, DOMINATING SET and complete problems at all levels of the W-hierarchy.

In this talk, we consider classes of graphs that are *sparse*. We consider a number of different measures of sparseness, including classes of bounded treewidth, bounded degree and bounded expansion. These enjoy the property that they also ensure that the satisfaction relation for first-order logic becomes fixed-parameter tractable. The methods deployed to establish these results essentially fall into three categories: those based on decompositions (used, for instance, on graphs of bounded treewidth [1]); those based on locality (for instance, the results on graphs of bounded degree [8]); and the recently established method of low-tree depth colourings (see [5]).

We survey these different methods and discuss how they combine in establishing the tractability of satisfaction on classes that locally exclude a minor [3] or those with locally bounded expansion [5]. Finally, we discuss how these results might be extended to classes that are *nowhere-dense*. This is a property introduced by Nešetřil and Ossona de Mendez in [6,7] and closely related to the property of quasi-wideness [2], which has been used to establish that certain domination problems are fixed-parameter tractable on these classes [4]. Nowhere-denseness seems to provide a natural limit for methods which rely on the sparsity of graphs.

References

1. Courcelle, B.: Graph rewriting: An algebraic and logic approach. In: van Leeuwen, J. (ed.) Handbook of Theoretical Computer Science, Volume B: Formal Models and Sematics (B), pp. 193–242. Elsevier, Amsterdam (1990)
2. Dawar, A.: Homomorphism preservation on quasi-wide classes. J. Computer and System Sciences 76, 324–332 (2009)
3. Dawar, A., Grohe, M., Kreutzer, S.: Locally excluding a minor. In: Proc. 22nd IEEE Symp. on Logic in Computer Science, pp. 270–279 (2007)
4. Dawar, A., Kreutzer, S.: Domination problems in nowhere-dense classes of graphs. In: FSTTCS 2009, pp. 157–168 (2009)
5. Dvořak, Z., Král, D., Thomas, R.: Deciding first-order properties for sparse graphs. In: FoCS 2010 (to appear, 2010)

V. Raman and S. Saurabh (Eds.): IPEC 2010, LNCS 6478, pp. 1–2, 2010.
© Springer-Verlag Berlin Heidelberg 2010

6. Nešetřil, J., Ossona de Mendez, P.: First-order properties of nowhere dense structures. Journal of Symbolic Logic 75, 868–887 (2010)
7. Nešetřil, J., Ossona de Mendez, P.: On nowhere dense graphs. European Journal of Combinatorics (to appear, 2010)
8. Seese, D.: Linear time computable problems and first-order descriptions. Math. Struct. in Comp. Science 6, 505–526 (1996)

Protrusions in Graphs and Their Applications

Fedor V. Fomin

Department of Informatics, Univeristy of Bergen, 5020 Bergen, Norway
`fomin@ii.uib.no`

A protrusion in a graph is a subgraph of constant treewidth that can be separated from the graph by removing a constant number of vertices. More precisely, given a graph $G = (V, E)$ and $S \subseteq V$, we denote by $\partial(S)$ the set of vertices in S that have a neighbour in $V \setminus S$.

Definition 1. *Given a graph $G = (V, E)$, we say that a set $X \subseteq V$ is an r-protrusion of G if $|\partial(X)| \leq r$ and the treewidth of the subgraph $G[X]$ of G induced by X is at most r.*

Protrusions were introduced in [1] as a tool for obtaining kernelization meta-theorems. Loosely speaking, if a parametrized problem can be expressed in a certain logic, then a sufficiently large protrusion can be replaced by a smaller "equivalent" one. Protrusions appear to be a handy tool for obtaining kernelization and approximation algorithms [2,3]. In this talk we discuss combinatorial properties of graphs implying existence of large protrusions and give a number of algorithmic applications of protrusions.

The talk is based on joint works with Hans L. Bodlaender, Daniel Lokshtanov, Elko Penninkx, Saket Saurabh, Venkatesh Raman, and Dimitrios M. Thilikos [1,2,3].

References

1. Bodlaender, H., Fomin, F.V., Lokshtanov, D., Penninkx, E., Saurabh, S., Thilikos, D.M.: (Meta) Kernelization. In: Proceedings of the 50th Annual Symposium on Foundations of Computer Science (FOCS 2009), pp. 629–638. IEEE, Los Alamitos (2009)
2. Fomin, F.V., Lokshtanov, D., Saurabh, S., Thilikos, D.M.: Bidimensionality and kernels. In: Proceedings of the 21th ACM-SIAM Symposium on Discrete Algorithms (SODA 2010), pp. 503–510 (2010)
3. Fomin, F.V., Lokshtanov, D., Raman, V., Saurabh, S.: Bidimensionality and EP-TAS, CoRR abs/1005.5449 (2010)

V. Raman and S. Saurabh (Eds.): IPEC 2010, LNCS 6478, p. 3, 2010.

Parameterized Complexity Results
in Symmetry Breaking*

Toby Walsh

NICTA and University of New South Wales
toby.walsh@nicta.com.au

Abstract. Symmetry is a common feature of many combinatorial problems. Unfortunately eliminating all symmetry from a problem is often computationally intractable. This paper argues that recent parameterized complexity results provide insight into that intractability and help identify special cases in which symmetry can be dealt with more tractably.

1 Introduction

Symmetry occurs in many constraint satisfaction problems. For example, in scheduling a round robin sports tournament, we may be able to interchange all the matches taking place in two stadia. Similarly, we may be able to interchange two teams throughout the tournament. As a second example, when colouring a graph (or equivalently when timetabling exams), the colours are interchangeable. We can swap red with blue throughout. If we have a proper colouring, any permutation of the colours is itself a proper colouring. Problems may have many symmetries at once. In fact, the symmetries of a problem form a group. Their action is to map solutions (a schedule, a proper colouring, etc.) onto solutions.

Symmetry is problematic when solving constraint satisfaction problems as we may waste much time visiting symmetric solutions. In addition, we may visit many (failing) search states that are symmetric to those that we have already visited. One simple but effective mechanism to deal with symmetry is to add constraints which eliminate symmetric solutions [1]. Unfortunately eliminating all symmetry is NP-hard in general [2]. However, recent results in parameterized complexity give us a good understanding of the source of that complexity. In this survey paper, I summarize results in this area. For more background, see [3–7].

2 An Example

To illustrate the ideas, we consider a simple problem from musical composition. The all interval series problem (prob007 in CSPLib.org [8]) asks for a permutation of the

---

* Supported by the Australian Government's Department of Broadband, Communications and the Digital Economy and the ARC. I wish to thank my co-authors in this area for their many contributions: Christian Bessiere, Brahim Hnich, Emmanuel Hebrard, George Katsirelos, Zeynep Kiziltan, Yat-Chiu Law, Jimmy Lee, Nina Narodytska Claude-Guy Quimper and Justin Yip.

V. Raman and S. Saurabh (Eds.): IPEC 2010, LNCS 6478, pp. 4–13, 2010.

numbers 0 to $n - 1$ so that neighbouring differences form a permutation of 1 to $n - 1$. For $n = 12$, the problem corresponds to arranging the half-notes of a scale so that all musical intervals (minor second to major seventh) are covered. This is a simple example of a graceful graph problem in which the graph is a path. We can model this as a constraint satisfaction problem in n variables with $X_i = j$ iff the ith number in the series is j. One solution for $n = 11$ is:

$$X_1, X_2, \ldots, X_{11} = 3, 7, 4, 6, 5, 0, 10, 1, 9, 2, 8 \tag{1}$$

The differences form the series: $4, 3, 2, 1, 5, 10, 9, 8, 7, 6$.

The all interval series problem has a number of different symmetries. First, we can reverse any solution and generate a new (but symmetric) solution:

$$X_1, X_2, \ldots, X_{11} = 8, 2, 9, 1, 10, 0, 5, 6, 4, 7, 3 \tag{2}$$

Second, the all interval series problem has a value symmetry as we can invert values. If we subtract all values in (1) from 10, we generate a second (but symmetric) solution:

$$X_1, X_2, \ldots, X_{11} = 7, 3, 6, 4, 5, 10, 0, 9, 1, 8, 2 \tag{3}$$

Third, we can do both and generate a third (but symmetric) solution:

$$X_1, X_2, \ldots, X_{11} = 2, 8, 1, 9, 0, 10, 5, 4, 6, 3, 7 \tag{4}$$

To eliminate such symmetric solutions from the search space, we can post additional constraints which eliminate all but one solution in each symmetry class. To eliminate the reversal of a solution, we can simply post the constraint:

$$X_1 < X_{11} \tag{5}$$

This eliminates solution (2) as it is a reversal of (1). To eliminate the value symmetry which subtracts all values from 10, we can post:

$$X_1 \leq 5, \ X_1 = 5 \Rightarrow X_2 < 5 \tag{6}$$

This eliminates solutions (2) and (3). Finally, eliminating the third symmetry where we both reverse the solution and subtract it from 10 is more difficult. We can, for instance, post:

$$[X_1, \ldots, X_{11}] \leq_{\text{lex}} [10 - X_{11}, \ldots, 10 - X_1] \tag{7}$$

Note that of the four symmetric solutions given earlier, only (4) with $X_1 = 2$, $X_2 = 8$ and $X_{11} = 7$ satisfies all three sets of symmetry breaking constraints: (5), (6) and (7). The other three solutions are eliminated.

3   Formal Background

We will need some formal notation to present some of the more technical results. A *constraint satisfaction problem* (CSP) consists of a set of variables, each with a finite domain of values, and a set of constraints [9]. Each *constraint* is specified by the allowed combinations of values for some subset of variables. For example, $X \neq Y$ is a binary constraint which ensures X and Y do not take the same values. A *global constraint* is one in which the number of variables is not fixed. For instance, the global constraint NVALUE($[X_1, \ldots, X_n], N$) ensures that n variables, X_1 to X_n, take N different values [10]. That is, $N = |\{X_i \mid 1 \leq i \leq n\}|$.

Constraint solvers typically use backtracking search to explore the space of partial assignments. After each assignment, constraint propagation algorithms prune the search space by enforcing local consistency properties like domain or bound consistency. A constraint is *domain consistent (DC)* iff when a variable is assigned any of the values in its domain, there exist compatible values in the domains of all the other variables of the constraint. Such values are called a *support*. A CSP is domain consistent iff every constraint is domain consistent.

Recently, Bessiere *et al.* have shown that a number of common global constraints are intractable to propagate [11, 12]. For instance, enforcing domain consistency on the NVALUE constraint is NP-hard [13, 14]. Parameterized complexity can provide a more fine-grained view of such results, identifying more precisely what makes a global constraint (in)tractable. We will say that a problem is *fixed-parameter tractable (FPT)* if it can be solved in $O(f(k)n^c)$ time where f is *any* computable function, k is some parameter, c is a constant, and n is the size of the input. For example, vertex cover ("Given a graph with n vertices, is there a subset of vertices of size k or less that cover each edge in the graph") is NP-hard in general, but fixed-parameter tractable with respect to k since it can be solved in $O(1.31951^k k^2 + kn)$ time [15]. Hence, provided k is small, vertex cover can be solved effectively.

4 Symmetry Breaking

As we have argued, symmetry is a common feature of many real-world problems that dramatically increases the size of the search space if it is not factored out. Symmetry can be defined as a bijection on assignments that preserves solutions. The set of symmetries form a group under the action of composition. We focus on two special types of symmetry. A *value symmetry* is a bijective mapping, σ of the values such that if $X_1 = d_1, \ldots, X_n = d_n$ is a solution then $X_1 = \sigma(d_1), \ldots, X_n = \sigma(d_n)$ is also. For example, in our all interval series example, there is a value symmetry σ that maps the value i onto $n - i$. A *variable symmetry*, on the other hand, is a bijective mapping, θ of the indices of variables such that if $X_1 = d_1, \ldots, X_n = d_n$ is a solution then $X_{\theta(1)} = d_1, \ldots, X_{\theta(n)} = d_n$ is also. For example, in our all interval series example, there is a variable symmetry θ that maps the index i onto $n + 1 - i$. This swaps the variable X_i with X_{n+1-i}.

A simple and effective mechanism to deal with symmetry is to add constraints to eliminate symmetric solutions [1, 2, 16–19]. The basic idea is very simple. We pick an ordering on the variables, and then post symmetry breaking constraints to ensure that the final solution is lexicographically less than any symmetric re-ordering of the variables. That is, we select the "lex leader" assignment. For example, to break the variable symmetry θ, we post the constraint:

$$[X_1, \ldots, X_n] \leq_{\text{lex}} [X_{\theta(1)}, \ldots, X_{\theta(n)}]$$

Efficient inference methods exist for propagating such constraints [20, 21]. The symmetry breaking constraints in our all interval series example can all be derived from such lex leader constraints.

In theory, the lex leader method solves the problem of symmetries, eliminating all symmetric solutions and pruning many symmetric states. Unfortunately, the set of symmetries might be exponentially large (for example, in a graph k-colouring, there are $k!$ symmetries). There may therefore be too many symmetry breaking constraints to post. In addition, decomposing symmetry breaking into many lex leader constraints typically hinders propagation. We focus on three special but commonly occurring cases where symmetry breaking is more tractable and propagation can be more powerful: value symmetry, interchangeable values, and row and column symmetry. In each case, we identify islands of tractability but show that the quick-sands of intractability remain close to hand.

5 Value Symmetry

Value symmetries are a commonly occurring symmetry that are more tractable to break [6]. For instance, Puget has proved that a linear number of symmetry breaking constraints will eliminate any number of value symmetries in polynomial time [22]. Given a set of value symmetries Σ, we can eliminate all value symmetry by posting the global constraint LEXLEADER$(\Sigma, [X_1, \ldots, X_n])$ [16]. This is a conjunction of lex leader constraints, ensuring that, for each $\sigma \in \Sigma$:

$$\langle X_1, \ldots, X_n \rangle \leq_{\text{lex}} \langle \sigma(X_1), \ldots, \sigma(X_n) \rangle$$

Unfortunately, enforcing domain consistency on this global constraint is NP-hard. However, this complexity depends on the number of symmetries. Breaking all value symmetry is fixed-parameter tractable in the number of symmetries.

Theorem 1. *Enforcing domain consistency on* LEXLEADER$(\Sigma, [X_1, \ldots, X_n])$ *is NP-hard in general but fixed-parameter tractable in* $k = |\Sigma|$.

Proof: NP-hardness is proved by Theorem 1 in [23], and fixed-parameter tractability by Theorem 7 in [24]. □

One situation where we may have only a small number of symmetries is when we focus on just the *generators* of the symmetry group [2, 25]. This is attractive as the size of the generator set is logarithmic in the size of the group, many algorithms in computational group theory work on generators, and breaking just the generator symmetries has been shown to be effective on many benchmarks [25]. In general, breaking just the generators may leave some symmetry. However, on certain symmetry groups (like that for interchangeable values considered in the next section), all symmetry is eliminated (Theorem 3 in [23]).

6   Interchangeable Values

By exploiting special properties of the value symmetry group, we can identify even more tractable cases. A common type of value symmetry with such properties is that due to interchangeable values. We can break all such symmetry using the idea of value *precedence* [26]. In particular, we can post the global symmetry breaking constraint PRECEDENCE($[X_1, \ldots, X_n]$). This ensures that for all $j < k$:

$$\min\{i \mid X_i = j \vee i = n + 1\} < \min\{i \mid X_i = k \vee i = n + 2\}$$

That is, the first time we use j is before the first time we use k for all $j < k$. For example, consider the assignment:

$$X_1, X_2, X_3, \ldots, X_n = 1, 1, 1, 2, 1, 3, \ldots, 2 \tag{8}$$

This satisfies value precedence as 1 first occurs before 2, 2 first occurs before 3, etc. Now consider the symmetric assignment in which we swap 2 with 3:

$$X_1, X_2, X_3, \ldots, X_n = 1, 1, 1, 3, 1, 2, \ldots, 3 \tag{9}$$

This does not satisfy value precedence as 3 first occurs before 2. A PRECEDENCE constraint eliminates all symmetry due to interchangeable values. In [27], we give a linear time propagator for enforcing domain consistency on the PRECEDENCE constraint. In [23], we argue that PRECEDENCE can be derived from the lex leader method (but offers more propagation by being a global constraint).

Another way to ensure value precedence is to map onto dual variables, Z_j which record the first index using each value j [22]. This transforms value symmetry into variable symmetry on the Z_j. We can then eliminate this variable symmetry with some ordering constraints:

$$Z_1 < Z_2 < Z_3 < \ldots < Z_m \tag{10}$$

In fact, Puget proves that we can eliminate *all* value symmetry (and not just that due to value interchangeability) with a linear number of such ordering constraints. Unfortunately, this decomposition into ordering constraints hinders propagation even for the tractable case of interchangeable values (Theorem 5 in [23]). Indeed, even with *just* two value symmetries, mapping into variable symmetry can hinder propagation. This is supported by the experiments in [23] where we see faster and more effective symmetry breaking with the global PRECEDENCE constraint. This global constraint thus appears to be a promising method to eliminate the symmetry due to interchangeable values.

A generalization of the symmetry due to interchangeable values is when values partition into sets, and values within each set (but not between sets) are interchangeable. The idea of value precedence can be generalized to this case [27]. The global GENPRECEDENCE constraint ensures that values in each interchangeable set occur in order. More precisely, if the values are divided into s equivalence classes, and the jth equivalence class contains the values $v_{j,1}$ to v_{j,m_j} then GENPRECEDENCE ensures $\min\{i \mid X_i = v_{j,k} \vee i = n + 1\} < \min\{i \mid X_i = v_{j,k+1} \vee i = n + 2\}$ for all $1 \le j \le s$ and $1 \le k < m_j$. Enforcing domain consistency on GENPRECEDENCE is NP-hard in general but fixed-parameter tractable in $k = s$ [23, 24].

7   Row and Column Symmetry

Another common type of symmetry where we can exploit special properties of the symmetry group is row and column symmetry [28]. Many problems can be modelled by a matrix model involving a matrix of decision variables [29–31]. Often the rows and columns of such matrices are fully or partially interchangeable [28]. For example, the Equidistant Frequency Permutation Array (EFPA) problem is a challenging combinatorial problem in coding theory. The aim is to find a set of v code words, each of length $q\lambda$ such that each word contains λ copies of the symbols 1 to q, and each pair of code words is at a Hamming distance of d apart. For example, for $v = 4$, $\lambda = 2$, $q = 3$, $d = 4$, one solution is:

$$
\begin{array}{cccccc}
0 & 2 & 1 & 2 & 0 & 1 \\
0 & 2 & 2 & 1 & 1 & 0 \\
0 & 1 & 0 & 2 & 1 & 2 \\
0 & 0 & 1 & 1 & 2 & 2
\end{array}
\tag{11}
$$

This problem has applications in communications, and is closely related to other combinatorial problems like finding orthogonal Latin squares. Huczynska *et al.* [32] consider a simple matrix model for this problem with a v by $q\lambda$ array of variables, each with domains 1 to q. This model has row and column symmetry since we can permute the rows and the columns of any solution. Although breaking all row and column symmetry is intractable in general, it is fixed-parameter tractable in the number of columns (or rows).

Theorem 2. *With a n by m matrix, checking lex leader constraints that break all row and column symmetry is NP-hard in general but fixed-parameter tractable in $k = m$.*

Proof: NP-hardness is proved by Theorem 3.2 in [2], and fixed-parameter tractability by Theorem 1 in [33].   □

Note that the above result only talks about *checking* a constraint which breaks all row and column symmetry. That is, we only consider the computational cost of deciding if a complete assignment satisfies the constraint. Propagation of such a global constraint is computationally more difficult.

   Just row or column symmetry on their own are tractable to break. To eliminate all row symmetry we can post lexicographical ordering constraints on the rows. Similarly, to eliminate all column symmetry we can post lexicographical ordering constraints on the columns. When we have both row and column symmetry, we can post a DOUBLELEX constraint that lexicographically orders both the rows and columns [28]. This does not eliminate all symmetry since it may not break symmetries which permute both rows and columns. Nevertheless, it is more tractable to propagate and is often highly effective in practice. Note that DOUBLELEX can be derived from a *strict* subset of the LEXLEADER constraints. Unfortunately propagating such a DOUBLELEX constraint completely is already NP-hard.

Theorem 3. *With a n by m matrix, enforcing domain consistency on DOUBLELEX is NP-hard in general.*

Proof: See Threorem 3 in [33].   □

There are two special cases of matrix models where row and column symmetry is more tractable to break. The first case is with an all-different matrix, a matrix model in which every value is different. If an all-different matrix has row and column symmetry then the lex-leader method ensures that the top left entry is the smallest value, and the first row and column are ordered [28]. Domain consistency can be enforced on such a global constraint in polynomial time [33]. The second more tractable case is with a matrix model of a function. In such a model, all entries are 0/1 and each row sum is 1. If a matrix model of a function has row and column symmetry then the lex-leader method ensures the rows and columns are lexicographically ordered, the row sums are 1, and the sums of the columns are in decreasing order [28, 34, 35]. Domain consistency can also be enforced on such a global constraint in polynomial time [33].

8   Related Work

The study of computational complexity in constraint programming has tended to focus on the structure of the constraint graph (e.g. especially measures like tree width [36, 37]) or on the semantics of the constraints (e.g. [38]). However, these lines of research are mostly concerned with constraint satisfaction problems as a whole, and do not say much about individual (global) constraints. For global constraints of bounded arity, asymptotic analysis has been used to characterize the complexity of propagation both in general and for constraints with a particular semantics. For example, the generic domain consistency algorithm of [39] has an $O(d^n)$ time complexity on constraints of arity n and domains of size d, whilst the domain consistency algorithm of [40] for the n-ary ALLDIFFERENT constraint has $O(n^{\frac{3}{2}}d)$ time complexity. Bessiere *et al.* showed that many global constraints like NVALUE are also intractable to propagate [11]. More recently, Samer and Szeider have studied the parameterized complexity of the EGCC constraint [41]. Szeider has also studied the complexity of symmetry in a propositional resolution calculus [42]. See Chapter 10 in [43] for more about symmetry of propositional systems.

9   Conclusions

We have argued that parameterized complexity is a useful tool with which to study symmetry breaking. In particular, we have shown that whilst it is intractable to break all symmetry completely, there are special types of symmetry like value symmetry and row and column symmetry which are more tractable to break. In these case, fixed-parameter tractability comes from natural parameters like the number of generators which tend to be small. In future, we hope that insights provided by such analysis will inform the design of new search methods. For example, we might build a propagator that propagates completely when the parameter is small, but only partially when it is large. In the longer term, we hope that other aspects of parameterized complexity like kernels will find application in the domain of symmetry breaking.

References

1. Puget, J.F.: On the satisfiability of symmetrical constrained satisfaction problems. In: Komorowski, J., Raś, Z.W. (eds.) ISMIS 1993. LNCS, vol. 689, pp. 350–361. Springer, Heidelberg (1993)

2. Crawford, J., Ginsberg, M., Luks, G., Roy, A.: Symmetry breaking predicates for search problems. In: Proceedings of the 5th International Conference on Knowledge Representation and Reasoning (KR 1996), pp. 148–159 (1996)
3. Walsh, T.: Symmetry within and between solutions. In: Zhang, B.-T., Orgun, M.A. (eds.) PRICAI 2010. LNCS, vol. 6230, pp. 11–13. Springer, Heidelberg (2010)
4. Katsirelos, G., Walsh, T.: Symmetries of symmetry breaking constraints. In: Coelho, H., Studer, R., Wooldridge, M. (eds.) Proceedings of 19th European Conference on Artificial Intelligence, ECAI 2010. Frontiers in Artificial Intelligence and Applications, vol. 215, pp. 861–866. IOS Press, Amsterdam (2010)
5. Heule, M., Walsh, T.: Symmetry in solutions. In: Fox, M., Poole, D. (eds.) Proceedings of the Twenty-Fourth AAAI Conference on Artificial Intelligence, AAAI 2010. AAAI Press, Menlo Park (2010)
6. Walsh, T.: Breaking value symmetry. In: Fox, D., Gomes, C. (eds.) Proceedings of the 23rd National Conference on AI, Association for Advancement of Artificial Intelligence, pp. 1585–1588 (2008)
7. Walsh, T.: Symmetry breaking. In: Sattar, A., Kang, B.-h. (eds.) AI 2006. LNCS (LNAI), vol. 4304, pp. 7–8. Springer, Heidelberg (2006)
8. Gent, I., Walsh, T.: CSPLib: a benchmark library for constraints. Technical report, Technical report APES-09-1999, A shorter version appears in the Proceedings of the 5th International Conference on Principles and Practices of Constraint Programming, CP 1999 (1999)
9. Rossi, F., van Beek, P., Walsh, T. (eds.): Handbook of Constraint Programming. Elsevier, Amsterdam (2006)
10. Pachet, F., Roy, P.: Automatic generation of music programs. In: Jaffar, J. (ed.) CP 1999. LNCS, vol. 1713, pp. 331–345. Springer, Heidelberg (1999)
11. Bessiere, C., Hebrard, E., Hnich, B., Walsh, T.: The complexity of global constraints. In: Proceedings of the 19th National Conference on AI, Association for Advancement of Artificial Intelligence (2004)
12. Bessière, C., Hebrard, E., Hnich, B., Walsh, T.: The tractability of global constraints. In: Wallace, M. (ed.) CP 2004. LNCS, vol. 3258, pp. 716–720. Springer, Heidelberg (2004)
13. Bessière, C., Hebrard, E., Hnich, B., Kiziltan, Z., Walsh, T.: Filtering algorithms for the NVALUE constraint. Constraints 11(4), 271–293 (2006)
14. Bessière, C., Hebrard, E., Hnich, B., Kiziltan, Z., Walsh, T.: Filtering algorithms for the NVALUE constraint. In: Integration of AI and OR Techniques in Constraint Programming for Combinatorial Optimization Problems, 2nd International Conference (CPAIOR 2005) (2005)
15. Downey, R.G., Fellows, M.R., Stege, U.: Parameterized complexity: A framework for systematically confronting computational intractability. In: Contemporary Trends in Discrete Mathematics: From DIMACS and DIMATIA to the Future. DIMACS Series in Discrete Mathematics and Theoretical Computer Science, vol. 49, pp. 49–99 (1999)
16. Walsh, T.: General symmetry breaking constraints. In: Benhamou, F. (ed.) CP 2006. LNCS, vol. 4204, pp. 650–664. Springer, Heidelberg (2006)
17. Katsirelos, G., Narodytska, N., Walsh, T.: Combining symmetry breaking and global constraints. In: Oddi, A., Fages, F., Rossi, F. (eds.) Recent Advances in Constraints. LNCS, vol. 5655, pp. 84–98. Springer, Heidelberg (2009)
18. Law, Y.C., Lee, J., Walsh, T., Yip, J.: Breaking symmetry of interchangeable variables and values. In: Bessière, C. (ed.) CP 2007. LNCS, vol. 4741, pp. 423–437. Springer, Heidelberg (2007)
19. Hnich, B., Kiziltan, Z., Walsh, T.: Combining symmetry breaking with other constraints: lexicographic ordering with sums. In: Proceedings of the 8th International Symposium on the Artificial Intelligence and Mathematics (2004)

20. Frisch, A., Hnich, B., Kiziltan, Z., Miguel, I., Walsh, T.: Global constraints for lexicographic orderings. In: Van Hentenryck, P. (ed.) CP 2002. LNCS, vol. 2470, p. 93. Springer, Heidelberg (2002)
21. Frisch, A., Hnich, B., Kiziltan, Z., Miguel, I., Walsh, T.: Propagation algorithms for lexicographic ordering constraints. Artificial Intelligence 170(10), 803–908 (2006)
22. Puget, J.F.: Breaking all value symmetries in surjection problems. In: van Beek, P. (ed.) CP 2005. LNCS, vol. 3709, pp. 490–504. Springer, Heidelberg (2005)
23. Walsh, T.: Breaking value symmetry. In: Bessière, C. (ed.) CP 2007. LNCS, vol. 4741, pp. 880–887. Springer, Heidelberg (2007)
24. Bessière, C., Hebrard, E., Hnich, B., Kiziltan, Z., Quimper, C.G., Walsh, T.: The parameterized complexity of global constraints. In: Fox, D., Gomes, C. (eds.) Proceedings of the 23rd National Conference on AI, Association for Advancement of Artificial Intelligence, pp. 235–240 (2008)
25. Aloul, F., Ramani, A., Markov, I., Sakallah, K.: Solving difficult SAT instances in the presence of symmetries. In: Proceedings of the Design Automation Conference, pp. 731–736 (2002)
26. Law, Y., Lee, J.: Global constraints for integer and set value precedence. In: Wallace, M. (ed.) CP 2004. LNCS, vol. 3258, pp. 362–376. Springer, Heidelberg (2004)
27. Walsh, T.: Symmetry breaking using value precedence. In: Brewka, G., Coradeschi, S., Perini, A., Traverso, P. (eds.) Proceedings of Including Prestigious Applications of Intelligent Systems (PAIS 2006), Riva del Garda, Italy, August 29-September 1. Frontiers in Artificial Intelligence and Applications, vol. 141, pp. 168–172. IOS Press, Amsterdam (2006)
28. Flener, P., Frisch, A., Hnich, B., Kiziltan, Z., Miguel, I., Pearson, J., Walsh, T.: Breaking row and column symmetry in matrix models. In: Van Hentenryck, P. (ed.) CP 2002. LNCS, vol. 2470, p. 462. Springer, Heidelberg (2002)
29. Flener, P., Frisch, A., Hnich, B., Kiziltan, Z., Miguel, I., Walsh, T.: Matrix modelling. Technical Report APES-36-2001, APES group, Presented at Formul, Workshop on Modelling and Problem Formulation, CP 2001 post-conference workshop (2001)
30. Flener, P., Frisch, A., Hnich, B., Kiziltan, Z., Miguel, I., Walsh, T.: Matrix modelling: Exploiting common patterns in constraint programming. In: Proceedings of the International Workshop on Reformulating Constraint Satisfaction Problems, held alongside CP 2002 (2002)
31. Walsh, T.: Constraint patterns. In: Rossi, F. (ed.) CP 2003. LNCS, vol. 2833, pp. 53–64. Springer, Heidelberg (2003)
32. Huczynska, S., McKay, P., Miguel, I., Nightingale, P.: Modelling equidistant frequency permutation arrays: An application of constraints to mathematics. In: Gent, I.P. (ed.) CP 2009. LNCS, vol. 5732, pp. 50–64. Springer, Heidelberg (2009)
33. Katsirelos, G., Narodytska, N., Walsh, T.: On the complexity and completeness of static constraints for breaking row and column symmetry. In: Cohen, D. (ed.) CP 2010. LNCS, vol. 6308, pp. 305–320. Springer, Heidelberg (2010)
34. Shlyakhter, I.: Generating effective symmetry-breaking predicates for search problems. Electronic Notes in Discrete Mathematics 9, 19–35 (2001)
35. Flener, P., Frisch, A., Hnich, B., Kiziltan, Z., Miguel, I., Pearson, J., Walsh, T.: Symmetry in matrix models. Technical Report APES-30-2001, APES group, Presented at SymCon 2001 (Symmetry in Constraints), CP 2001 post-conference workshop (2001)
36. Freuder, E.: A sufficient condition for backtrack-free search. Journal of the Association for Computing Machinery 29(1), 24–32 (1982)
37. Dechter, R., Pearl, J.: Tree clustering for constraint networks. Artificial Intelligence 38, 353–366 (1989)
38. Cooper, M., Cohen, D., Jeavons, P.: Characterizing tractable constraints. Artificial Intelligence 65, 347–361 (1994)

39. Bessiere, C., Régin, J.: Arc consistency for general constraint networks: Preliminary results. In: Proceedings of the 15th International Conference on AI, International Joint Conference on Artificial Intelligence, pp. 398–404 (1997)
40. Régin, J.C.: A filtering algorithm for constraints of difference in CSPs. In: Proceedings of the 12th National Conference on AI, Association for Advancement of Artificial Intelligence, pp. 362–367 (1994)
41. Samer, M., Szeider, S.: Tractable cases of the extended global cardinality constraint. In: Proceedings of CATS 2008, Computing: The Australasian Theory Symposium (2008)
42. Szeider, S.: The complexity of resolution with generalized symmetry rules. Theory of Computing Systems 38(2), 171–188 (2005)
43. Biere, A., Heule, M., van Maaren, H., Walsh, T. (eds.): Handbook of Satisfiability. IOS Press, Amsterdam (2009)

On the Kernelization Complexity of Colorful Motifs

Abhimanyu M. Ambalath[1], Radheshyam Balasundaram[2],
Chintan Rao H.[3], Venkata Koppula[4],
Neeldhara Misra[5], Geevarghese Philip[5], and M.S. Ramanujan[5]

[1] National Institute of Technology, Calicut
[2] Birla Institute of Technology and Science, Pilani
[3] Indian Institute of Science, Bangalore
[4] Indian Institute of Technology, Kanpur
[5] The Institute of Mathematical Sciences, Chennai
abhimanyu.m.a@nitc.ac.in, adyarshyam@gmail.com,
chintanraoh@csa.iisc.ernet.in, kvenkata@iitk.ac.in,
{neeldhara,gphilip,msramanujan}@imsc.res.in

Abstract. The COLORFUL MOTIF problem asks if, given a vertex-colored graph G, there exists a subset S of vertices of G such that the graph induced by G on S is connected and contains every color in the graph exactly once. The problem is motivated by applications in computational biology and is also well-studied from the theoretical point of view. In particular, it is known to be NP-complete even on trees of maximum degree three [Fellows et al, ICALP 2007]. In their pioneering paper that introduced the color-coding technique, Alon et al. [STOC 1995] show, *inter alia*, that the problem is FPT on general graphs. More recently, Cygan et al. [WG 2010] showed that COLORFUL MOTIF is NP-complete on *comb graphs*, a special subclass of the set of trees of maximum degree three. They also showed that the problem is not likely to admit polynomial kernels on forests.

We continue the study of the kernelization complexity of the COLORFUL MOTIF problem restricted to simple graph classes. Surprisingly, the infeasibility of polynomial kernelization persists even when the input is restricted to comb graphs. We demonstrate this by showing a simple but novel composition algorithm. Further, we show that the problem restricted to comb graphs admits polynomially many polynomial kernels. To our knowledge, there are very few examples of problems with many polynomial kernels known in the literature. We also show hardness of polynomial kernelization for certain variants of the problem on trees; this rules out a general class of approaches for showing many polynomial kernels for the problem restricted to trees. Finally, we show that the problem is unlikely to admit polynomial kernels on another simple graph class, namely the set of all graphs of diameter two. As an application of our results, we settle the classical complexity of CONNECTED DOMINATING SET on graphs of diameter two — specifically, we show that it is NP-complete.

V. Raman and S. Saurabh (Eds.): IPEC 2010, LNCS 6478, pp. 14–25, 2010.
© Springer-Verlag Berlin Heidelberg 2010

1   Introduction and Motivation

Algorithms that are designed to reduce the size of an instance in polynomial time are widely referred to as preprocessing algorithms. It is natural to study such algorithms in the context of problems that are NP-hard — preprocessing techniques are used in almost every practical computer implementation that deals with an NP-hard problem. We study *kernelization algorithms* — these are preprocessing algorithms that have provable performance bounds, both in the running time and in the extent of reduction in instance size. In particular, we are interested in polynomial time algorithms that reduce the size of a parameterized problem (cf. Section 2 and [8,11] for definitions) to an instance whose size is a polynomial in the parameter. Such a reduced instance is called a *polynomial kernel*.

It is natural to examine the possibility of preprocessing strategies when a problem is notoriously intractable. In this work, our interests center around the COLORFUL MOTIF problem. The problem is intractable, in the classical sense, even on seemingly "simple" classes of graphs. Using a recent framework for showing lower bounds on polynomial kernelization, we establish that not only are these problems unlikely to admit a polynomial time algorithm that solves them — they are unlikely to admit polynomial time algorithms that reduce them to instances whose size is bounded by a polynomial in the parameter.

The GRAPH MOTIF problem concerns a vertex-colored undirected graph G and a *multiset* M of colors. We are asked whether there is a set S of vertices of G such that the subgraph induced on S is connected and there is a color-preserving bijective mapping from S to M. That is, the problem is to find if there is a connected subgraph H of G such that the multiset of colors of H is identical to M.

The GRAPH MOTIF problem has immense utility in bioinformatics, especially in the context of metabolic network analysis (eg. motif search in metabolic reaction graphs with vertices representing reactions and edges connecting successive reactions) [4,14] . The problem is NP-complete even in very restricted cases, such as when G is a tree with maximum degree 3, or when G is a bipartite graph with maximum degree 4 and M is a multiset over just two colors. When parameterized by $|M|$, the problem is FPT, and it is W[2]-hard when parameterized by the *number of colors* in M, even when G is a tree [9].

The COLORFUL MOTIF problem is a simpler version of the GRAPH MOTIF problem, where M is a set (and not a multiset). Even this problem is NP-hard on simple classes of graphs, such as when G is a tree with maximum degree 3 [9]. The problem is FPT on general graphs when parameterized by $|M|$, and the current fastest FPT algorithm, by Guillemot and Sikora, runs in $\mathcal{O}^*(2^{|M|})$ time[1] and polynomial space [13].

---

[1] Given $f : \mathbb{N} \to \mathbb{N}$, we define $\mathcal{O}^*(f(n))$ to be $O(f(n) \cdot p(n))$, where $p(\cdot)$ is some polynomial function. That is, the $\mathcal{O}^*$ notation suppresses polynomial factors in the expression.

We now turn to an example of a seemingly simple graph class on which the problem continues to be intractable. A graph is called a *comb graph* if (i) it is a tree, (ii) all vertices are of degree at most 3, (iii) all the vertices of degree 3 lie on a single simple path. Cygan et al. [6] recently showed that the problem is NP-hard even on comb graphs. Further, they show that the parameterized version of the problem is unlikely to admit a polynomial kernel on forests unless $NP \subseteq coNP/Poly$, which would in turn imply an unlikely collapse of the Polynomial Hierarchy [5].

We begin by pushing the borders of classical tractability. We show that while the problem is polynomial time on *caterpillars* (trees where the removal of all leaf vertices results in a *path*, called the *spine* of the caterpillar), it is NP-hard on *lobsters* (trees where the removal of all leaf vertices results in a *caterpillar*). In fact, we show that even more is true: the problem is NP-hard even on rooted trees of height two, or equivalently, on trees of diameter at most four.

Next, we extend the known results on the hardness of kernelization for this problem [6]. Specifically, we show that the lower bound can be tightened to hold for comb graphs as well. This is established by demonstrating a simple but unusual composition algorithm for the problem restricted to comb graphs. The composition is unusual because it is not the trivial composition (via disjoint union), and yet, it does not employ gadgets to encode the identity of the instances. To the best of our knowledge, this is an uncommon style of composition. On the positive side, we show a straightforward argument that yields polynomially many polynomial kernels for the problem on comb graphs, *a la* the many polynomial kernels obtained for k-Leaf Out Branching [10]. Again, to the best of our knowledge, this is one of the very few examples of many polynomial kernels for a parameterized problem for which polynomial kernelization is infeasible.

However, in our attempts to obtain many polynomial kernels for the more general case of trees, we learn that some natural approaches fail. Specifically, we show that two natural variants of the problem — Rooted Colorful Motif[2], Subset Colorful Motif[3] — do not admit polynomial kernels unless $NP \subseteq coNP/Poly$. This shows, for instance, that the "guess" for obtaining many polynomial kernels has to be more than, or different from, a subset of vertices.

While we show that Colorful Motif is NP-hard on trees of diameter at most four, the kernelization complexity of the problem on this class of graphs is still open. However, we show that Colorful Motif is NP-hard on general graphs of diameter *three*, and the same reduction also shows that polynomial kernels are unlikely for graphs of diameter three. We employ a reduction from Colorful Motif on general graphs. Using similar techniques, we show that the problem is NP-hard on general graphs of diameter *two*. This turns out to be useful to show the NP-hardness of Connected Dominating Set on the same class of graphs.

---

[2] Does there exist a colorful subtree that contains a specific vertex?

[3] Does there exist a colorful subtree that contains a specific subset of vertices?

The results we obtain in this paper contribute to the rapidly growing collection of problems for which polynomial kernels do not exist under reasonable complexity-theoretic assumptions. Given that many of our results pertain to very special graph classes, we hope these hardness results — which make these special problems available as starting points for further reductions — will be useful in settling the kernelization complexity of many other problems. In fact, we demonstrate the utility of the NP-completeness of COLORFUL MOTIF on graphs of diameter two, by showing that CONNECTED DOMINATING SET on graphs of diameter two is NP-complete. The classical complexity of CONNECTED DOMINATING SET on graphs of diameter two was hitherto unknown, although it was known to be NP-complete on graphs of diameter three, and trivial on graphs of diameter one. Also, since COLORFUL MOTIF is both well-motivated and well-studied, we believe that these results are of independent interest.

2 Preliminaries

A parameterized problem is denoted by a pair $(Q, k) \subseteq \Sigma^* \times \mathbb{N}$. The first component Q is a classical language, and the number k is called the parameter. Such a problem is *fixed–parameter tractable* (FPT) if there exists an algorithm that decides it in time $\mathcal{O}(f(k)n^{\mathcal{O}(1)})$ on instances of size n. A *many-to-one kernelization algorithm* (or, simply, a kernelization algorithm) for a parameterized problem takes an instance (x, k) of the problem as input, and in time polynomial in $|x|$ and k, produces an equivalent instance (x', k') such that $|x'|$ is a function purely of k. The output x' is called a *kernel* for the problem instance, and $|x'|$ is called the *size* of the kernel. A kernel is said to be a *polynomial kernel* if its size polynomial in the parameter k. We refer the reader to [8,16] for more details on the notion of fixed-parameter tractability.

The notion of *Turing kernelization* was introduced to formalize the idea of a "cheat kernel", wherein, given an instance of a parameterized problem, an algorithm outputs polynomially many polynomial kernels rather than a single kernel [15]. Formally, a t-oracle for a parameterized problem Π is an oracle that takes as input (I, k) with $|I| \leq t, |k| \leq t$ and decides whether $(I, k) \in \Pi$ in constant time. Π is said to have a $g(k)$-sized *Turing kernel* if there is an algorithm which, given input (I, k) and a $g(k)$-oracle for Π, decides whether $(I, k) \in \Pi$ in time polynomial in $|I + k|$. The Turing kernel is polynomial if $g()$ is a polynomial function.

To prove our lower bounds on polynomial kernels, we need a few notions and results from the recently developed theory of kernel lower bounds [2,3,7,12]. We use a notion of reductions, similar in spirit to those used in classical complexity to show NP-hardness results, to show this kernelization lower bound. We begin by associating a classical decision problem with a parameterized problem in a natural way as follows:

Definition 1. [Derived Classical Problem] [3] *Let* $\Pi \subseteq \Sigma^* \times \mathbb{N}$ *be a parameterized problem, and let* $1 \notin \Sigma$ *be a new symbol. We define the* derived classical problem *associated with* Π *to be* $\left\{ x1^k \mid (x, k) \in \Pi \right\}$.

The notion of a composition algorithm plays a key role in the lower bound argument.

Definition 2. [Composition Algorithm, Compositional Problem] [2] *A composition algorithm for a parameterized problem $\Pi \subseteq \Sigma^* \times \mathbb{N}$ is an algorithm that*

- *takes as input a sequence $\langle (x_1, k), (x_2, k), \ldots, (x_t, k) \rangle$ where each $(x_i, k) \in \Sigma^* \times \mathbb{N}$,*
- *runs in time polynomial in $\sum_{i=1}^{t} |x_i| + k$,*
- *and outputs an instance $(y, k') \in \Sigma^* \times \mathbb{N}$ with*
    1. *$(y, k') \in L \iff (x_i, k) \in L$ for some $1 \leq i \leq t$, and*
    2. *k' is polynomial in k.*

We say that a parameterized problem is compositional *if it has a composition algorithm.*

Theorem 1. *[2, Lemmas 1 and 2] Let L be a compositional parameterized problem whose derived classical problem is* NP-*complete. If L has a polynomial kernel, then $NP \subseteq coNP/Poly$.*

Now we define the class of reductions which lead to the kernel lower bound.

Definition 3. [3] *Let P and Q be parameterized problems. We say that P is polynomial time and parameter reducible to Q, written $P \leq_{Ptp} Q$, if there exists a polynomial time computable function $f : \Sigma^* \times \mathbb{N} \to \Sigma^* \times \mathbb{N}$, and a polynomial $p : \mathbb{N} \to \mathbb{N}$, and for all $x \in \Sigma^*$ and $k \in \mathbb{N}$, if $f((x, k)) = (x', k')$, then $(x, k) \in P$ if and only if $(x', k') \in Q$, and $k' \leq p(k)$. We call f a polynomial parameter transformation (or a PPT) from P to Q.*

This notion of a reduction is useful in showing kernel lower bounds because of the following theorem:

Theorem 2. [3, Theorem 3] *Let P and Q be parameterized problems whose derived classical problems are P^c, Q^c, respectively. Let P^c be* NP-*complete, and $Q^c \in$* NP. *Suppose there exists a PPT from P to Q. Then, if Q has a polynomial kernel, then P also has a polynomial kernel.*

We use $[n]$ to denote $\{1, 2, \ldots, n\} \subseteq \mathbb{N}$. The operation of *subdividing* an edge (u, v) involves replacing the edge (u, v) with two new edges (u, x_{uv}) and (x_{uv}, v), where x_{uv} is a new vertex (the *subdivided vertex*). For any two vertices u and v, the *distance* between u and v, denoted $d(u, v)$, is the length of a shortest path between u and v. The k-neighborhood of a vertex u in a graph G is the set of all vertices in G that are at a distance of at most k from u. A *rooted tree* is a pair (T, r) where T is a tree and $r \in V(T)$. A leaf node in a rooted tree (T, r) is said to be a *lowest leaf* if it is a leaf at the maximum distance from the root r. The *diameter* of a graph G is defined to be $max_{u,v \in V(G)} d(u, v)$. In other words, diameter of a graph is the length of a "longest shortest" path in the graph. A *superstar graph* is a tree with diameter at most 4. Note that in any superstar G, there exists a vertex v such that G rooted at v has height at most two.

The problem at the heart of this work is the following:

COLORFUL MOTIF

Input:	A graph $G = (V, E)$, $k \in \mathbb{N}$, and a coloring function $c : V \to [k]$.
Parameter:	k
Question:	Does G contain a subtree T on k vertices such that c restricted to T is bijective?

3 Hardness on Superstar Graphs

We begin by observing that the COLORFUL MOTIF problem is NP-complete even on simple classes of graphs. It is already known that the problem is NP-complete on comb graphs [6]. In this section, we show that the problem is NP-complete on superstars — or equivalently, on rooted trees of height at most two. To begin with, consider COLORFUL MOTIF on paths. A solution corresponds to a colorful subpath, which, if it exists, we can find in polynomial time by guessing its end points. It is easy to see that this approach can be extended to a polynomial time algorithm for COLORFUL MOTIF on caterpillars, in which case we are looking for a colorful "subcaterpillar": We may guess the end points of the spine of the subcaterpillar, and for any given guess, if the subpath on the spine does not span the entire set of colors, we check if they can be found on the leaves. The details are omitted as the argument is straightforward.

Recall that a lobster is a tree where the removal of all leaf vertices results in a caterpillar. Lobsters are a natural generalization of caterpillars, and we show that the COLORFUL MOTIF problem is NP-hard on lobsters. In fact, we show that the problem is NP-hard on lobsters whose spine has just *one* vertex. Observe that every such graph is a superstar graph; thus we show that the problem is NP-hard on superstars. To show these hardness results, we reduce from the following variant of the well-known SET COVER problem:

COLORFUL SET COVER

Input:	A finite universe U, a finite family $\mathcal{F} = \{F_1, F_2, \ldots, F_n\}$ of subsets of U that is such that there is no i, j for which $F_i \cup F_j = U$, and a function $C : \mathcal{F} \twoheadrightarrow U$ such that $C(F_i) \subset F_i$.
Question:	Does there exist $\mathcal{R} \subseteq \mathcal{F}$ such that $\bigcup_{S \in \mathcal{R}} S = U$ and C is injective on $\mathcal{R}$?

We will need the fact that SET COVER is NP-complete even when no two sets in the family span the universe. Formally, if the input to SET COVER is restricted to families that have the property that no two subsets in the family are such that their union is the universe, it remains NP-complete:

AT-LEAST-THREE SET COVER

Input:	A finite universe U, a finite family $\mathcal{F} = \{F_1, F_2, \ldots, F_n\}$ of subsets of U, such that there is no i, j for which $F_i \cup F_j = U$.
Question:	Does there exist $\mathcal{R} \subseteq \mathcal{F}$ such $\bigcup_{S \in \mathcal{R}} S = U$?

Proposition 1. [⋆] AT-LEAST-THREE SET COVER *is NP-complete.*

Lemma 1. [⋆] COLORFUL SET COVER *is NP-hard.*

Theorem 3. [⋆]COLORFUL MOTIF *on superstar graphs is NP-hard.*

Proposition 2. [⋆] *Let* (T, C) *be an instance of* COLORFUL MOTIF, *where* T *is a superstar graph. Let* $u_1, \ldots u_r$ *be the children of the root of* T. *Let* V_i *denote the set of leaves adjacent to* u_i, *and let* U_i *denote* $V_i \cup \{u_i\}$. *For* $X \subseteq V(T)$, *let* $c(X)$ *denote the set of colors used on* X, *that is:*

$$c(X) = \{d \mid \exists x \in X, c(x) = d\}.$$

The COLORFUL MOTIF *problem is hard on superstar graphs even on instances where no two subtrees are colored with the entire set of colors: that is, for any* $i \neq j$,

$$(c(U_i) \cup c(U_j)) \setminus C \neq \phi.$$

4   Colorful Motifs on Graphs of Diameter Two and Three

In this section, we consider the COLORFUL MOTIF problem restricted to graphs of diameter two and three. We show that the COLORFUL MOTIF problem on superstars reduces to COLORFUL MOTIF on graphs of diameter two, thereby establishing that the problem is NP-complete. Also, we show that the COLORFUL MOTIF problem on general graphs reduces to COLORFUL MOTIF on graphs of diameter three, thereby establishing that the problem is NP-complete, and that polynomial kernels are infeasible. These reductions are quite similar, with only subtle differences.

Lemma 2. [⋆][4] *The* COLORFUL MOTIF *problem on superstars with parameter* k *reduces to* COLORFUL MOTIF *with parameter* k *on graphs of diameter two.*

Lemma 3. [⋆] *The* COLORFUL MOTIF *problem with parameter* k *reduces to* COLORFUL MOTIF *with parameter* $(k + 1)$ *on graphs of diameter three. The reduction serves to show both NP-hardness and infeasibility of polynomial kernelization (being a polynomial parameter transformation).*

5   Colorful Motifs on Comb Graphs

In [6], Cygan et al. show that COLORFUL MOTIF is NP-complete on *comb graphs*, defined as follows:

---

[4] Due to space constraints, proofs marked with a ⋆ have been deferred to a full version [1] of the paper.

Definition 4. *A graph $G = (V, E)$ is called a* comb graph *if (i) it is a tree, (ii) all vertices are of degree at most 3, (iii) all the vertices of degree 3 lie on a single simple path. The maximal path, which starts and ends in degree 3 vertices is called the* spine *of a comb graph. One of the two endpoint vertices of the spine is arbitrarily chosen as the first vertex, and the other as the last. A path from a degree 3 vertex to a leaf which contains exactly one degree 3 vertex, is called a* tooth.

In this section, we present a composition algorithm for COLORFUL MOTIF on comb graphs. Note that in [6], it is observed that COLORFUL MOTIF is unlikely to admit polynomial kernels on forests. This simple composition obtained using disjoint union does not work "as is" when we restrict our attention to comb graphs, since the graph resulting from the disjoint union of comb graphs is not a comb graph, as it is not connected.

5.1   A Composition Algorithm

We begin by introducing some notation that will be useful in the proof of the main result of this section. Let (T, k, c) be an instance of COLORFUL MOTIF restricted to comb graphs, that is, let T denote a comb graph, and let $c : V(T) \to [k]$ be a coloring function.

Let T_p and T_q be two comb graphs, and let $l_p \in V(T_p)$ be the last vertex on the spine of T_p, and let $f_q \in V(T_q)$ be the first vertex on the spine of T_q. We define $T_p \odot T_q$ as follows:

(i) $V(T_p \odot T_q) = V(T_p) \uplus V(T_q) \cup \{v_p, v_q\}$, where $\{v_p, v_q\}$ are "new" vertices, and

(ii) $E(T_p \odot T_q) = E(T_p) \uplus E(T_q) \cup \{(l_p, v_p), (v_p, v_q), (v_q, f_q)\}$

We are now ready to present the main result of this section:

Lemma 4. *The COLORFUL MOTIF problem does not admit a polynomial kernel on comb graphs unless $NP \subseteq coNP/Poly$.*

Proof. **(Sketch)** Let $(T_1, c_1, k), (T_2, c_2, k), \dots (T_t, c_t, k)$ be the instances that are input to the composition algorithm. Let $\mathcal{T}$ denote the graph:

$$\mathcal{T} = T_1 \odot T_2 \odot \cdots \odot T_t.$$

Let N denote the set of all new vertices introduced by the $\odot$ operations. Notice that any vertex of $\mathcal{T}$ that does not belong to N is a vertex from one of the instances T_i. We will refer to such vertices in $\mathcal{T}$ as being from $\mathcal{T}(T_i)$.

We refer to the pair of vertices in N adjacent to the endpoints of the spine of a T_i as the *guards* of T_i (notice that any T_i has at most two guard vertices). We define the coloring function c on $\mathcal{T}$ as follows. For every vertex $u \in T_i$, $c(u) = c_i(u)$. For every vertex u that is a guard of T_i, $c(u) = c(v)$, where v is the vertex of T_i adjacent to u. We claim that $(\mathcal{T}, k)$ is the composed instance. The details of this proof are defered to a full version due to space constraints.   □

Corollary 1. [⋆] *The* COLORFUL MOTIF *problem on lobsters does not admit a polynomial kernel unless* $NP \subseteq coNP/Poly$.

5.2   Many Polynomial Kernels

Although a parameterized problem may not necessarily admit a polynomial kernel, it may admit many of them, with the property that the instance is in the language if and only if at least one of the kernels corresponds to an instance that is in the language. We now show that the COLORFUL MOTIF problem admits n kernels of size $\mathcal{O}(k^2)$ each on comb graphs. This is established by showing that a closely related variant, the ROOTED COLORFUL MOTIF problem, admits a polynomial kernel. The ROOTED COLORFUL MOTIF problem is the following:

ROOTED COLORFUL MOTIF
Input: A graph $G = (V, E)$, $k \in \mathbb{N}$, a coloring function $c : V \to [k]$, and $r \in V$.
Parameter: k
Question: Does G contain a subtree T on k vertices, containing r, such that c restricted to T is bijective?

Lemma 5. [⋆] *The* COLORFUL MOTIF *problem admits many polynomial kernels on comb graphs.*

6   Colorful Motifs on Trees

In this section, we demonstrate the infeasibility of some strategies for showing many polynomial kernels for COLORFUL MOTIF restricted to trees. Observe that the COLORFUL MOTIF problem is unlikely to admit a polynomial kernel on trees, since a polynomial kernel on trees would imply a polynomial kernelization procedure for comb graphs, which is infeasible (see Lemma 4).

6.1   Hardness with a Fixed Root

In the case of comb graphs, we were able to establish that the problem of finding a colorful subtree with a fixed root admits a $\mathcal{O}(k^2)$ kernel. Unfortunately, this approach does not extend to trees, as we establish that ROOTED COLORFUL MOTIF is compositional on trees.

Proposition 3. [⋆] *The* ROOTED COLORFUL MOTIF *problem restricted to trees is NP-hard.*

Proposition 4. [⋆] *The* ROOTED COLORFUL MOTIF *problem when restricted to trees does not admit a polynomial kernel unless* $NP \subseteq coNP/Poly$.

Hardness on Trees of Bounded Degree. Since the problem of COLOR-
FUL MOTIF is fixed-parameter tractable on trees with running time $\mathcal{O}^*(2^k)$, we
may assume, without loss of generality[5], that the number of instances input to
the composition algorithm is at most $\mathcal{O}^*(2^k)$. Consider the COLORFUL MOTIF
problem restricted to rooted binary trees. We establish in this section that this
problem is compositional as well:

Lemma 6. *The* COLORFUL MOTIF *problem is compositional on rooted binary
trees, and does not admit a polynomial kernel unless* $NP \subseteq coNP/Poly$.

Proof. We build the composed instance $\mathcal{T}$ by combining the input instances with
a complete binary tree on t leaves, where t is the number of instances input to
the composition algorithm. Let (T_i, r_i, c_i, k), for $i \in [t]$ be the input instances to
the composition algorithm, where T_i is a rooted binary tree rooted at r_i. Let $\mathcal{B}$
be the complete binary tree on t leaves. Note that the depth of $\mathcal{B}$ is $\log t$, and
since we have assumed that $t \leq 2^k$, the depth of $\mathcal{B}$ is at most k. Denote the
root of this tree by r, and let $D(i)$ the set of all vertices at a distance i from
r. Assume that the leaves are ordered in some arbitrary but fixed fashion. We
identify the vertices l_i and r_i, where l_i is the ith leaf of $\mathcal{B}$. We define the coloring
function

$$c : V(\mathcal{T}) \rightarrow [2k],$$

as follows: $c(u) = k+i+1$ for every $u \in D(i)$, and $c(u) = c_i(u)$ for any $u \in T_i$. We
claim that $(\mathcal{T}, r, c, 2k)$ is the composed instance. The correctness follows from
reasons similar to those in the proof of Proposition 4, and the straightforward
details are omitted.   □

6.2   Hardness with a Fixed Subset of Vertices

Now, we have seen that "fixing" one vertex does not help the cause of kernelization for trees in general. In fact, more is true: fixing any constant number of
vertices does not help. The problem we study in this section is the following:

SUBSET COLORFUL MOTIF
Input: A graph $G = (V, E)$, a coloring function $c : V \rightarrow [k]$, and a set
                   of vertices $U \subseteq V$, $|U| = s = \mathcal{O}(1)$.
Parameter: k
Question: Does G contain a subtree T on k vertices, such that $U \subseteq V(T)$,
                   and c restricted to T is bijective?

Proposition 5. [⋆] *The* SUBSET COLORFUL MOTIF *problem restricted to trees
does not admit a polynomial kernel unless* $NP \subseteq coNP/Poly$.

---

[5] If the number of instances is greater than $f(k)n^c$, where $f(k)n^c$ is the time taken
by a FPT algorithm to solve the problem, then the composition algorithm can solve
every problem and trivially return an appropriate answer within the required time
bounds.

7   Connected Dominating Set on Graphs of Diameter Two

In this section, we show that CONNECTED DOMINATING SET on graphs of diameter two is NP-complete. The classical complexity of CONNECTED DOMINATING SET on graphs of diameter two was hitherto unknown, although it was known to be NP-complete on graphs of diameter three, and trivial on graphs of diameter one. We establish this by a non-trivial reduction from COLORFUL MOTIF on graphs of diameter two, which is NP-complete by Lemma 2.

Theorem 4. [⋆] *The* CONNECTED DOMINATING SET *problem, when restricted to graphs of diameter two, is* NP-*complete.*

8   Summary, Conclusions and Further Directions

We studied the problem of COLORFUL MOTIF on various graph classes. We proved that the problem of COLORFUL MOTIF restricted to superstars is NP-complete. We also showed NP-completeness on graphs of diameter two. We applied this result towards settling the classical complexity of CONNECTED DOMINATING SET on graphs of diameter two — specifically, we show that it is NP-complete. Further, we showed that on graphs of diameter two, the problem is NP-complete *and* is unlikely to admit a polynomial kernel.

Next, we showed that obtaining polynomial kernels for COLORFUL MOTIF on comb graphs is infeasible, but we show the existence of n polynomial kernels. Further, we study the problem of COLORFUL MOTIF on trees, where we observe that the natural strategies for many polynomial kernels are not successful. For instance, we show that "guessing" a root vertex, which helped in the case of comb graphs, fails as a strategy because the ROOTED COLORFUL MOTIF problem has no polynomial kernels on trees. In fact, this lower bound holds even on rooted binary trees. We summarize our results about COLORFUL MOTIF in special graph classes in the following theorem:

Theorem 5

1. *On the class of comb graphs,* COLORFUL MOTIF *is* NP-*complete and* ROOTED COLORFUL MOTIF *has an* $\mathcal{O}(k^2)$ *kernel. Equivalently,* COLORFUL MOTIF *has* n *kernels of size* $\mathcal{O}(k^2)$ *each.*
2. ROOTED COLORFUL MOTIF *does not admit a polynomial kernel on binary rooted trees, unless* $NP \subseteq coNP/Poly$.
3. SUBSET COLORFUL MOTIF *does not admit a polynomial kernel on trees unless* $NP \subseteq coNP/Poly$.

Finally, we leave open the questions of whether the COLORFUL MOTIF problem admits polynomial kernels on superstars, and many polynomial kernels when restricted to trees.

Acknowledgements. We thank our anonymous reviewers for their painstakingly detailed comments for improving the paper. We especially thank two reviewers who pointed out to us a significantly simpler composition algorithm for the COLORFUL MOTIF problem on comb graphs.

References

1. Ambalath, A.M., Balasundaram, R., Chintan Rao, H., Koppula, V., Misra, N., Philip, G., Ramanujan, M.S.: On the kernelization complexity of colorful motifs (2010), http://www.imsc.res.in/~gphilip/publications/cm.pdf [Full Version]
2. Bodlaender, H.L., Downey, R.G., Fellows, M.R., Hermelin, D.: On problems without polynomial kernels. Journal of Computer and System Sciences 75(8), 423–434 (2009)
3. Bodlaender, H.L., Thomassé, S., Yeo, A.: Kernel bounds for disjoint cycles and disjoint paths. In: Fiat, A., Sanders, P. (eds.) ESA 2009. LNCS, vol. 5757, pp. 635–646. Springer, Heidelberg (2009)
4. Bruckner, S., Hüffner, F., Karp, R.M., Shamir, R., Sharan, R.: Topology-free querying of protein interaction networks. In: Batzoglou, S. (ed.) RECOMB 2009. LNCS, vol. 5541, pp. 74–89. Springer, Heidelberg (2009)
5. Cai, J.-y., Chakaravarthy, V.T., Hemaspaandra, L.A., Ogihara, M.: Competing provers yield improved Karp-Lipton collapse results. Information and Computation 198(1), 1–23 (2005)
6. Cygan, M., Pilipczuk, M., Pilipczuk, M., Wojtaszczyk, J.O.: Kernelization hardness of connectivity problems in d-degenerate graphs. Accepted at WG 2010 (2010)
7. Dom, M., Lokshtanov, D., Saurabh, S.: Incompressibility through Colors and IDs. In: Proceedings of ICALP 2009. LNCS, vol. 5555, pp. 378–389. Springer, Heidelberg (2009)
8. Downey, R.G., Fellows, M.R.: Parameterized Complexity. Springer, Heidelberg (1999)
9. Fellows, M.R., Fertin, G., Hermelin, D., Vialette, S.: Sharp tractability borderlines for finding connected motifs in vertex-colored graphs. In: Arge, L., Cachin, C., Jurdziński, T., Tarlecki, A. (eds.) ICALP 2007. LNCS, vol. 4596, pp. 340–351. Springer, Heidelberg (2007)
10. Fernau, H., Fomin, F.V., Lokshtanov, D., Raible, D., Saurabh, S., Villanger, Y.: Kernel(s) for problems with no kernel: On out-trees with many leaves. In: Proceedings of STACS 2009, pp. 421–432 (2009)
11. Flum, J., Grohe, M.: Parameterized Complexity Theory. Springer, Heidelberg (2006)
12. Fortnow, L., Santhanam, R.: Infeasibility of instance compression and succinct PCPs for NP. In: Proceedings of STOC 2008, pp. 133–142. ACM, New York (2008)
13. Guillemot, S., Sikora, F.: Finding and counting vertex-colored subtrees. In: Hliněný, P., Kučera, A. (eds.) MFCS 2010. LNCS, vol. 6281, pp. 405–416. Springer, Heidelberg (2010)
14. Lacroix, V., Fernandes, C.G., Sagot, M.-F.: Motif search in graphs: Application to metabolic networks. IEEE/ACM Transactions on Computational Biology and Bioinformatics 3(4), 360–368 (2006)
15. Lokshtanov, D.: New Methods in Parameterized Algorithms and Complexity. PhD thesis, University of Bergen, Norway (2009)
16. Niedermeier, R.: Invitation to Fixed Parameter Algorithms. Oxford Lecture Series in Mathematics and Its Applications. Oxford University Press, Oxford (2006)

Partial Kernelization for Rank Aggregation: Theory and Experiments

Nadja Betzler*, Robert Bredereck*, and Rolf Niedermeier

Institut für Informatik, Friedrich-Schiller-Universität Jena,
Ernst-Abbe-Platz 2, D-07743 Jena, Germany
{nadja.betzler,robert.bredereck,rolf.niedermeier}@uni-jena.de

Abstract. RANK AGGREGATION is important in many areas ranging from web search over databases to bioinformatics. The underlying decision problem KEMENY SCORE is NP-complete even in case of four input rankings to be aggregated into a "median ranking". We study efficient polynomial-time data reduction rules that allow us to find *optimal* median rankings. On the theoretical side, we improve a result for a "partial problem kernel" from quadratic to linear size. On the practical side, we provide encouraging experimental results with data based on web search and sport competitions, e.g., computing optimal median rankings for real-world instances with more than 100 candidates within milliseconds.

1 Introduction

We investigate the effectiveness of data reduction for computing *optimal* solutions of the NP-hard RANK AGGREGATION problem. Kemeny's corresponding voting scheme can be described as follows. An *election* (V, C) consists of a set V of n votes and a set C of m candidates. A vote or a *ranking* is a total order of all candidates. For instance, in case of three candidates a, b, c, the order $c > b > a$ means that candidate c is the best-liked one and candidate a is the least-liked one. For each pair of votes v, w, the *Kendall-Tau distance* (also known as the number of inversions) between v and w is defined as

$$\text{KT-dist}(v, w) = \sum_{\{c,d\} \subseteq C} d_{v,w}(c, d),$$

where $d_{v,w}(c, d)$ is set to 0 if v and w rank c and d in the same order, and is set to 1, otherwise. The *score* of a ranking l with respect to an election (V, C) is defined as $\sum_{v \in V} \text{KT-dist}(l, v)$. A ranking l with a minimum score is called a *Kemeny ranking* of (V, C) and its score is the *Kemeny score* of (V, C). The central problem considered in this work is as follows:

RANK AGGREGATION: Given an election (V, C), find a Kemeny ranking of (V, C).

Its decision variant KEMENY SCORE asks whether there is a Kemeny ranking of (V, C) with score at most some additionally given positive integer k. The RANK AGGREGATION problem has numerous applications, ranging from building meta-search engines for the web or spam detection [7] over databases [8] to the construction of genetic maps

---

* Supported by the DFG, research project PAWS, NI 369/10.

V. Raman and S. Saurabh (Eds.): IPEC 2010, LNCS 6478, pp. 26–37, 2010.

in bioinformatics [9]. Kemeny rankings are also desirable in classical voting scenarios such as the determination of a president (see, for example, www.votefair.org) or the selection of the best qualified candidates for job openings. The wide range of applications is due to the fulfillment of many desirable properties from the social choice point of view [18], including the *Condorcet property*: if there is a candidate (*Condorcet winner*) who is better than every other candidate in more than half of the votes, then this candidate is also ranked first in every Kemeny ranking.

Previous work. First computational complexity studies of KEMENY SCORE go back to Bartholdi et al. [2], showing its NP-hardness. Dwork et al. [7] showed that the problem remains NP-hard even in the case of four votes. Moreover, they identified its usefulness in aggregating web search results and provided several approximation and heuristic algorithms. Recent papers showed constant-factor approximability [1,17] and an (impractical) PTAS [11]. Schalekamp and van Zuylen [15] provided a thorough experimental study of approximation and heuristic algorithms. Due to the importance of computing optimal solutions, there have been some experimental studies in this direction [5,6]: An integer linear program and a branch-and-bound approach were applied to random instances generated under a noise model (motivated by the interpretation of Kemeny rankings as maximum likelihood estimators [5]). From a parameterized complexity perspective, the following is known. First fixed-parameter tractability results have been shown with respect to the single parameters number of candidates, Kemeny score, maximum range of candidate positions, and average KT-distance d_a [3]. The *average KT-distance*

$$d_a := \sum_{v,w \in V, v \neq w} \text{KT-dist}(v,w))/(n(n-1))$$

will also play a central role in this work. Moreover, KEMENY SCORE remains NP-hard when the average range of candidate positions is two [3], excluding hope for fixed-parameter tractability with respect to this parameterization. Simjour [16] further introduced the parameter "Kemeny score divided by the number of votes" (also showing fixed-parameter tractability) and improved the running times for the fixed-parameter algorithms corresponding to the parameterizations by average KT-distance and Kemeny score. Recently, Karpinski and Schudy [10] devised subexponential-time fixed-parameter algorithms for the parameters Kemeny score, d_a, and Kemeny score divided by the number of votes. Mahajan et al. [12] studied above guarantee parameterization with respect to the Kemeny score. Introducing the new concept of partial kernelization, it has been shown that with respect to the average KT-distance d_a one can compute in polynomial time an equivalent instance where the number of candidates is at most $162d_a^2 + 9d_a$ [4]. This equivalent instance is called *partial kernel*[1] with respect to the parameter d_a because it only bounds the number of candidates but not the number of votes instead of bounding the total instance size (as one has in classical problem kernels).

Our contributions. On the theoretical side, we improve the previous partial kernel [4] from $162d_a^2 + 9d_a$ candidates to $11d_a$ candidates. Herein, the central point is to exploit "stronger majorities", going from ">_{2/3}-majorities" as used before [4] to "$\geq_{3/4}$-majorities". In this line, we also prove that the consideration of "$\geq_{3/4}$-majorities" is

---

[1] A formal definition of partial kernels appears in [4].

optimal in the sense that "$\geq_s$-majorities" with $s < 3/4$ do not suffice. Moreover, our "linear partial kernel" result with respect to d_a is based on results for related parameterizations based on "dirty pairs according to $\geq_s$-majorities" which might be considered as parameterizations of independent interest.

On the practical side, we provide strong empirical evidence for the usefulness of data reduction rules associated with the above mentioned kernelization. An essential property of our data reduction rules is that they can break instances into several subinstances to be handled independently, that is, the relative order between the candidates in two different subinstances in a Kemeny ranking is already determined. This also means that for hard instances which we could not completely solve we were still able to compute "partial rankings" of the top and bottom ranked candidates. Finally, we employ some of the known fixed-parameter algorithms and integer linear programming to solve sufficiently small parts of the instances remaining after data reduction. Our experiments showed that the data reduction rules allow for drastically decreased running times and are crucial to solve larger instances. For example, using data reduction rules decreased the running time for a test instance from winter sports with 33 candidates from 103 seconds to few milliseconds. Furthermore, several instances with more than 100 candidates, which could not be solved without data reduction, were solved in a few seconds.

2   Majority-Based Data Reduction Rules

We start with some definitions and sketch some relevant previous results [4]. Then we show how to extend the previous results to obtain a linear partial kernel for the average KT-distance by providing a new data reduction rule. We also show the "limitations" of our new reduction rule. Finally, we provide two more reduction rules of practical relevance.

Definitions and previous results. The data reduction framework from previous work [4] introduces a "dirtiness concept" and shows that one can delete some "non-dirty candidates" by a data reduction rule leading to a partial kernel with respect to the average KT-distance. The "dirtiness" of a pair of candidates is measured by the amount of agreement of the votes for this pair. To this end, we introduce the following notation. For an election (V, C), two candidates $c, c' \in C$, and a rational number $s \in \,]0.5, 1]$, we write

$$c \geq_s c'$$

if at least $\lceil s \cdot |V| \rceil$ of the votes prefer c to c'. A candidate pair $\{c, c'\}$ is *dirty according to the $\geq_s$-majority* if neither $c \geq_s c'$ nor $c' \geq_s c$. All remaining pairs are *non-dirty according to the $\geq_s$-majority*. This directly leads to the parameter n_d^s denoting the number of dirty pairs according to the $\geq_s$-majority. Previous work only considered $>_{2/3}$-majorities[2] and provided a reduction rule such that the number of candidates in a reduced instance is at most quadratic in $\min\{n_d^{>2/3}, d_a\}$. In this work, we provide a linear partial kernel with respect to n_d^s according to the $\geq_s$-majority for $s \geq 3/4$ and show that this leads to a linear partial kernel with respect to d_a.

---

[2] To simplify matters, we write "$>_{2/3}$" instead of "$\geq_s$ with $s > 2/3$", and if the value of s is clear from the context, then we speak of "dirty pairs" and omit "according to the $\geq_s$-majority".

Table 1. Partial kernelization and polynomial-time solvability. The term dirty refers to the $\geq_s$-majority for the respective values of s. The number of dirty pairs is n_d^s. A linear partial kernel w.r.t. the average KT-distance follows directly from the linear partial kernel w.r.t. n_d^s (Theorem 1).

value of s	partial kernel result	special case: no dirty pairs
$2/3 < s < 3/4$	quadratic partial kernel w.r.t. n_d^s [4, Theorem 2]	polynomial-time solvable
$3/4 \leq s \leq 1$	linear partial kernel w.r.t. n_d^s (Theorem 1)	[4, Proposition 3]

We say that c and c' are *ordered according to the $\geq_s$-majority* in a preference list l if $c \geq_s c'$ and $c > c'$ in l. If all candidate pairs are non-dirty with respect to the $\geq_s$-majority for an $s > 2/3$, then there exists a $\geq_s$-*majority order*, that is, a preference list in which all candidate pairs are ordered according to the $\geq_s$-majority [4]. Furthermore, the corresponding $>_{2/3}$-majority order can be found in polynomial time and is a Kemeny ranking [4]. Candidates appearing only in non-dirty pairs according to a $\geq_s$-majority are called *non-dirty candidates* and all remaining candidates are *dirty candidates* according to the $\geq_s$-majority. Note that with this definition a non-dirty pair can also be formed by two dirty candidates. See Table 1 for an overview of partial kernelization and polynomial-time solvability results.

We end with some notation needed to state our data reduction rules. For a candidate subset $C' \subseteq C$, a ranking fulfills the condition $C' > C \setminus C'$ if every candidate from C' is preferred to every candidate from $C \setminus C'$. A subinstance of (V, C) *induced* by a candidate subset $C' \subseteq C$ is given by (V', C') where every vote in V' one-to-one corresponds to a vote in V keeping the relative order of the candidates from C'.

2.1 New Results Exploiting $\geq_{3/4}$-Majorities

We improve the partial kernel upper bound [4] for the parameter d_a from quadratic to linear, presenting a new data reduction rule. The crucial idea for the new reduction rule is to consider $\geq_{3/4}$-majorities instead of $>_{2/3}$-majorities. We further show that the new reduction rule is tight in the sense that it does not work for $>_{2/3}$-majorities.

Reduction Rule. The following lemma allows us to formulate a data reduction rule that deletes all non-dirty candidates and additionally may break the remaining set of dirty candidates into several subsets to be handled independently from each other.

Lemma 1. *Let $a \in C$ be a non-dirty candidate with respect to the $\geq_{3/4}$-majority and $b \in C \setminus \{a\}$. If $a \geq_{3/4} b$, then in every Kemeny ranking one must have "$a > b$"; if $b \geq_{3/4} a$, then in every Kemeny ranking one must have "$b > a$".*

Proof. Let the *partial score* of a candidate subset C' be the Kemeny score of the subinstance induced by C'. We consider the case $a \geq_{3/4} b$; the case $b \geq_{3/4} a$ follows in complete analogy. The proof is by contradiction. Assume that there is a Kemeny ranking l with "$\ldots > b > D > a > \ldots$" for some $D \subseteq C \setminus \{a, b\}$. Since a is non-dirty, for every candidate $d \in D$, it must either hold that $a \geq_{3/4} d$ or $d \geq_{3/4} a$. Let $D_1 := \{d \in D \mid a \geq_{3/4} d\}$ and $D_2 := \{d \in D \mid d \geq_{3/4} a\}$. Consider the ranking l' obtained from l through replacing

$$b > D > a \quad \text{by} \quad D_2 > a > b > D_1,$$

where the positions of all other candidates remain unchanged and the candidates within D_1 and D_2 have the same relative order as within D.

We show that the score of l is greater than the score of l' contradicting that l is a Kemeny ranking. The only pairs of candidates that have different orders in l and l' and thus can contribute with different partial scores to the scores of l and l' are $\{a, d\}$ and $\{d_2, d\}$ for all $d \in D_1 \cup \{b\}$ and all $d_2 \in D_2$. Consider any $d \in D_1 \cup \{b\}$ and $d_2 \in D_2$. Since $|\{v \in V \mid d_2 \geq_{3/4} a\}| \geq 3/4 \cdot |V|$ and $|\{v \in V \mid a \geq_{3/4} d\}| \geq 3/4 \cdot |V|$, the intersection of these two sets must contain at least $|V|/2$ elements, that is, there must be at least $|V|/2$ votes with "$\ldots > d_2 > \ldots > a > \ldots > d > \ldots$". Thus, the partial score of $\{d_2, d\}$ in l is at least as high as its partial score in l'. The partial score of every pair $\{a, d\}$ with $d \in D_1 \cup \{b\}$ in l' is strictly less than the partial score in l. Since $|D_1 \cup \{b\}| \geq 1$, the score of l' is smaller than the score of l and thus l cannot be a Kemeny ranking, a contradiction. $\square$

As a direct consequence of Lemma 1 we can partition the candidates of an election (V, C) as follows. Let $N := \{n_1, \ldots, n_t\}$ denote the set of non-dirty candidates with respect to the $\geq_{3/4}$-majority such that $n_i \geq_{3/4} n_{i+1}$ for $1 \leq i \leq t-1$. Then, $D_0 := \{d \in C \setminus N \mid d \geq_{3/4} n_1\}$, $D_i := \{d \in C \setminus N \mid n_i \geq_{3/4} d$ and $d \geq_{3/4} n_{i+1}\}$ for $1 \leq i \leq t-1$, and $D_t := \{d \in C \setminus N \mid n_t \geq_{3/4} d\}$.

3/4-Majority Rule. *Let (V, C) be an election and N and $D_0, \ldots, D_t$ be the sets of non-dirty and dirty candidates as specified above. Replace the original instance by the $t + 1$ subinstances induced by D_i for $i \in \{0, \ldots, t\}$.*

The soundness of the 3/4-Majority Rule follows directly from Lemma 1 and it is straightforward to verify its running time $O(nm^2)$. An instance reduced by the 3/4-Majority Rule contains only dirty candidates with respect to the original instance. Making use of a simple relation between the number of dirty candidates and the average KT-distance as also used previously [4], one can state the following.

Theorem 1. *Let d_a be the average KT-distance and n_d^s be the number of dirty pairs according to the $\geq_s$-majority for an $s \geq 3/4$. Then,* KEMENY SCORE *admits a partial kernel with less than $\min\{11d_a, 2n_d^s\}$ candidates which can be computed in $O(nm^2)$ time.*

Proof. It is easy to adapt the 3/4-Majority Rule to work for the decision problem: An instance is reduced by deleting all candidates from N, reordering every vote such that $D_0 > D_1 > \cdots > D_t$ where within D_i, $0 \leq i \leq t$, the order of the candidates remains unchanged, and decreasing the Kemeny score appropriately. It is not hard to verify that the 3/4-Majority Rule can be executed in $O(m^2 n)$ time by using a table storing the outcomes for all pairs of candidates. After applying the adapted reduction rule, only dirty candidates remain. Let their number be i. Since every dirty candidate must be involved in at least one candidate pair that is not ordered according to the $\geq_{3/4}$-majority, there must be at least $i/2$ candidate pairs that contribute with at least $|V|/4 \cdot 3|V|/4$ to the average KT-distance of the original input instance. By definition of the average KT-distance, it follows that

$$d_a \geq \frac{1}{|V|(|V|-1)} \cdot \frac{i}{2} \cdot \frac{|V|}{4} \cdot \frac{3|V|}{4} > \frac{3}{32} \cdot i \Rightarrow 11 \cdot d_a > i. \qquad \square$$

Table 2. Properties "induced" by $\geq_s$-majorities for different values of s

value of s	properties
$1/2 \leq s \leq 2/3$	a $\geq_s$-majority order does not necessarily exist (Example 1)
$2/3 < s < 3/4$	a $\geq_s$-majority order exists (follows from [4, Theorem 4])
	but a non-dirty candidate and a dirty candidate do not have to be ordered according to the $\geq_s$-majority in a Kemeny ranking (Theorem 2)
$3/4 \leq s \leq 1$	a $\geq_s$-majority order exists (follows from [4, Theorem 4])
	and in every Kemeny ranking every non-dirty candidate is ordered according to the $\geq_s$-majority with respect to all other candidates (Lemma 1)

Tightness results. We investigate to which $\geq_s$-majorities the results obtained for $\geq_{3/4}$-majorities extend. An overview of properties for a Kemeny ranking for different values of s is provided in Table 2.

For the $>_{2/3}$-majority, instances without dirty candidates are polynomial-time solvable [4]. More precisely, the $>_{2/3}$-majority order is a Kemeny ranking. A simple example shows that, for any $s \leq 2/3$, a $\geq_s$-majority order does not always exist (**Example 1**): Consider the election consisting of the three candidates a, b, and c and the three votes "$a > b > c$", "$b > c > a$", and "$c > a > b$". Here, $a \geq_{2/3} b$, $b \geq_{2/3} c$, and $c \geq_{2/3} a$. No linear order fulfills all three relations.

The existence of a data reduction rule analogously to the 3/4-Majority Rule for $\geq_s$-majorities for $s < 3/4$ would be desirable since such a rule might be more effective: There are instances for which a candidate is dirty according to the $\geq_{3/4}$-majority but non-dirty according to a $\geq_s$-majority with $s < 3/4$. Hence, for many instances, the number of dirty pairs according to the $\geq_{3/4}$-majority assumes higher values than it does according to smaller values of s. In the following, we discuss why an analogous s-Majority Rule with $s < 3/4$ cannot exist. The decisive point of the 3/4-Majority Rule is that, in a Kemeny ranking, every non-dirty candidate must be ordered according to the $\geq_{3/4}$-majority with respect to *every* other candidate. The following theorem shows that this is not true for $\geq_s$-majorities with $s < 3/4$. We omit the corresponding intricate construction.

Theorem 2. *Consider a $\geq_s$-majority for any rational $s \in \,]2/3, 3/4[$. For a non-dirty candidate x and a dirty candidate y, $x \geq_s y$ does not imply $x > y$ in a Kemeny ranking.*

2.2 Exploiting the Condorcet Property

We present a well-known data reduction rule of practical relevance and show that it reduces an instance at least as much as the 3/4-Majority Rule. The reduction rule is based on the following easy-to-verify observation.

Observation 1. *Let $C' \subseteq C$ be a candidate subset with $c' \geq_{1/2} c$ for every $c' \in C'$ and every $c \in C \setminus C'$. Then there must be a Kemeny ranking fulfilling $C' > C \setminus C'$.*

To turn Observation 1 into a reduction rule, we need a polynomial-time algorithm to identify appropriate "winning subsets" of candidates. We use the following simple strategy, called *winning subset routine*: For every candidate c, compute a minimal winning subset M_c by iteratively adding every candidate c' with $c' >_{1/2} c''$, $c'' \in M_c$, to M_c. After this, we choose a smallest winning subset.

Condorcet-Set Rule. *If the winning subset routine returns a subset C' with $C' \neq C$, then replace the original instance by the two subinstances induced by C' and $C \setminus C'$.*

It is easy to see that the Condorcet-Set Rule can be carried out in $O(nm^3)$ time. The following proposition shows that the Condorcet-Set Rule is at least as powerful as the 3/4-Majority Rule, implying that the Condorcet-Set Rule provides a partial kernel with less than $11d_a$ candidates.

Proposition 1. *An instance where the Condorcet-Set Rule has been exhaustively applied cannot be further reduced by the 3/4-Majority Rule.*

Proposition 1 shows that the 3/4-Majority Rule cannot lead to a "stronger" reduction of an instance than the Condorcet-Set Rule does. However, since the Condorcet-Set Rule has a higher running time, that is $O(nm^3)$ compared to $O(nm^2)$, applying the 3/4-Majority Rule before the Condorcet-Set Rule may lead to an improved running time in practice. This is also true for the consideration of the following "special case" of the Condorcet-Set Rule also running in $O(nm^2)$ time.

Condorcet Rule. *If there is a candidate $c \in C$ with $c \geq_{1/2} c'$ for every $c' \in C \setminus \{c\}$, then delete c.*

Our experiments will show that combining the Condorcet-Set Rule with the other rules significantly speeds up the practical running times for many instances. On the contrary, it is easy to see that other than the Condorcet-Set Rule the Condorcet Rule alone does not yield a partial kernel.

3   Experimental Results

To solve sufficiently small remaining parts of the instances left after the application of our data reduction rules, we implemented three exact algorithms. First, an extended version of the search tree algorithm showing fixed-parameter tractability with respect to the Kemeny score [3]. Second, a dynamic programming algorithm running in $O(2^m \cdot nm^2)$ time for m candidates and n votes [3,14]. Third, the integer linear program [5, Linear Program 3] which was the fastest exact algorithm in previous experimental studies [5,15]. We use the freely available ILP-solver GLPK[3] to solve the ILP.

Our algorithms are implemented in C++ using several libraries of the boost package. Our implementation consists of about 4000 lines of code. All experiments were carried out on a PC with 3 GHz and 4 GB RAM (CPU: Intel Core2Quad Q9550) running under Ubuntu 9.10 (64 bit) Linux. Source code and test data are available under the GPL Version 3 license under http://theinf1.informatik.uni-jena.de/kconsens/.

In the following, first we describe our results for two different types of web search data, and second we investigate instances obtained from sport competitions.

3.1   Search Result Rankings

A prominent application of RANK AGGREGATION is the aggregation of search result rankings obtained from different web search engines. We queried the same 37 search

---

[3] http://www.gnu.org/software/glpk/

Table 3. We describe the results for different combinations of the data reduction rules for the three instances corresponding to the search terms "blues", "gardening", and "classical guitar". The first column encodes the combination of reduction rules used: the first digit is "1" if the Condorcet-Set Rule is applied, the second if the Condorcet Rule is applied and the last digit is "1" if the $3/4$-Majority Rule is applied. We give the running times of data reduction in seconds and the profiles describing the result of the data reduction process. The profiles are to read as follows. Every "1" stands for a position for which a candidate was determined in a Kemeny ranking and higher numbers for groups of candidates whose "internal" order could not be determined by the data reduction rules. Sequences of i ones are abbreviated by 1^i. For example, for the search term "gardening" and the combination 111, we know the order of the best 54 candidates, we know the set of candidates that must assume positions 55-64 without knowledge of their relative orders, and so on.

	blues		gardening		classical guitar	
	time	profile	time	profile	time	profile
001	0.03	$1^2 > 5 > 1 > 101 > 1 > 2$	0.01	$1 > 2 > 1 > 102$	0.03	$1 > 114$
010	0.10	$1^{74} > 9 > 1^{29}$	0.05	$1^{54} > 43 > 1^9$	0.06	$1^6 > 92 > 1^{17}$
011	0.10	$1^{74} > 9 > 1^{29}$	0.05	$1^{54} > 43 > 1^9$	0.07	$1^6 > 92 > 1^{17}$
100	0.84	$1^{74} > 9 > 1^{29}$	0.95	$1^{54} > 20 > 1^3 > 9 > 1^{10} > 4 > 1^6$	1.89	$1^6 > 7 > 1^{50} > 35 > 1^{17}$
101	0.10	$1^{74} > 9 > 1^{29}$	1.03	$1^{54} > 20 > 1^3 > 9 > 1^{10} > 4 > 1^6$	2.03	$1^6 > 7 > 1^{50} > 35 > 1^{17}$
110	0.10	$1^{74} > 9 > 1^{29}$	0.10	$1^{54} > 20 > 1^3 > 9 > 1^{10} > 4 > 1^6$	0.19	$1^6 > 7 > 1^{50} > 35 > 1^{17}$
111	0.10	$1^{74} > 9 > 1^{29}$	0.11	$1^{54} > 20 > 1^3 > 9 > 1^{10} > 4 > 1^6$	0.18	$1^6 > 7 > 1^{50} > 35 > 1^{17}$

terms as Dwork et al. [7] and Schalekamp and van Zuylen [15] to generate rankings. We used the search engines Google, Lycos, MSN Live Search, and Yahoo! to generate rankings of 1000 candidates. We consider two search results as identical if their URL is identical up to some canonical form (cutting after the top-level domain). Results not appearing in all rankings are ignored. Ignoring the term "zen budism" with only 18 candidates, this results in 36 instances having between 55 and 163 candidates. We start with a systematic investigation of the performance of the individual reduction rules followed by describing our results for the web instances.

We systematically applied all combinations of reduction rules, always sticking to the following rule ordering: If applied, the Condorcet-Set Rule is applied last and the $3/4$-Majority Rule is applied first. After a successful application of the Condorcet-Set Rule, we "jump" back to the other rules (if "activated"). Examples for the combinations are given in Table 3. This led to the following observations.

First, surprisingly, the Condorcet Rule alone led to a stronger reduction than the $3/4$-Majority Rule in most of the instances whereas the $3/4$-Majority Rule never led to a stronger reduction than the Condorcet Rule. Second, for several instances the Condorcet-Set Rule led to a stronger reduction than the other two rules, for example, for gardening and classical guitar (see Table 3). It led to a stronger reduction for 14 out of the 36 instances and restricted to the 15 instances with more than 100 candidates (given in Table 4), it led to a stronger reduction for eight of them. The running times for the Condorcet-Set Rule in combination with the other rules are given in the left part of Fig. 1. Applying the Condorcet Rule before the Condorcet-Set Rule led to a significant speed-up. Additionally applying the $3/4$-Majority Rule changes the running time only marginally. Note that jumping back to the "faster" rules after applying the Condorcet-Set Rule is crucial to obtain the given running times. In the following, by "our reduction rules", we refer to all three rules applied in the order: $3/4$-Majority Rule, Condorcet Rule, and Condorcet-Set Rule.

Table 4. Web data instances with more than 100 candidates. The first column denotes the search term, the second the number of candidates, the third the running time of data reduction in seconds, and the last column the "profiles" remaining after data reduction.

search term	# cand.	time	structure of reduced instance
affirmative action	127	0.21	1^{27} $> 41 >$ 1^{59}
alcoholism	115	0.10	1^{115}
architecture	122	0.16	1^{36} $> 12 > 1^{30} > 17 >$ 1^{27}
blues	112	0.10	1^{74} $> 9 >$ 1^{29}
cheese	142	0.20	1^{94} $> 6 >$ 1^{42}
classical guitar	115	0.19	1^{6} $> 7 > 1^{50} > 35 >$ 1^{17}
Death+Valley	110	0.11	1^{15} $> 7 > 1^{30} > 8 >$ 1^{50}
field hockey	102	0.17	1^{37} $> 26 > 1^{20} > 4 >$ 1^{15}
gardening	106	0.10	1^{54} $> 20 > 1 > 1 > 9 > 1^{8} > 4 > 1^{9}$
HIV	115	0.13	1^{62} $> 5 > 1^{7} > 20 >$ 1^{21}
lyme disease	153	3.08	1^{25} $> 97 >$ 1^{31}
mutual funds	128	2.08	1^{9} $> 45 > 1^{9} > 5 > 1 > 49 >$ 1^{10}
rock climbing	102	0.07	1^{102}
Shakespeare	163	0.26	1^{100} $> 10 > 1^{25} > 6 >$ 1^{22}
telecommuting	131	1.60	1^{9} $> 109 >$ 1^{13}

For all instances with more than 100 candidates, the results of our reduction rules are displayed in Table 4: the data reduction rules are not only able to reduce candidates at the top and the last positions but also partition some instances into several smaller subinstances. Out of the 36 instances, 22 were solved directly by the reduction rules and one of the other algorithms in less than five minutes. Herein, the reduction rules always contributed with less than four seconds to the running time. For all other instances we still could compute the "top" and the "flop" candidates of an optimal ranking. For example, for the search term "telecommuting" there remains a subinstance with 109 candidates but we know the best nine candidates (and their order). The effectiveness in terms of top candidates of our reduction rules combined with the dynamic programming algorithm is illustrated in Fig. 1. For example, we were able to compute the top seven candidates for all instances and the top 40 candidates for 70 percent of the instances.

3.2   Impact Rankings

We generated rankings that measure the "impact in the web" of different search terms. For a search engine, a list of search terms is ranked according to the number of the hits of each single term. We used Ask, Google, MSN Live Search, and Yahoo! to generate rankings for all capitals, all nations, and the 103 richest people of the world.[4] Our biggest instance is built from a list of 1349 mathematicians.[5]

As for the capitals, in less than a second, our algorithms (reduction rules and any of the other algorithms for solving subinstances up to 11 candidates) computed the following "profile" of a Kemeny ranking: $1^{45} > 34 > 1^{90} > 43 > 1^{26}$ (see Table 3 for a

---

[4] http://en.wikipedia.org/wiki/List_of{capitals_by_countries, richest_people}

[5] http://aleph0.clarku.edu/~djoyce/mathhist/chronology.html

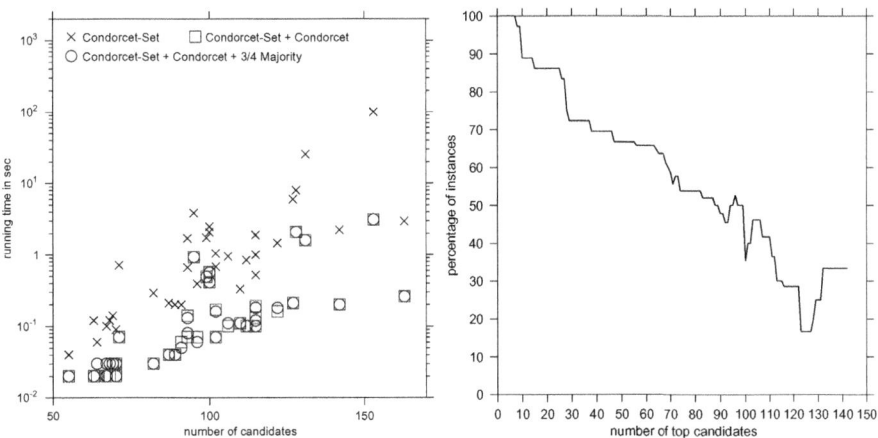

Fig. 1. Left: Running times of different combinations of reduction rules. To improve readability, we omitted the data points for the Condorcet-Set Rule combined with the 3/4-Majority Rule which was usually worse and in no case outperformed the best running times for the other combinations. Right: Percentage of the web search instances for which the x top candidates could be determined by data reduction and dynamic programming within five minutes. For a given number x of top positions, we only considered instances with at least x candidates.

description of the profile concept). The final Kemeny ranking starts as follows: London > Paris > Madrid > Singapore > Berlin > $\cdots$. For aggregating the nation rankings, our algorithms were less successful. However, we could still compute the top 6 and the flop 12 candidates. Surprisingly, the best represented nation in the web seems to be Indonesia, followed by France, the United States, Canada, and Australia. The instance consisting of the 103 richest persons could be solved exactly in milliseconds by the data reduction rules. In contrast, for the mathematicians we could only compute the top 31 and flop 31 candidates but could not deal with a subinstance of 1287 candidates between. For the mathematicians instance, the search strategy for minimal subsets for the Condorcet-Set Rule as given in Section 2 led to a running time of more than a day. Hence, we used a cutoff of 20 candidates for the size of the minimal subsets. This decreased the running time to less than one hour.

3.3   Winter Sport Competitions

For ski jumping and cross skiing, we considered the world cup rankings from the seasons 2005/2006 to 2008/2009,[6] ignoring candidates not appearing in all four rankings. Without data reduction, the ski jumping instance, consisting of 33 candidates, was solved by the ILP-solver GLPK in 103 seconds whereas the search tree and dynamic programming algorithms did not find a solution within five minutes. In contrast, the instance was solved in milliseconds by only applying the reduction rules. The cross skiing instance, consisting of 69 candidates, could not be solved without data reduction within five minutes by any of our algorithms but was reduced in 0.04 seconds such that one

---

[6] Obtained from http://www.sportschau.de/sp/wintersport/

component with 12 and one component with 15 candidates were left while all other positions could be determined by the reduction rules. The remaining parts could be solved, for example by the dynamic programming algorithm, within 0.12 and 0.011 seconds.

4   Conclusion

Our experiments showed that the described data reduction rules allow for the computation of optimal Kemeny rankings for real-world instances of non-trivial sizes within seconds. We could solve large instances with more than 50 candidates by means of data reduction while the corresponding instances could not be directly solved by the ILP, the previously fastest exact algorithm [5], or the two other implemented fixed-parameter algorithms. A key feature of the data reduction rules is to break instances into smaller, independent instances. A crucial observation in the experiments with the different data reduction rules regards certain cascading effects, that is, jumping back to the faster-to-execute rules after a successful application of the Condorcet-Set Rule significantly improves the running time. This shows that the order of applying data reduction rules is important. We could not observe a specific behavior of our data reduction rules for the different types of data under consideration. However, a further extension of the data sets and experiments in this direction seem interesting.

On the theoretical side, we improved the previous partial kernel [4] with respect to the parameter average KT-distance from quadratic to linear size. It is open whether there is a linear partial kernel with respect to the $\geq_s$-majority for any $s < 3/4$. A natural step in answering this question seems to investigate whether for two *non-dirty* candidates a, b, there must be a Kemeny ranking with $a > b$ if $a \geq_s b$. An important extension of RANK AGGREGATION is to consider "constraint rankings", that is, the problem input additionally contains a prespecified order of some candidate pairs in the consensus list [17]. Here, our data reduction rules cannot be applied anymore. New reduction rules to be developed for this scenario could also be used in "combination" with the search tree algorithm [3] in an "interleaving mode" [13].

Acknowledgements. We are grateful to the anonymous referees of *IPEC-2010* and of the *Third International Workshop on Computational Social Choice (COMSOC-2010)* for constructive feedback helping to improve this work.

References

1. Ailon, N., Charikar, M., Newman, A.: Aggregating inconsistent information: ranking and clustering. Journal of the ACM 55(5), Article 23 (2008)
2. Bartholdi III, J., Tovey, C.A., Trick, M.A.: Voting schemes for which it can be difficult to tell who won the election. Social Choice and Welfare 6, 157–165 (1989)
3. Betzler, N., Fellows, M.R., Guo, J., Niedermeier, R., Rosamond, F.A.: Fixed-parameter algorithms for Kemeny rankings. Theoretical Computer Science 410(45), 4554–4570 (2009)
4. Betzler, N., Guo, J., Komusiewicz, C., Niedermeier, R.: Average parameterization and partial kernelization for computing medians. Journal of Computer and System Sciences (2010), doi:10.1016/j.jcss.2010.07.005
5. Conitzer, V., Davenport, A., Kalagnanam, J.: Improved bounds for computing Kemeny rankings. In: Proc. of 21st AAAI, pp. 620–626. AAAI Press, Menlo Park (2006)

6. Davenport, A., Kalagnanam, J.: A computational study of the Kemeny rule for preference aggregation. In: Proc. of 19th AAAI, pp. 697–702. AAAI Press, Menlo Park (2004)
7. Dwork, C., Kumar, R., Naor, M., Sivakumar, D.: Rank aggregation methods for the Web. In: Proc. of 10th WWW, pp. 613–622 (2001)
8. Fagin, R., Kumar, R., Sivakumar, D.: Efficient similarity search and classification via rank aggregation. In: Proc. of 22nd ACM SIGMOD, pp. 301–312. ACM, New York (2003)
9. Jackson, B.N., Schnable, P.S., Aluru, S.: Consensus genetic maps as median orders from inconsistent sources. IEEE/ACM Transactions on Computational Biology and Bioinformatics 5(2), 161–171 (2008)
10. Karpinski, M., Schudy, W.: Approximation schemes for the betweenness problem in tournaments and related ranking problems. In: Proc. of 21st ISAAC (to appear, 2010)
11. Kenyon-Mathieu, C., Schudy, W.: How to rank with few errors. In: Proc. of 39th STOC, pp. 95–103. ACM, New York (2007)
12. Mahajan, M., Raman, V., Sikdar, S.: Parameterizing above or below guaranteed values. Journal of Computer and System Sciences 75, 137–153 (2009)
13. Niedermeier, R.: Invitation to Fixed-Parameter Algorithms. Oxford University Press, Oxford (2006)
14. Raman, V., Saurabh, S.: Improved fixed parameter tractable algorithms for two "edge" problems: MAXCUT and MAXDAG. Information Processing Letters 104(2), 65–72 (2007)
15. Schalekamp, F., van Zuylen, A.: Rank aggregation: Together we're strong. In: Proc. of 11th ALENEX, pp. 38–51. SIAM, Philadelphia (2009)
16. Simjour, N.: Improved parameterized algorithms for the Kemeny aggregation problem. In: Proc. of 4th IWPEC. LNCS, vol. 5917, pp. 312–323. Springer, Heidelberg (2009)
17. van Zuylen, A., Williamson, D.P.: Deterministic pivoting algorithms for constrained ranking and clustering problems. Mathematics of Operations Research 34, 594–620 (2009)
18. Young, H., Levenglick, A.: A consistent extension of Condorcet's election principle. SIAM Journal on Applied Mathematics 35(2), 285–300 (1978)

Enumerate and Measure: Improving Parameter Budget Management

Daniel Binkele-Raible and Henning Fernau

Univ.Trier, FB 4—Abteilung Informatik, 54286 Trier, Germany
{raible,fernau}@informatik.uni-trier.de

Abstract. Measure & Conquer (M&C) is the prominent technique for analyzing exact algorithms for computationally hard problems. It tries to balance worse and better situations within the algorithm analysis.

Several obstacles prevent the application of this technique in parameterized algorithmics, making it rarely applied in this area. However, these difficulties can be handled in some situations. We will exemplify this with two problems related to VERTEX COVER, namely CONNECTED VERTEX COVER and EDGE DOMINATING SET. For both problems, several parameterized algorithms have been published, all based on the idea of first enumerating minimal vertex covers and then producing solutions to the requested problem. Using M&C in this context will improve on the hitherto published running times, offering some unifying view. In contrast to some of the earlier suggested algorithms, ours will use polynomial space.

1 Introduction

Measure & Conquer (M&C) is nowadays the standard approach for developing and analyzing exact exponential-time algorithms for computationally hard problems, in particular, for graph problems, see [9] for an overview. It balances worse and better situations within the algorithm analysis. For example, there are now algorithms for MINIMUM DOMINATING SET with search tree sizes estimated only slightly larger than 1.5^n, while it was believed for quite some time that the naïve enumerative algorithm that considers 2^n cases cannot be improved.

M&C is less frequent in parameterized algorithmics. The reasons for this are various; let us make these clear with the example of VERTEX COVER in mind. (1) M&C works if one can gain a "part" of a vertex upon "first visiting" it, because the "next visit" would be in some nicer situation. This works fine as long as we measure the run time in, say, the number of vertices, n. However, if we measure in some parameter k upperbounding the size of a vertex cover, then it is hard to justify why or when we can "gain something" of some vertices which are not put into the cover. This would be like a credit on the future, and we should pay it back if we do not put this vertex later into the cover. This leads to problems since we should never increase the measure. (2) Having decided on the first issue, we have to be careful that there is a polynomial-time answer when this measure drops below zero. (3) We must ensure that the measure will not increase upon applying reduction rules (or restrict their use).

V. Raman and S. Saurabh (Eds.): IPEC 2010, LNCS 6478, pp. 38–49, 2010.

Hence, M&C techniques are rarely found in parameterized algorithmics. Some earlier papers combine a small problem kernel with an exact algorithm analyzed by M&C, but this approach is not feasible for CONNECTED VERTEX COVER, since no polynomial-size kernel exists for this problem, unless the polynomial hierarchy collapses to the third level, see [3]. Only few examples for such an amortized genuinely parameterized search-tree analysis are known, the first one being one on VERTEX COVER on subcubic graphs [1]. As it is well-known, VERTEX COVER possesses a linear vertex kernel. This is an important observation, since S. Gaspers formulated the following challenge in the *Parameterized Complexity News* from September 2009 (p.5): "However, it would be very interesting to see a parameterized M&C analysis for a problem that does not have a linear kernel." Our contribution delivers the first examples of a successful application of parameterized M&C to problems (most likely) having no linear kernels.[1]

We will overcome these difficulties with two problems related to VERTEX COVER: CONNECTED VERTEX COVER and EDGE DOMINATING SET. For both problems, several parameterized algorithms have been already developed, all based on the idea of first enumerating minimal vertex covers and then producing solutions to the requested problem. Using M&C in this context will improve on the hitherto published running times, offering some unifying view. In contrast to some of the earlier suggested algorithms, ours will use polynomial space.

We will freely use standard terminology from graph theory and from parameterized algorithms in what follows. Recall that we aim at obtaining a valid solution set (here: a connected vertex cover or an edge dominating set) of size at most k with an algorithm that is based on a search tree with at most c^k leaves, where $c > 1$ is some hopefully small constant. The amount of work to be done in the nodes of the search tree is polynomial-time only. This results in running times of $\mathcal{O}^*(c^k)$, referring to $\mathcal{O}(n^d \cdot c^k)$ (with d being a constant).

We should mention that for both problems, a parameterized algorithms' race[2] has been going on now for years, competing for the fastest run times. In [11], an $\mathcal{O}^*(6^k)$ algorithm for CONNECTED VERTEX COVER (CVC) has been established. This was improved in [6] to $\mathcal{O}^*(2.9316^k)$ and in [13] further to $\mathcal{O}^*(2.7606^k)$. Using the M&C approach, we will further lower the run time of parameterized algorithms for CVC to $\mathcal{O}^*(2.4882^k)$. All the mentioned algorithms consume polynomial space only. To the knowledge of the authors, no previous results have been published that solve the natural question whether MIN-CVC can be solved in time less than $\Theta^*(2^n)$, with n being the number of vertices of the input graph (which would correspond to a trivial enumeration algorithm). We solve this question affirmatively, exhibiting an alternative analysis of our algorithm that (also) runs in time $\mathcal{O}^*(1.8698^n)$.

For EDGE DOMINATING SET (EDS), E. Prieto in her PhD thesis [15] obtained a kernel of quadratic size for MINIMUM MAXIMAL MATCHING (i.e., INDEPENDENT EDGE DOMINATING SET), a problem that is basically the same as EDS,

---

[1] Regarding EDGE DOMINATING SET, D. Lokshtanov said: "Resolving whether edge dominating set has an $o(k^2)$ (vertex) kernel gives 100$ from me." according to the *Parameterized Complexity News* from September 2009 (p.1).

[2] Also testified by the *Parameterized Newsletter* edited by F. Rosamond.

see [20], which leads to an algorithm of run time $\mathcal{O}^*(c^{k^2})$ by naïve enumeration, following kernelization. This was considerably improved in [4] to $\mathcal{O}^*(2.6162^k)$ (using polynomial space). A further improvement for EDS was exhibited in [19], providing a parameterized analysis of an algorithm of Fomin *et al.* [8], yielding a run time of $\mathcal{O}^*(2.4178^k)$, with the same amount of space consumption. We will here describe another polynomial-space algorithm that runs in time $\mathcal{O}^*(2.3819^k)$.

Besides these technical contributions, this paper also offers some new methodological insights that reach beyond. Namely, in order to cope with the mentioned problems that seem to prevent the application of M&C techniques, previous papers used quite a descriptive meaning of the "credit" that could be taken "from the future": It was either bound to certain vertices or edges that would be re-visited later (examples are the algorithms for finding maximum forests in subcubic graphs or for finding spanning trees with a maximum number of internal vertices, see [7,5]), or it is clear (for example, in algorithms for finding spanning tree with a maximum number of leaves) that certain kinds of vertices (here, leaves) will show up in any valid solution later somewhere in the graph, see [2,12,16]. The present approach is different, as this localization of the credit is no longer possible.

2   Parameterized Branching and Measure & Conquer

The Measure & Conquer approach requires the definition of an appropriate measure that should express the "gain" we obtain by branching and by applying reduction rules. For example, consider the classical VERTEX COVER problem. Here, we are given a graph and a parameter k, and we are asked to find a set of at most k vertices whose removal will leave only isolated vertices. The usual way to handle a simple search tree algorithm that preferably branches on high-degree vertices and also exhaustively applies the well-known Degree-0, Degree-1 and Degree-2 Rules can be expressed as follows in the terminology of a *measure*:

$$\phi(G, k) = k - |C| - |F|$$

Here, k is the original parameter value (in contrast to the classical approach, k does not change during the run of the algorithm), C is the (partial) vertex cover solution found and fixed so far, and F is the set of Folding Rule applications (part of the Degree-2 Rule). At the very beginning, $C = F = \emptyset$, so $\phi = k$. During the run of the algorithm, C and F will increase, decreasing the measure accordingly. Clearly, when ϕ drops below zero, we face a NO-instance. This is far less clear if we use a more elaborate measure that might already reduce when, e.g., vertices of degree three are created, because "soon" reduction rules trigger or "finally" graphs of maximum degree two evolve, which are solvable in polynomial time. What if the measure drops below zero now?

We will discuss this in the following with two related problems: CONNECTED VERTEX COVER and EDGE DOMINATING SET. Due to space restriction, some proofs are omitted.

3 First Case Study: Connected Vertex Cover CVC

In this section we will present a parameterized algorithm for CONNECTED VERTEX COVER. Given a graph $G = (V, E)$, a set $V' \subseteq V$ of cardinality at most k is sought such that V' is a vertex cover which is connected at the same time. All previous approaches are based on listing all minimal vertex covers of size up to k. Then in a second phase, more vertices are added to the vertex cover to achieve connectedness. The same strategy is followed here. In addition, we use a more flexible measure which enables us to better balance the two phases.

As in the course of the forthcoming recursive algorithm vertices will be fixed to belong to the future solution we focus on an annotated version of our problem:

ANNOTATED CONNECTED VERTEX COVER (ACVC)
Given: $G = (V, E)$, a subset $Q \subseteq V$, and the parameter k.
We ask: Find $V' \subseteq V$ such that for all $e \in E$ we have $V' \cap e \neq \emptyset$, $|V'| \leq k$, $Q \subseteq V'$ and $G[V']$ is connected.

The next (easy) lemma will help us to justify some of the branching rules of the forthcoming algorithm.

Lemma 1. *Let V' be a connected vertex cover. Assume that $t \in V'$ is not a cut-vertex. If $G_t := G[V' \setminus \{t\}]$ consists of exactly two components V'_1 and V'_2 and $N(t) \subseteq V'$, then there is a connected vertex cover V'' with $t \notin V''$ and $|V''| \leq |V|$.*

Reduction Rules. We introduce the following reduction rules:

Deg1: Let $u \in V \setminus Q$ such that $N(u) = \{v\}$. Then delete u and set $Q := Q \cup \{v\}$.
CutVertex: Let $u \in V \setminus Q$ such that $G[V \setminus \{u\}]$ contains at least two components. Then set $Q := Q \cup \{u\}$.
Deg2a: Let $u \in V \setminus Q$ such that $\deg(u) = 2$. Then delete u and set $Q := Q \cup N(u)$.
Contract: If there are $u, v \in Q$ such that $\{u, v\} \in E$ then contract $\{u, v\}$ (substituting double edges by simple ones).

Deg2b: Let $u \in Q$ such that $N(u) = \{x_1, x_2\} \cap Q = \emptyset$. Then delete u and introduce the edge $\{x_1, x_2\}$.

As in the introductory VERTEX COVER example with the folding rule, the last two listed reduction rules actually decrease the parameter k of the ACVC instance with the classical understanding of an analysis of a parameterized algorithm; here, this effect is modeled by keeping k fixed (in the measure) and subtracting the number of these rule applications. With this understanding, the following is not hard to derive.

Lemma 2. *The reduction rules are sound.*

Let $\mathcal{B}_{ij} := \{u \in V \setminus Q \mid |N(u) \cap Q| = i, |N(u)| = j\}$ and $\mathcal{B}^{\geq}_{ij} := \bigcup\limits_{i' \geq i, j' \geq j} \mathcal{B}_{i'j'}$.

Note that in a reduced instance $\mathcal{B}_{01} = \mathcal{B}_{11} = \mathcal{B}_{02} = \mathcal{B}_{12} = \mathcal{B}_{22} = \emptyset$ due to reduction rules **Deg1**, **CutVertex** and **Deg2a**. Now, we can present Algorithm 1. Observe that we can solve CVC by calling Algorithm 1 for all $h \in V$ with the sets $Q = \{h\}$ and $I = \emptyset$ (I is only used for reasons of bookkeeping).

Algorithm 1. An Algorithm for ANNOTATED CONNECTED VERTEX COVER

Input: A connected graph $G = (V, E)$, a set $\emptyset \subsetneq Q \subseteq V$, an integer k and a set $I \subseteq V$.
Output: A set $V' \subset V$ such that $Q \subseteq V'$, $G[V']$ is connected and V' is a vertex cover.
Procedure: SolveCVC(G, Q, k, I)

1: Apply **Deg2b** on every $u \in I$ (delete u and add $\{x_1, x_2\}$ where $N(u) = \{x_1, x_2\}$).
2: $I := \emptyset$.
3: Apply the reduction rules **CutVertex**, **Deg1**, **Deg2a** and **Contract** exhaustively with priorities corresponding to the given order.
4: **if** $|Q| + \#\{$**Contract** applications$\} + \#\{$**Deg2b** applications$\} > k$ **or** $G(V, E)$ is disconnected **then**
5:     **return** NO
6: **else if** $V = Q$ **then**
7:     **return** YES.
8: **else if** $\exists v \in \mathcal{B}_{14}^{\geq}$ such that $N(v) \setminus Q \neq \emptyset$ **then** {Begin Phase I}
9:     Branch binary by a) setting $Q := Q \cup \{v\}$
        b) deleting v and setting $Q := Q \cup N(v)$.
10: **else if** $\mathcal{B}_{13} \cup \mathcal{B}_{23} \neq \emptyset$ **then**
11:     Choose $v \in (\mathcal{B}_{13} \cup \mathcal{B}_{23})$ according to the next priorities:
12:         1. I. $\exists \{x_1, x_2\} \in E$ (s.t. $N(v) \setminus Q = \{x_1, x_2\}$) or II. $\exists u \in (N(v) \setminus Q) : \deg(u) = 3$.
13:         Branch binary by a) setting $Q := Q \cup \{v\}$;   b) deleting v, setting $Q := Q \cup N(v)$ and $I := \{u \mid u \in N(v) \wedge |N(u) \setminus \{v\}| = 2 \wedge N(u) \cap Q = \emptyset\}$.
14:         2. $v \in \mathcal{B}_{23}$
15:             Choose $u \in (N(v) \setminus Q)$ and branch binary by
                a) setting $Q := Q \cup \{u\}$ b) deleting u and setting $Q := Q \cup N(u)$.
16:         3. $v \in \mathcal{B}_{13}$ and thus $N(v) \setminus Q = \{b, c\}$ with $b \neq c$.
17:             3.1 $\exists r \in N(b) \cap N(c)$ where $r \neq v$.
18:             a) Delete v, $Q := Q \cup \{b, c\}$;   b) Delete b, $Q := Q \cup \{v\} \cup N(b)$;
                c) Delete c, $Q := Q \cup \{v, b\} \cup N(c)$ d) Delete r, $Q := Q \cup \{v, b, c\} \cup N(r)$
19:             3.2 $N(b) \cap N(c) = \{v\}$.
20:             a) Delete v, $Q := Q \cup \{b, c\}$;   b) $Q := Q \cup \{v, b\}$;
                c) Delete b, $Q := Q \cup \{v, c\} \cup N(b)$;   d) Delete b, c, $Q := Q \cup \{v\} \cup N(b) \cup N(c)$.
21: **else** {Begin Phase II}
22:     Apply the Steiner-Tree-Algorithm of [14] with Q as the terminal set.
23: **end if**

Correctness. The lemma below is needed as an intermediate step towards demonstrating the correctness of Algorithm 1.

Lemma 3. *In step 8 of Alg. 1, we have that a) $Q \neq \emptyset$ and b) G is connected.*

Lemma 4. *Algorithm 1 solves ACVC correctly.*

Proof. By Lemma 2, the reduction rules are sound. We now show that if no case in Phase I applies, then the Steiner-Tree-Algorithm in Phase II can be used. Firstly, if the first part of the or-statement in step 4 applies then the size of the vertex cover is greater than k. Note that any **Contract** or **Deg2b**-application implicitly puts a vertex (which is not present in the current graph) in the final vertex cover. Thus, we correctly answer NO. Secondly, if $V = Q$ then by **Contract** and Lemma 3.b) we have $|V| = 1$. Thus, we have a trivial instance and can return YES.

Otherwise, there is a vertex $h \in V \setminus Q$. Additionally we can require that $N(h) \cap Q \neq \emptyset$ by Lemma 3. If there is at least one such vertex h where $N(h) \not\subseteq Q$ then we have $h \in \bigcup_{z,\ell:\ell>\max\{2,z\}} \mathcal{B}_{z\ell}$ due to the reduction rules. Thus, Phase I applies. If for all such h we have $N(h) \subseteq Q$, then Q is an independent vertex cover by **Contract**. The final task is to find some minimum cardinality set $B \subseteq V \setminus Q$ such that $G[Q \cup B]$ is connected. This is now the Steiner-Tree problem and therefore we apply an appropriate existing algorithm for this task (Phase II).

Any branching is exhaustive except the one in step 18.

We turn to step 18. Here in the last recursive call d) we delete r and therefore any solution C with $v, b, c, r \in C$ is not considered. But notice that by Lemma 1 a solution $\tilde{C}$ no greater in size is guaranteed with $v \notin \tilde{C}$ as $G(C \setminus \{v\})$ consists of two components. $\tilde{C}$ is found in the recursive call a) in step 18.   **Q.E.D.**

Run Time Analysis. We use a rather simple measure:

$$\phi(G, Q) := k - \omega \cdot |Q| - c \text{ where } \omega \leq 0.5$$

Here $c = \#\{\textbf{Contract} \text{ applications}\} + \#\{\textbf{Deg2b} \text{ applications}\}$, i.e., it counts the number of **Contract** and **Deg2b** applications. Observe that c counts the number of vertices of the original instance fixed to be in Q which are not present anymore in the current instance. Therefore, a decrease of the initial budget k by an amount of c is justified. The following lemma is important yet clear.

Lemma 5. *The reduction rules do not increase the measure* $\phi(G, Q)$.

Here we like to point out the following observation: Once we have $\phi(G, Q) < 0$ during our algorithm, line 4 of Algorithm 1 applies and YES is returned. Thus, $\phi(G, Q)$ indeed can be used to derive a run time of the form $\mathcal{O}^*(c^k)$.

Phase II. J. Nederlof [14] showed that the STEINER TREE problem can be solved in time $\mathcal{O}^*(2^\ell)$, where $\ell = |Q|$ is the number of given terminals, using polynomial space only. But note that by

$$\phi(G, Q) := k - \omega \cdot \ell - c \geq k - \omega \cdot \ell - (k - \ell) = (1 - \omega) \cdot \ell$$

we can upper bound step 22 in Algorithm 1 by $\mathcal{O}^*(2^{\frac{1}{1-\omega}\phi(G,Q)})$. Observe that we used the fact that $k - \ell \geq c$, which is due to step 4.

Phase I. We further have to find the correct branching vectors in case we are branching in Algorithm 1:

Step 9. For $v \in \mathcal{B}_{ij}$, we derive the branching vector $(i \cdot (1 - \omega) + \omega, (j - i) \cdot \omega)$, where $i \geq 1, j \geq 4$ and $i < j$. Observe that we only need to consider branching vectors up to $j = 4$. Any branching vector where $j > 4$ is dominated by one of the latter.

Step 13. I. If $N(v) \setminus Q = \{x_1, x_2\}$ and $\{x_1, x_2\} \in E$, then in the second branch **Contract** is triggered on x_1 and x_2, resulting in a reduction of $1 + w$ with respect to $\phi(G, Q)$. Thus, a $(1, 1 + w)$ branching vector evolves (note $I = \emptyset$). II(a). Assume $v \in \mathcal{B}_{13}$ and (due to the previous case) $\{x_1, x_2\} \notin E$. Let $x_1 \in \mathcal{B}_{s3}$ where $0 \le s \le 2$. Then the following branching vectors are entailed if $s \ge 1$: $((1 - w) + w, 2w + (1 - w))$ and if $s = 0$: $((1 - w) + w, w + 1)$. Note that in the last branching vector in the second case **Deg2b** applies in the next recursive call as $I \ne \emptyset$.
    II(b). Assume $v \in \mathcal{B}_{23}$. Consider $\{x_1\} = N(v) \setminus Q$, $x_1 \in \mathcal{B}_{s3}$. By the same analysis as in the previous case we arrive at the following branching vectors: If $s \ge 1$, $(2 \cdot (1 - w) + w, w + (1 - w))$ and if $s = 0$, $(2 \cdot (1 - w) + w, 1)$.

Step15. Note that $\deg(u) \ge 4$ due to step 13 and that $N(u) \cap Q = \emptyset$ due to step 9. Hence, the branching vector $(w, 4w + 2 \cdot (1 - w))$ is entailed. Note the application of **Contract** in the second part of the branch.

Step 18. The previous branching cases imply $\deg(b) \ge 4$, $\deg(c) \ge 4$ and $(N(b) \cup N(c)) \cap Q = \emptyset$. Firstly, suppose $N(r) \cap Q \ne \emptyset$. Then a branching vector $(2w, 2 + 2w, 3 + 2w, 3)$ follows due to the application of **Contract** with respect to r in the second and third part of the branching. Secondly, if $N(r) \cap Q = \emptyset$ we get a branching vector $(2w, 1 + 3w, 2 + 3w, 3 + w)$. Note that the additional amount of w in the last entry is due to the fact that $|N(r) \setminus (\{c, b\} \cup Q)| \ge 1$ because of **Deg2a**.

Step 20. This case yields a straight-forward $(2w, 2, 2 + 3w, 1 + 6w)$ branching vector as $N(b) \cap N(c) = \{v\}$.

On the basis of the above discussion, we get the main result of this section.

Theorem 1. CONNECTED VERTEX COVER *can be solved in time* $\mathcal{O}^*(2.4882^k)$ *and* $\mathcal{O}^*(1.8658^n)$, *using polynomial space.*

Proof. By choosing $w = 0.23956$ the maximum branching number of the above branching vectors is 2.4882. Additionally, by the choice of w the run time of Phase II can be upper bounded by $\mathcal{O}^*(2.4882^k)$. To achieve the run time upper bound in terms of n we use a method of [17]: By invoking Algorithm 1 for every $1 \le k' \le n/2 + \alpha$ where $\alpha = 0.1842n$ and by iterating over any non-separating independent set with maximum size $n/2 - \alpha$ we get the bound of $\mathcal{O}^*(1.8658^n)$. **Q.E.D.**

In this case study we have seen that with some moderate effort, we could achieve an improved run time for an already broadly studied problem. This was possible even though the basic method did not change substantially. We first branched towards creating a vertex cover and then in a second phase additional vertices had to be added for the sake of connectivity. This case study showed what an impact a proper chosen measure can have for the understanding of the problem and the designed algorithm. Let us also mention the cases which determine the run time: 1. Step 20 and 2. Phase II.

4   Second Case Study: Edge Dominating Set

An *edge dominating set* is a set of edges D with the property that every other edge shares at least one endpoint with one edge from D. Note that any such set $E' \subset E$ has the property that $V(E') = (\bigcup_{e' \in E'} e')$ is a vertex cover of size at most $2|E'|$. This fact was used by H. Fernau [4] to design an $\mathcal{O}^*(2.6181^k)$-algorithm. This algorithm is based on enumerating all minimal vertex covers of size no more than $2k$. A similar exponential time algorithm was developed by J. M. M. van Rooij and H. L. Bodlaender [19] consuming $\mathcal{O}^*(1.3226^n)$ time. We basically analyze a parameterized variant of their algorithm enriched by two special branching cases. Thus, during the course of the algorithm we handle a set L of vertices which is supposed to be covered by the final edge dominating set. Therefore, we present an annotated version of the problem.

k-ANNOTATED EDGE DOMINATING SET (k-AEDS)
Given: $G = (V, E)$, a set $L \subseteq V$, and the parameter k.
We ask: Find $E' \subseteq E$, $|E'| \leq k$, $L \subseteq V(E')$, with $e \cap V(E') \neq \emptyset$ for any $e \in E$.

We call a given annotated instance *redundant* if $G[V \setminus L]$ exclusively consists of components $C_1, \ldots, C_\ell$ which are either single vertices or edges, i.e., these components are K_1's and K_2's.

Lemma 6. *[4] A minimum cardinality solution to a redundant instance of k-AEDS can be found in polynomial time.*

We now present Algorithm 2, which lists all minimal vertex covers VC of size at most $2k$ until a redundant instance is reached. This redundant instance is solved in polynomial time using Lemma 6, yielding a solution $S \subseteq E$. Note that U and U^s form the set L given in the definition of k-AEDS.

Correctness. Notice that any recursive branching step in Algorithm 2 is exhaustive except the one in step 7. Here case a) should be considered. In additional to p_1, also p_3 is adjoined to U which has to be justified. Note that if $p_3 \notin U$ then it follows that p_2, p_4 have to be adjoined yielding a non-minimal solution. If both $p_1, p_3 \subset U$ then p_2, p_4 have to be deleted for the sake of minimality.
    Notice also that vertices that are not taken into the vertex cover during the enumeration phase (and hence go into the complement, i.e., an independent set) can be deleted whenever it is guaranteed that all their neighbors (in the current graph) are moved into the vertex cover.
    Let $X^\ell(B) := \{\{p_1, \ldots, p_\ell\} \mid p_1, \ldots, p_\ell$ is a path component in $G[V \setminus B]\}$. For example, $X^1(B)$ collects the isolated vertices in $G[V \setminus B]$. In a sense, $\ell = 1$ is the smallest meaningful value of ℓ; however, the case $\ell = 0$ will be useful as a kind of boundary case in the following. The measure is defined as follows:

$$\lambda(G, U, U^s) := 2k - \omega \cdot |U| - (2 - \omega)|U^s| \tag{1}$$
$$- \chi(\forall v \in V \setminus (U \cup U^s) : d_{G[V \setminus (U \cup U^s)]}(v) \leq 2) \cdot \left(\sum_{i \geq 0} \gamma_i \cdot |X^i(U \cup U^s)| \right)$$

Algorithm 2. An Algorithm for k-Annotated Edge Dominating Set

---

Input: A graph $G = (V, E)$ and the parameter k.
Output: YES if a solution to k-AEDS exists, NO otherwise.

SolveEDS($G,L,\emptyset$); this procedure is described in the following.

Procedure: SolveEDS(G, U, U^s)

1: **if** $\lambda(G, U, U^s) < 0$ (defined in Eq. (1)) **then**
2:     **return** NO
3: **else if** $\exists v \in V \setminus (U \cup U^s)$ such that $d_{G[V \setminus (U \cup U^s)]}(v) \geq 3$ **then** {Begin Phase I}
4:     Branch by a) $U := U \cup \{v\}$ and b) Delete v, $U := U \cup N(v)$.
5: **else if** $G[V \setminus (U \cup U^s)]$ has a cycle comp. $p_1, \ldots, p_\ell$, $\ell \geq 3$ **then** {Begin Phase II}
6:     **if** $\ell = 4$ **then**
7:         Branch a) Delete p_2, p_4, $U := U \cup \{p_1, p_3\}$; b) Delete p_1, $U := U \cup \{p_2, p_4\}$.
8:     **else if** $\ell = 6$ **then**
9:         Branch a) $U := U \cup \{p_1, p_4\}$; b) Delete p_4, $U := U \cup \{p_1, p_3, p_5\}$; c) Delete p_1,
        $U := \{p_2, p_6\}$.
10:     **else**
11:         Branch a) $U := U \cup \{p_1\}$; b) Delete p_1, $U := U \cup \{p_2, p_\ell\}$.
12:     **end if**
13: **else if** $\exists$ a path component $p_1, \ldots, p_\ell$, $\ell \geq 4$ in $G[V \setminus (U \cup U^s)]$ **then**
14:     Branch a) $U := U \cup \{p_3\}$; b) Delete p_3, $U := U \cup \{p_2, p_4\}$.
15: **else if** $\exists$ a path component $p_1, \ldots, p_3$ in $G[V \setminus (U \cup U^s)]$ **then** {Begin Phase III}
16:     Branch a) $U^s := U^s \cup \{p_2\}$; b) Delete p_2, $U^s := U^s \cup \{p_1, p_3\}$.
17: **else** {Begin Phase IV}
18:     Solve the problem in polynomial time according to Lem. 6.
19: **end if**

---

where $\omega = 0.8909$, $\gamma_0 = \gamma_1 = 0$, $\gamma_2 = 1$, $\gamma_3 = 0$, $\gamma_4 = 0.2365$ and $\gamma_j = 0.3505$ for $j \geq 5$ and χ is the indicator function. Note that the initial budget $2k$ is only decreased by adjoining vertices to U or U^s (and k remains fixed). Note that once $G[V \setminus U]$ consist only of three-vertex-paths the vertices adjoined to the vertex cover are collected in U^s (and no longer in U). Observe that any vertex in U is only counted by a fraction of ω instead of 1. On the other hand, the vertices in U^s are counted by $2 - \omega > 1$. Thus, the U^s-vertices are overestimated. But it is also true that the vertices from a three-vertex-path are surrounded by U-vertices. In the analysis we show that the overestimation of the U^s-vertices can be compensated by the underestimation of the U-vertices due to this fact.

In contrast to the previous case study, it is now not so clear how to answer if the measure drops below zero. This is handled by the following lemma.

Lemma 7. *If in step 1 of Algorithm 2 $\lambda(G, U, U^s) < 0$, then there is no annotated edge dominating set ED extending $L := U \cup U^s$ with $|ED| \leq k$.*

Proof. Suppose the contrary and let ED be a solution to k-AEDS with c components. Let $V_{ED} = (\bigcup_{e' \in ED} e')$ and note that $L \subseteq V_{ED}$ (where here $L = U \cup U^s$). Observe that, w.l.o.g., $G[ED]$ is acyclic and hence $|V_{ED}| \leq k + c$. Let $B_1, \ldots, B_c$ be the components of $G[ED]$. Let $C^s := \{B_i \mid V(B_i) \cap U^s \neq \emptyset\}$

and $m : V_{ED} \rightarrow (V_{ED} \cup \{-1\})$ be defined as follows: 1. For any $B_i \in C^s$ pick exactly one $x \in V(B_i) \cap U^s$ and let its image $m(x)$ be an arbitrary neighbor of x in $G[ED]$. 2. The image of any other vertex is -1 (where -1 is not a name of any element of V_{ED}). This notation easily extends to sets by setting $m(B) = \{m(v) \mid v \in B\}$. Let $\mathcal{M} = \{u, v \in V_{ED} \mid m(u) = v\}$ and observe that $|C^s| = |\frac{\mathcal{M}}{2}|$. If $\lambda(G, U, U^s) < 0$ happens after a recursive call of Phase II (but not Phase III) we have: Since ED is an edge dominating set, we know that for each $i \geq 2$ and for all paths $p_1, \ldots, p_i \in X^i(U \cup U^s)$, there exists some $1 \leq j \leq i$ such that $p_j \in V_{ED}$. Hence, $|X^i(U \cup U^s)| \leq |V(X^i(U \bigcup U^s)) \cap V_{ED}|$ ($\bigstar_1$). Notice that, since we are in Phase II, the term $\chi(\forall v \in V \setminus (U \cup U^s) : d_{G[V \setminus (U \cup U^s)]}(v) \leq 2)$ equals one. Then we deduce: $2k < \omega \cdot |U| + (\bigcup_{i \geq 0} \gamma_i \cdot |X^i(U \cup U^s)|) \overset{\bigstar_1, \bigstar_4}{\leq} |U| +$
$(\bigcup_{i \geq 2} \gamma_i \cdot |V(X^i(U \cup U^s)) \cap V_{ED}|) \leq |U| + (\bigcup_{i \geq 2} |V(X^i(U \cup U^s)) \cap V_{ED}|) \overset{\bigstar_5}{\leq} |V_{ED}|$.
We were also using that $\gamma_0 = \gamma_1 = 0$ ($\bigstar_4$) and that ED covers $U \cup U^s$ ($\bigstar_5$).

So we should discuss in details if $\lambda(G, U, U^s) < 0$ happens after a recursive call of the third Phase (but not in Phase IV). Consider now $u \in U^s$ with $m(u) \neq -1$; then there exists a path p_1, p_2, p_3 in $G[V \setminus U]$. Firstly, suppose $u = p_1$. Thus, u has been adjoined to U^s in case $b)$ of step 16. Since p_2 got deleted, $p_2 \notin V_{ED}$. By $p_2 \notin V_{ED}$ we have $m(u) \in U$. Thus, summing up for u and $m(u)$ the vertices are counted by two in V_{ED} whereas we decremented λ by $\omega + (2 - \omega) = 2$. ($\bigstar_2$). Secondly, if $u = p_2$ then if $m(p_2) \in U$ the same reasoning applies. Otherwise, w.l.o.g., $m(p_2) = p_1$ (note that $(N(p_2) \setminus \{p_1, p_3\}) \subseteq U$). Then, in λ both vertices together were counted by a fraction of $(2 - \omega)$ where in V_{ED} they both together have been counted by two ($\bigstar_3$). This seems to be an invalid argument at first glance, but will be justified below. All this has to be considered more formally. Let $\mathcal{R} := \{x \in U^s \mid m(x) = -1\}$ and note that $\mathcal{R} = U^s \setminus \{z \in U^s \mid m(z) \in U \text{ or } m(z) \in (V(X^1(U \cup U^s) \cap V_{ED})\}$ and that $\mathcal{M} \setminus U^s = ((m(U^s) \cap U) \cup (m(U^s) \cap (V(X^1(U \cup U^s) \cap V_{ED}))$ as we are facing Phase III and because of $\bigstar_2$ and $\bigstar_3$. Also note that $\mathcal{R} \cap \mathcal{M} = \emptyset$ and therefore $|\mathcal{R}| \leq (|E(G[C^s])| + |C^s| - |\mathcal{M}|) = |E(G[C^s])| - |C^s| \leq k - c$ ($\bigstar_7$).

$$2k \quad < \quad \omega \cdot |U| + (2 - \omega)|U^s| + (\bigcup_{i \geq 0} \gamma_i \cdot |X^i(U \cup U^s)|)$$

$$\overset{\bigstar_1, \bigstar_4, \bigstar_6}{\leq} \quad \omega \cdot |U| + (2 - \omega)|U^s| + |\overbrace{(V(X^2(U \cup U^s)) \cap V_{ED})}^{\mathcal{C}}|$$

$$\overset{\bigstar_2, \bigstar_3}{\leq} \quad |U| + |\overbrace{m(U^s) \cap U}^{\mathcal{A}}| + |\overbrace{m(U^s) \cap (V(X^1(U \cup U^s) \cap V_{ED})}^{\mathcal{B}}| + |V(X^1(U \cup U^s) \cap V_{ED}| + 2 \cdot |\mathcal{R}| + |\mathcal{C}|$$

$$\overset{\bigstar_7}{\leq} \quad |U| + |\mathcal{A}| + |\mathcal{B}| + |\mathcal{C}| + |\mathcal{R}| + |V(X^1(U \cup U^s) \cap V_{ED}| + (k - c)$$

$$\overset{\bigstar_9}{\leq} \quad |V_{ED}| + (k - c)$$

In Phase III, we have $\bigcup_{i \geq 4} X^i(U \cup U^s) = \emptyset$ and $\gamma_0 = \gamma_1 = \gamma_3 = 0$ ($\bigstar_6$). Note that $V_{ED} \supset U \cup \mathcal{R} \cup m^{-1}(\mathcal{A} \cup \mathcal{B})$ and $m^{-1}(\mathcal{A})$, $m^{-1}(\mathcal{B})$ and $\mathcal{R}$ are pairwise disjoint (✦$_9$). We explain the third last inequality more detailed. Observe that any vertex t from $m^{-1}(\mathcal{A} \cup \mathcal{B})$ is counted by an amount of one in $|V_{ED}|$ but by an amount of $(2 - \omega)$ in λ. This presents a potential danger. Note that $m(t) \in (\mathcal{A} \cup \mathcal{B})$ and $U^s = m^{-1}(\mathcal{A} \cup \mathcal{B}) \cup \mathcal{R}$. Either t finds its partner in U or in $V(X^1(U \cup U^s)) \cap V_{ED}$. The drop of $(1 - \omega)$ with respect to t is compensated by the increase due to $m(t)$.

$$(1-\omega)|m^{-1}(\mathcal{A} \cup \mathcal{B})| = (1-\omega)(|\mathcal{A}| + |\mathcal{B}|) \le (1-\omega)(|U| + |V(X^1(U \cup U^s) \cap V_{ED}|).$$

By the inequality above we finally derive $k + c < |V_{ED}|$, a contradiction. **Q.E.D.**

Run Time Analysis. This analysis is rather straightforward in this case study and hence omitted.

Theorem 2. EDGE DOMINATING SET *can be solved in time* $\mathcal{O}^*(1.5433^{2k}) \subseteq \mathcal{O}^*(2.3819^k)$ *consuming polynomial space.*

The last run time analysis showed once more that with an appropriate measure a better run time can be proven. This method also allowed us to identify a cycle of length six in $G[V \setminus (U \cup U^s)]$ (step 9 in Algorithm 2) as the critical case in the spirit of *design by M&C* [18]. Therefore, a special branching rule for this case has been invented. This is a further point where we gained on the run time.

5   Conclusions

We have focused on the development of parameterized algorithms for two well-studied problems, based on the M&C paradigm. We achieved run time upper-bounds considerably better compared to previous attempts, although we did not change the principal ideas of the underlying algorithms.

Let us point to the similarity of both considered problems, CONNECTED VERTEX COVER and EDGE DOMINATING SET, not only from the algorithms that have been presented for both problems. Namely, if G admits a connected vertex cover of size k, then it also admits a connected edge dominating set (also known as a *tree cover*) of size $k - 1$ and vice versa. This was observed before, see, e.g., [10]. It allows us to compare the run time of both algorithms: Namely, it can be assumed that the size of a minimum edge dominating set is considerably smaller than that of a minimum tree cover, so that our algorithm for EDGE DOMINATING SET will show results much faster than our algorithm for CONNECTED VERTEX COVER, assuming that both are applied to the same graph and used to find a minimum solution (by gradually increasing the parameter).

The authors of [19] showed how to solve EDGE DOMINATING SET in time $\mathcal{O}^*(2.4178^k)$ with the same space consumption. Their approach can be used to obtain a further slight run time improvement.

Theorem 3. *Provided exponential space* EDGE DOMINATING SET *can be solved in time* $\mathcal{O}^*(2.373^k)$.

Our EDGE DOMINATING SET algorithm can be modified to cope with rational weights $w_e \ge 1$ associated to an edge e, as described in [4], again using polynomial space only. Namely, weighted redundant instances remain polynomial time solvable and $\lambda(G, U, U^s)$ can be also used for its run time analysis.

References

1. Chen, J., Kanj, I.A., Xia, G.: Labeled search trees and amortized analysis: improved upper bounds for NP-hard problems. Algorithmica 43, 245–273 (2005)
2. Daligault, J., Gutin, G., Kim, E.J., Yeo, A.: FPT algorithms and kernels for the directed k-leaf problem. J. Comput. Syst. Sci. 76, 144–152 (2010)
3. Dom, M., Lokshtanov, D., Saurabh, S.: Incompressibility through colors and IDs. In: ICALP 2009, Part I. LNCS, vol. 5555, pp. 378–389. Springer, Heidelberg (2009)
4. Fernau, H.: Edge dominating set: efficient enumeration-based exact algorithms. In: Bodlaender, H.L., Langston, M.A. (eds.) IWPEC 2006. LNCS, vol. 4169, pp. 142–153. Springer, Heidelberg (2006)
5. Fernau, H., Gaspers, S., Raible, D.: Exact and parameterized algorithms for max internal spanning tree. In: Paul, C., Habib, M. (eds.) Graph-Theoretic Concepts in Computer Science. LNCS, vol. 5911, pp. 100–111. Springer, Heidelberg (2010)
6. Fernau, H., Manlove, D.F.: Vertex and edge covers with clustering properties: Complexity and algorithms. J. Disc. Alg. 7, 149–167 (2009)
7. Fernau, H., Raible, D.: Exact algorithms for maximum acyclic subgraph on a superclass of cubic graphs. In: Nakano, S.-i., Rahman, M. S. (eds.) WALCOM 2008. LNCS, vol. 4921, pp. 144–156. Springer, Heidelberg (2008)
8. Fomin, F.V., Gaspers, S., Saurabh, S., Stepanov, A.A.: On two techniques of combining branching and treewidth. Algorithmica 54, 181–207 (2009)
9. Fomin, F.V., Grandoni, F., Kratsch, D.: A measure & conquer approach for the analysis of exact algorithms. J. ACM 56(5) (2009)
10. Fujito, T., Doi, T.: A 2-approximation NC algorithm for connected vertex cover and tree cover. Inform. Process Lett. 90, 59–63 (2004)
11. Guo, J., Niedermeier, R., Wernicke, S.: Parameterized complexity of vertex cover variants. Theory Comput. Syst. 41, 501–520 (2007)
12. Kneis, J., Langer, A., Rossmanith, P.: A new algorithm for finding trees with many leaves. In: Hong, S.-H., Nagamochi, H., Fukunaga, T. (eds.) ISAAC 2008. LNCS, vol. 5369, pp. 270–281. Springer, Heidelberg (2008)
13. Mölle, D., Richter, S., Rossmanith, P.: Enumerate and expand: Improved algorithms for connected vertex cover and tree cover. Theory Comput. Syst. 43, 234–253 (2008)
14. Nederlof, J.: Fast polynomial-space algorithms using Möbius inversion: Improving on Steiner tree and related problems. In: ICALP 2009, Part I. LNCS, vol. 5555, pp. 713–725. Springer, Heidelberg (2009)
15. Prieto, E.: Systematic Kernelization in FPT Algorithm Design. PhD thesis, The University of Newcastle, Australia (2005)
16. Raible, D., Fernau, H.: An amortized search tree analysis for k-Leaf Spanning Tree. In: SOFSEM. LNCS, vol. 5901, pp. 672–684. Springer, Heidelberg (2010)
17. Raman, V., Saurabh, S., Sikdar, S.: Improved exact exponential algorithms for vertex bipartization and other problems. In: Coppo, M., Lodi, E., Pinna, G.M. (eds.) ICTCS 2005. LNCS, vol. 3701, pp. 375–389. Springer, Heidelberg (2005)
18. van Rooij, J.M.M., Bodlaender, H.L.: Design by measure and conquer, a faster exact algorithm for dominating set. In: STACS, Schloss Dagstuhl — Leibniz-Zentrum für Informatik, Germany. LIPIcs, vol. 1, pp. 657–668 (2008)
19. van Rooij, J.M.M., Bodlaender, H.L.: Exact algorithms for edge domination. In: Grohe, M., Niedermeier, R. (eds.) IWPEC 2008. LNCS, vol. 5018, pp. 214–225. Springer, Heidelberg (2008)
20. Yannakakis, M., Gavril, F.: Edge dominating sets in graphs. SIAM J. Appl. Math. 38, 364–372 (1980)

On the Exact Complexity of Evaluating Quantified k-CNF

Chris Calabro[1], Russell Impagliazzo[2,*], and Ramamohan Paturi[2,**]

[1] Google Inc.
[2] Department of Computer Science and Engineering
University of California, San Diego
La Jolla, CA 92093-0404, USA

Abstract. We relate the exponential complexities $2^{s(k)n}$ of k-SAT and the exponential complexity $2^{s(\Pi_2 3\text{-SAT})n}$ of $\Pi_2 3$-SAT (evaluating formulas of the form $\forall \boldsymbol{x} \exists \boldsymbol{y} \phi(\boldsymbol{x}, \boldsymbol{y})$ where ϕ is a 3-CNF in $\boldsymbol{x}$ variables and $\boldsymbol{y}$ variables) and show that $s(\infty)$ (the limit of $s(k)$ as $k \to \infty$) is at most $s(\Pi_2 3\text{-SAT})$. Therefore, if we assume the Strong Exponential-Time Hypothesis, then there is no algorithm for $\Pi_2 3$-SAT running in time 2^{cn} with $c < 1$. On the other hand, a nontrivial exponential-time algorithm for $\Pi_2 3$-SAT would provide a k-SAT solver with better exponent than all current algorithms for sufficiently large k. We also show several syntactic restrictions of $\Pi_2 3$-SAT have nontrivial algorithms, and provide strong evidence that the hardest cases of $\Pi_2 3$-SAT must have a mixture of clauses of two types: one universal literal and two existential literals, or only existential literals. Moreover, the hardest cases must have at least $n - o(n)$ universal variables, and hence only $o(n)$ existential variables. Our proofs involve the construction of efficient minimally unsatisfiable k-CNFs and the application of the Sparsification Lemma.

1 Introduction

From the viewpoint of exponential complexity of **NP**-complete problems, the most studied and best understood problems are probably the restricted versions of the general satisfiability problem (SAT), in particular, k-SAT, the restriction of SAT to k-CNFs, and CNF-SAT, the restriction to general CNFs. There has been a sequence of highly nontrivial and interesting algorithmic approaches to these problems [Sch99, PPZ99, PPSZ05, Sch05, DW05, Rol05], where the best known constant factor improvements in the exponent are of the form $1 - 1/\Theta(k)$ for k-SAT and $1 - 1/\Theta(\lg c)$ for CNF-SAT with at most cn clauses. More recently,

* This research is supported by supported by the Simonyi Fund, the Bell Company Fellowship and the Fund for Math, and NSF grants DMS-083573, CNS-0716790 and CCF-0832797.
** This research is supported by NSF grant CCF-0947262 from the Division of Computing and Communication Foundations. Any opinions, findings and conclusions or recommendations expressed in this material are those of the authors and do not necessarily reflect the views of the National Science Foundation.

V. Raman and S. Saurabh (Eds.): IPEC 2010, LNCS 6478, pp. 50–59, 2010.

the algorithmic approaches have been generalized to obtain improved exponential time algorithms for bounded-depth linear size circuits [CIP09]. Also, a sequence of papers ([IPZ98, IP01, CIKP08, CIP06]) has shown many nontrivial relationships between the exponential complexities of these problems, and helped characterize their hardest instances (under the assumption that they are indeed exponentially hard).

We consider the complexity of evaluating a quantified k-CNF ϕ. A quantified k-CNF is a quantified expression of the following form $Q_1 \boldsymbol{x_1} \cdots Q_i \boldsymbol{x_i} \, \phi$ where $\phi \in$ k-CNF, each $\boldsymbol{x_j}$ is a tuple of Boolean variables, and each $Q_j \in \{\forall, \exists\}$. Similarly, we define a quantified circuit to be a quantified expression $Q_1 \boldsymbol{x_1} \cdots Q_i \boldsymbol{x_i} \, \phi$ where ϕ is a Boolean circuit. For $k = 2$, the problem of evaluating quantified k-CNF can be solved in linear time [APT79], but for $k \geq 3$, it is PSPACE-complete. Note that the exponential complexity of evaluating arbitrary quantified circuits C, which is at least as hard, is at most $2^n \text{poly}(|C|)$: evaluation of a circuit can be carried out in time polynomial in its size, and backtracking need explore at most 2^n paths, each of length at most n, where n is the number of variables of the input (and this only requires polynomial space as well). We ask whether there exist nontrivial exponential-time algorithms for evaluating quantified k-CNF for $k \geq 3$.

Since we cannot hope for a nontrivial, unconditional result about the exponential complexity of qantified k-CNF, we will relate the exponential complexity of evaluating quantified k-CNF to the exponential complexities of k-SAT. As in [IP01], we define the *exponential complexity* of an evaluation problem Φ to be $s(\Phi) = \inf\{c | \exists$ a randomized algorithm for evaluating an instance ϕ of Φ with time complexity $\text{poly}(m) 2^{cn}$ where m is the size of ϕ and n is the number of variables$\}$. While [IP01] did not explicitly use the words 'randomized algorithm' in its definition of $s(k)$ (it uses the word 'algorithm' without qualification), it meant to employ a broader class of algorithms (randmozed algorithms) so as to define the constants $s(k)$ robustly.

Let **ETH** denote the Exponential-Time Hypothesis: $s(3) > 0$. Let $s(\infty)$ denote the limit of the sequence $\{s(k)\}$. [IP01] proposed the open question whether $s(\infty) = 1$, which we will call the *Strong Exponential-Time Hypothesis* (**SETH**). Since the best known upper bounds for $s(k)$ are all of the form $1 - 1/\Theta(k)$, **SETH** is plausible.

As with CNF-SAT, we will consider various syntactic restrictions of quantified k-SAT to arrive at a *minimally complex* set of formulas which are as hard as the standard existentially quantified 3-CNFs as far as exponential complexity is concerned. However, merely bounding k does not seem to be enough. Our main restriction is to bound the number of alternations of quantifier type. Define $\Pi_i k$-SAT ($\Sigma_i k$-SAT) as the problem of evaluating a quantified k-CNF $Q_1 \boldsymbol{x_1} \cdots Q_i \boldsymbol{x_i} \, \phi$ where $\phi \in k$-CNF, each $\boldsymbol{x_j}$ is a tuple of Boolean variables, and each $Q_j \in \{\forall, \exists\}$ is $\forall$ iff j is odd (even). Using the earlier notation, $s(\Pi_i k$-SAT$)$ and $s(\Sigma_i k$-SAT$)$ denote the exponential comlexities of the evaluation problems $\Pi_i k$-SAT and $\Sigma_i k$-SAT respectively. Note that k-SAT is the same problem as

$\Sigma_1 k$-SAT (also, $s(k) = s(\Sigma_1 k$-SAT$)$) and that $\Pi_1 k$-SAT is the problem of evaluating universally quantified k-CNF, which can be done in polynomial time.

Our main result is $s(\infty) \leq s(\Pi_2 3$-SAT$)$. Thus **SETH** would imply that for $\Pi_2 3$-SAT, virtually no improvement over the 2^n exhaustive search algorithm is possible. Since $\Pi_i k$-SAT is at least as hard when $i \geq 2$, $k \geq 3$, the complexity of problems in the polynomial hierarchy would seem to top off rather early. Of course, SETH is a very strong assumption, so this evidence should be considered weak. Conversely, a single nontrivial algorithm for $\Pi_2 3$-SAT would provide a k-SAT solver better than all current algorithms for sufficiently large k.

We also show several syntactic restrictions of $\Pi_2 3$-SAT that have nontrivial algorithms, concluding that the hardest cases of $\Pi_2 3$-SAT must have a mixture of clauses of two types under the assumption **SETH**: one universal literal and two existential literals, or only existential literals. Algorithmic design may benefit by concentrating on the hard cases, in much the same way that much of the progress in k-SAT solvers has been driven by focusing on the hard cases – when there are a linear number of clauses, only one solution, expanding variable-dependence graph, etc.

Lastly, we relate the exponential complexities $s(\Pi_i k$-SAT$)$ ($s(\Sigma_i k$-SAT$)$) for various i, k when the clause density is not too high, although it may well be that these are all simply 1.

[Wil02] explored similar problems, namely exponential upper bounds for solving quantified constraint satisfaction problems (QCSP) with domain size d and cn constraints, each of width $\leq k$. When $d = 2$ and the formula is in k-CNF, we'll call this k-QCNF. [Wil02] solves QCSP for $d = 3$ in time $O(2^{.669cn})$, QCNF in $O(2^{.536cn})$ time and space, 3-QCNF in time $O(2^{.610cn})$, and the special case of 3-QCNF with $c = 1$ in time $O(2^{.482n})$. Each of these techniques outperforms exhaustive search only when c is very small, certainly no more than 2. This is probably not an accident: our results suggest that finding a substantially improved algorithm for moderate c, even for low quantifier depth, is unlikely.

The main step of our lower bound proof is to encode a k-CNF as a $\Pi_2 3$-CNF using only $o(n)$ additional variables. For this encoding, we construct a minimally unsatisfiable CNF formula with many clauses while using few additional variables. Another key technique in our proofs is the Sparsification Lemma [IPZ98, CIP06]:

Lemma 1 (Sparsification Lemma). $\exists$ *algorithm* A $\forall k \geq 2, \epsilon \in (0,1], \phi \in k$-CNF *with* n *variables,* $A_{k,\epsilon}(\phi)$ *outputs a sequence* $\phi_1, \ldots, \phi_s \in k$-CNF *with* $n^{O(k)}$ *delay such that*

1. $s \leq 2^{\epsilon n}$
2. $\mathrm{sol}(\phi) = \bigcup_i \mathrm{sol}(\phi_i)$, *where* $\mathrm{sol}(\phi)$ *is the set of satisfying assignments of* ϕ
3. $\forall i \in [s]$ *each disjunction of* h *literals occurs* $\leq O(\frac{k}{\epsilon})^{3(k-h)}$ *times in* ϕ_i, *which also implies*
   (a) *each literal occurs* $\leq O(\frac{k}{\epsilon})^{3k}$ *times, or* $\leq O(\frac{1}{\epsilon} \lg \frac{1}{\epsilon})$ *times for* $k = 2$, *in* ϕ_i
   (b) $|\phi_i| \leq O(\frac{k}{\epsilon})^{3k} n$.

1.1   Notation

A CNF is a set of clauses and a clause is a set of literals. A CNF F is regarded as a formula at times and as a set of clauses at other times. Intended meaning should be clear from the context. For example, $F \subseteq G$ indicates that the CNFs F and G should ve viewed as sets of clauses. $|F|$ denotes the number of clauses in the CNF F. A k-clause contains at most k literals. A k-CNF is a CNF which contains only k-clauses. We also use CNF to denote the set of all CNFs. If ϕ is a formula, then var(ϕ) denotes set of variables occurring in ϕ. If a is a partial assignment to the variables of ϕ, $\phi|a$ denotes the restriction of ϕ to a. We use SAT to denote the set of satisfiable formulas.

2 $\Pi_2 3$-SAT Lower Bound

Our proof uses a construction similar to [Che09], proposition 3.2, that allows us to reduce the clause width of a k-CNF using a small number of new quantified variables. For this, we need for all k' a "minimally unsatisfiable" k'-CNF which includes a large number of $k' - 1$ clauses. Moreover, we would like to employ as few additional variables as possible, certainly no more than $o(n)$ additional variables. The following construction suffices for our purposes. However, it is an open question whether there are other constructions that employ fewer variables. Our construction differs from [Che09] primarily in its efficiency, since we are more concerned with exponential complexity rather than hardness for a class closed under polynomial time reductions, such as NP or coNP.

Say a pair (F, G) of formulas in CNF is *minimally unsatisfiable* iff $\forall F' \subseteq F$ $((F' \wedge G) \notin \text{SAT} \leftrightarrow F' = F)$. In other words, pair (F, G) is minimally unsatisfiable iff $F \wedge G$ is unsatisfiable, but for any proper subset $F' \subsetneq F$, $F' \wedge G$ is satisfiable. Say that $\phi \in$ CNF is *combinable* with (F, G) iff $|\phi| \leq |F|$ and var$(\phi) \cap \text{var}(F \wedge G) = \emptyset$. In this case, letting $f : \phi \to F$ be an arbitrary injection, we define

$$\text{combine}(\phi, F, G) = G \cup \{\{\bar{l}\} \cup f(C) \mid l \in C \in \phi\} \cup (F - \text{range}(f)),$$

i.e., starting with G, for each clause C of ϕ and literal l of C, we add a clause meaning "l implies the clause of F corresponding to C"; and any clauses of F that are left over get added as-is, without adjoining a literal.

Lemma 2. *Let ϕ be combinable with minimally unsatisfiable (F, G) and $\phi' =$ combine(ϕ, F, G). Then $\forall a \in \{0,1\}^{\text{var}(\phi)}$ $(\phi(a) = 1 \leftrightarrow \phi'|a \notin \text{SAT})$. In other words, for every assignment to the variables of ϕ, $\phi|a$ evaluates to true iff $\phi'|a$ is unsatisfiable.*

Proof. This is simply because in ϕ', after assigning to var(ϕ), each clause $f(C)$ appears in the resulting formula iff some literal $l \in C$ is assigned true. Note that this does not depend on the specific injection $f : \phi \to F$.   □

Lemma 3. *$\forall k \geq 3$, given $m \in \mathbb{N}$, one can construct in time $\text{poly}(m)$, $F \in$ $(k - 1)$-CNF and $G \in k$-CNF such that*

- (F, G) *is minimally unsatisfiable*
- $|F| \geq m$
- $|\operatorname{var}(F \wedge G)| \leq \frac{(k-1)^2}{k-2} \lceil m^{\frac{1}{k-1}} \rceil$.

Proof. Let $l = \lceil m^{\frac{1}{k-1}} \rceil$. Consider a $(k-1) \times l$ matrix A of Boolean variables $x_{i,j}$ and the contradictory statements

1. "every row of A contains a 1",
2. "every choice of one entry from each row contains a 0".

We can express 1. by using at most $\frac{k-1}{k-2} l$ new variables $y_{i,j}$ and k-clauses G as follows. For each i, partition $x_{i,1}, \ldots, x_{i,l}$ into $b = \lceil \frac{l}{k-2} \rceil$ blocks of size at most $k - 2$. Add to G k-clauses expressing that $y_{i,1}$ is the OR of the variables of the first block; $y_{i,j}$ is the OR of $y_{i,j-1}$ and the variables of the jth block, for $j \in [2, b-1]$; and that the OR of $y_{i,b-1}$ and the variables of the bth block is true. Note that $y = z_1 \vee z_2 \vee \cdots \vee z_l$ can be expressed as an $(l+1)$-CNF since every Boolean function in n variables can be expressed as a CNF where each clause has width at most n. We can express 2. by using a set F of $(k-1)$-clauses where $|F| = l^{k-1} \geq m$. Each clause in F is a disjunction of the negations of the variables in the matrix where each row contributes exactly one variable. F and G have the desired properties. □

Corollary 1. $\forall k, k' \geq 3$, *given* $\phi \in k$-CNF *with* n *variables* $\boldsymbol{x}$ *and* m *clauses, one can construct in time* $\operatorname{poly}(n)$, $\phi' \in k'$-CNF *with* n *variables* $\boldsymbol{x}$ *and* $\frac{(k'-1)^2}{k'-2} \lceil m^{\frac{1}{k'-1}} \rceil$ *variables* $\boldsymbol{y}$ *such that* $\forall a \in \{0,1\}^{\operatorname{var}(\phi)}$ $(\phi(a) = 1 \leftrightarrow \phi'|a \notin \text{SAT})$. *In particular,* $\phi \in \text{SAT}$ *iff* $\forall \boldsymbol{x} \exists \boldsymbol{y}$ ϕ' *is false.*

Proof. Use lemma 3 to construct minimally unsatisfiable (F, G) with $F \in (k' - 1)$-CNF, $G \in k'$-CNF, $|F| = m$, $|\operatorname{var}(F \wedge G)| \leq \frac{(k'-1)^2}{k'-2} \lceil m^{\frac{1}{k'-1}} \rceil$, and $\operatorname{var}(\phi) \cap \operatorname{var}(F \wedge G) = \emptyset$. Let $\phi' = \operatorname{combine}(\phi, F, G)$ and apply lemma 2. □

Theorem 1. $s(\infty) \leq s(\Pi_2 3\text{-SAT})$.

Proof. Fix k and $\epsilon > 0$. We will show that $s(k) \leq s(\Pi_2 3\text{-SAT}) + 4\epsilon$. Let ϕ be a k-CNF with n variables. Use Lemma 1 to sparsify ϕ into at most $2^{\epsilon n}$ many subformulas ϕ_i so that each has at most $m = \lfloor \frac{\epsilon n}{4} \rfloor^2$ clauses for all sufficiently large n. By corollary 1, with $k' = 3$, for each ϕ_i, we can construct in time $\operatorname{poly}(n)$ a $\psi_i \in \Pi_2 3\text{-CNF}$ with $(1 + \epsilon)n$ variables such that $\phi_i \in \text{SAT}$ iff ψ_i is false. So $\phi \in \text{SAT}$ iff ψ_i is false for some i. Evaluating each ψ_i with some $\Pi_2 3\text{-SAT}$ solver with exponential complexity $\leq s(\Pi_2 3\text{-SAT}) + \epsilon$, we see that

$$s(k) \leq (s(\Pi_2 3\text{-SAT}) + \epsilon)(1 + \epsilon) + \epsilon \leq s(\Pi_2 3\text{-SAT}) + 4\epsilon. \qquad \square$$

The previous result puts us in an interesting situation. [IP01] showed that **ETH**, namely that $s(3) > 0$, implies $s(k)$ increases infinitely often as a function of k. Consider the following much weaker analogue.

Conjecture 1. **ETH** $\Rightarrow \exists i \geq 2, k \geq 3 \qquad s(\Pi_i k\text{-SAT}) < s(\Pi_{i+1} k + 1\text{-SAT})$.

Conjecture 1, together with Theorem 1, would imply that $s(\infty) < 1$. To see this, assume that for some i and k, $s(\Pi_i k\text{-SAT}) < s(\Pi_{i+1} k + 1\text{-SAT})$. Since $s(\Pi_{i+1}k + 1\text{-SAT}) \leq 1$, it follows that $s(\Pi_2 3\text{-SAT}) \leq s(\Pi_i k\text{-SAT}) < 1$, which implies $s(\infty) < 1$ by Theorem 1.

3   Algorithms for Two Special Cases

In this section, we show that Quantified CNF-SAT has a poly time algorithm if each clause has at most one existential literal and that $\Pi_2 3$-SAT has a nontrivial algorithm if each clause has at least one universal literal. The purpose of such theorems is to find where the hard cases of $\Pi_i k$-SAT lie.

In a quantified CNF, we say that two clauses A, B *disagree* on variable x iff one of A, B contains x and the other contains $\neg x$. A, B are *resolvable* iff they differ in exactly 1 variable x, and the *resolvent* is resolve$(A, B) = A \cup B - \{x, \neg x\}$. Also, define eliminate$(A)$ to be A after removing any universal literal l of A for which no existential literal l' of A occurs after l in the order of quantification. A, B are *Q-resolvable* iff they are resolvable on an existential variable, and the *Q-resolution* operation is qresolve$(A, B) = $ eliminate$($resolve$(A, B))$. Büning, Karpinski and Flögel [BKF95] introduced the Q-resolution proof system and showed that it is sound and complete, in the sense that a quantified CNF is false iff the empty clause can be derived by first replacing each clause C by eliminate(C) and then repeatedly applying qresolve to generate new clauses.

Theorem 2. *Let $\psi = \forall x_1 \exists x_2 \cdots \forall x_n \, \phi$ be a quantified Boolean formula where each clause of $\phi \in$ CNF has at most one existential literal. Then ψ can be evaluated in polynomial time.*

Proof. Since each clause contains at most one existential literal, any application of the Q-resolution operator will produce the empty clause. Letting ψ' be ψ after replacing each clause C by eliminate(C), ψ is true iff ψ' does not contain the empty clause and no two clauses are Q-resolvable in ψ'. □

Theorem 3. *Let $\psi = \forall x \, \exists y \, \phi$, where $\phi \in$ 3-CNF. If each clause of ϕ has at least one universal literal, then ψ can be evaluated in time 2^{cn} for some $c < 1$.*

Proof. Let $G = (V, E)$ where

$$V = \{\text{existential literals of } \phi\}$$
$$E = \{(a, b) \text{ labeled with } c \mid (\bar{a} \vee b \vee \bar{c}) \text{ is a clause of } \phi$$
$$\text{and } a, b \text{ are existential and } c \text{ is universal}\}.$$

While we define the graph G assuming that each clause has exactly two existential literals and one universal literal, other cases can be handled as well. For example, if $(a \vee b \vee c)$ is a clause with a as the sole existential variable and the remaining two literals $\bar{b}$ and $\bar{c}$ are universal, we introduce the edge $(\bar{a}, a)$ with the label $b \wedge c$. 2-clauses can be treated similarly. Branching rules used in

[Wil02] can also be applied to reduce the fomulas with these special types of clauses (clauses with more than one universal literal or 2-clauses) to formulas with clauses containing exactly one universal literal and two existential literals.

A *consistent path* in G is a path such that no two edge labels disagree, a *consistent cycle* is a consistent path that is a cycle. Consistency of a path can be checked in polynomial time since all labels are products of literals. Then ψ is false iff there is an existential variable z such that there is a consistent cycle in G containing nodes $z, \bar{z}$.

Notice that a simple cycle containing nodes $z, \bar{z}$ cannot use more edge labels than $4\epsilon n$, where ϵn is the number of existential variables. So we can test whether there is a consistent cycle in G containing $z, \bar{z}$ in time at most $\text{poly}(|\psi|)\binom{2n}{4\epsilon n} \leq \text{poly}(|\psi|)2^{2h(2\epsilon)n}$ where h is the binary entropy function.

If ϵ is large, we can exhaustively search over all settings of the universal variables and then use a 3-SAT algorithm on the rest. If our 3-SAT algorithm runs in time $2^{s(3)n}$, then the combined algorithm runs in time at most $\min\{2^{(1-\epsilon)n+\epsilon s(3)n}, 2^{2h(2\epsilon)n}\}$, which is maximized at about $\epsilon = .05$ for $s(3) = .403$ and yields a run time of at most $2^{.97n}$. If $s(3)$ is smaller, this value would be even smaller. Also, there is already an algorithm for 3-SAT with $s(3) \leq .403$ [Rol05]. □

It is unlikely that we can find a polynomial time algorithm for the variant where each clause has at least one universal literal because of the following.

Theorem 4. *The language of true $\Pi_2$3-CNF formulas where each clause has at least one universal literal is coNP-complete.*

Proof. The language is in coNP because a witness for falsehood is a consistent cycle containing some existential variable z and its negation $\bar{z}$. To show coNP-hardness, we reduce from 3-UNSAT. Let $\phi \in$ 3-CNF have n variables $\boldsymbol{x}$ and m clauses. Assume without loss of generality that m is even, otherwise, just repeat some clause twice. Let $n' = \frac{m}{2}$, $\boldsymbol{y} = (y_1, \ldots, y_{n'})$ be new variables, and consider the following contradictory 2-CNF:

$$F = \{(y_i \to y_{i+1}), (y_{i+1} \to y_i) \mid i \in [n'-1]\} \cup \{(y_{n'} \to \bar{y}_1), (\bar{y}_1 \to y_{n'})\}.$$

Then the pair of formulas $(F, \emptyset)$ is minimally unsatisfiable and $|F| = m$. By Lemma 2, $\psi = \forall \boldsymbol{x} \exists \boldsymbol{y}$ combine$(\phi, F, \emptyset)$ is a $\Pi_2$3-CNF with at least one universal variable in each clause and such that $\phi \in$ SAT iff ψ is false. □

The coNP-hardness of theorem 4 also follows from [Che09], proposition 3.2, if one observes that the number of universal variables in each constraint constructed there is 1.

4   Reduction to a Canonical Form

Theorems 1 and 3 together suggest that pure existential clauses (those with only existentially quantified literals) and clauses with a mixture of universal and existential literals are both essential to make an instance of $\Pi_2 3$-SAT hard. Below we present further evidence of this.

Let $\Pi_2 3$-SAT$'$ be the special case of $\Pi_2 3$-SAT where each clause is one of just two types:

(1) one universal literal and two existential literals
(2) pure existential.

We then obtain the following theorem, which uses the same idea as in [Wil02] and follows from the algorithm therein.

Theorem 5. $s(\Pi_2 3$-SAT$') < 1 \Rightarrow s(\Pi_2 3$-SAT$) < 1$.

Proof. Let A be an algorithm with run time $\leq O(2^{cn})$ with $c < 1$ for $\Pi_2 3$-SAT$'$. Given a $\Pi_2 3$-CNF, do this: while there is a clause other than types (1) or (2), i.e., a clause C with at least one universal variable and at most one existential variable, reject if there is no existential variable in C, but otherwise branch on the universal variables. One of the branches will force the existential variable. Each formula at a leaf has only clauses of type (1) or (2), so apply algorithm A. This solves the general case in time at most $O(2^{dn})$ where $d = \max\{c, \frac{\lg 7}{3}\} < 1$. □

We may also assume, without loss of generality, that in a $\Pi_2 k$-SAT the number of existential variables is $o(n)$, since otherwise we could branch on every possible setting of the universal variables and then invoke a k-SAT solver to obtain a nontrivial algorithm.

5   Parameter Trade-Off at Higher Levels of the Hierarchy

In the next theorem, we show that when confronted with a $\Pi_i k$-SAT instance, one may be able to reduce k to $k' < k$ if one is willing to increase i by 1, and the input clause density is not too high. If m is a function, let $\Sigma_i k$-SAT$_m$ ($\Pi_i k$-SAT$_m$) be $\Sigma_i k$-SAT ($\Pi_i k$-SAT) but where the number of clauses is promised to be at most $m(n)$.

Theorem 6. *Let $k, k' \geq 3, m \in o(n^{k'-1})$. $\forall$ odd $i \geq 1$,*

$$s(\Sigma_i k\text{-SAT}_m) \leq s(\Pi_{i+1} k'\text{-SAT})$$
$$s(\Pi_i k\text{-SAT}_m) \leq s(\Sigma_i k'\text{-SAT})$$

(which is uninteresting for $i = 1$ since $\Pi_1 k$-SAT is in P). $\forall$ even $i \geq 2$,

$$s(\Sigma_i k\text{-SAT}_m) \leq s(\Pi_i k'\text{-SAT})$$
$$s(\Pi_i k\text{-SAT}_m) \leq s(\Sigma_{i+1} k'\text{-SAT})$$

Proof. Consider a quantified formula $Q_1 \boldsymbol{x_1} \cdots Q_i \boldsymbol{x_i} \phi$ where each Q_j is a quantifier, each $\boldsymbol{x_j}$ is a tuple of Boolean variables, and ϕ is a k-CNF with n variables $\boldsymbol{x}$ and $\leq m(n)$ clauses. Using corollary 1, $\forall \epsilon > 0$ and for sufficiently large n, we can construct in polynomial time a k'-CNF ϕ' such that

- ϕ' has $(1 + \epsilon)n$ vars: $\boldsymbol{x}$ and ϵn new ones $\boldsymbol{y}$
- for each assignment a to $\boldsymbol{x}$, $\phi(a) = 1$ iff $\phi'|a \notin \mathrm{SAT}$.

So

$$Q_1 \boldsymbol{x_1} \cdots Q_i \boldsymbol{x_i} \ \phi$$
$$\Leftrightarrow Q_1 \boldsymbol{x_1} \cdots Q_i \boldsymbol{x_i} \ \neg \exists \boldsymbol{y} \ \phi'$$
$$\Leftrightarrow \neg Q_1' \boldsymbol{x_1} \cdots Q_i' \boldsymbol{x_i} \exists \boldsymbol{y} \ \phi'$$

where Q_j' is existential iff Q_j is universal. $\square$

It would be nice to eliminate the requirement that ϕ have at most $o(n^{k'-1})$ clauses. One might try to do this by first sparsifying ϕ, but the possibly complex quantification of the variables prevents this.

Acknowledgments. We thank the reviewers for their helpful comments.

References

[APT79]   Aspvall, B., Plass, M.F., Tarjan, R.E.: A linear-time algorithm for testing the truth of certain quantified boolean formulas. Inf. Process. Lett. 8(3), 121–123 (1979)

[BKF95]   Büning, H.K., Karpinski, M., Flögel, A.: Resolution for quantified boolean formulas. Information and Computation 117(1), 12–18 (1995)

[Che09]   Chen, H.: Existentially restricted quantified constraint satisfaction. Information and Computation 207(3), 369–388 (2009)

[CIKP08] Calabro, C., Impagliazzo, R., Kabanets, V., Paturi, R.: The complexity of unique k-SAT: An isolation lemma for k-CNFs. Journal of Computer and Systems Sciences 74(3), 386–393 (2008); Preliminary version in Proceedings of the Eighteenth IEEE Conference on Computational Complexity, pp. 135–144 (2003)

[CIP06]   Calabro, C., Impagliazzo, R., Paturi, R.: A duality between clause width and clause density for SAT. In: CCC 2006: Proceedings of the 21st Annual IEEE Conference on Computational Complexity, Washington, DC, USA, pp. 252–260. IEEE Computer Society, Los Alamitos (2006)

[CIP09]   Calabro, C., Impagliazzo, R., Paturi, R.: The complexity of satisfiability of small depth circuits. In: Parameterized and Exact Computation: 4th International Workshop, IWPEC 2009, Revised Selected Papers, Copenhagen, Denmark, September 10-11, pp. 75–85. Springer, Heidelberg (2009)

[DW05]   Dantsin, E., Wolpert, A.: An improved upper bound for SAT. In: Bacchus, F., Walsh, T. (eds.) SAT 2005. LNCS, vol. 3569, pp. 400–407. Springer, Heidelberg (2005)

[IP01]   Impagliazzo, R., Paturi, R.: The complexity of k-SAT. Journal of Computer and Systems Sciences 62(2), 367–375 (2001); Preliminary version in 14th Annual IEEE Conference on Computational Complexity, pp. 237–240 (1999)

[IPZ98]   Impagliazzo, R., Paturi, R., Zane, F.: Which problems have strongly exponential complexity? Journal of Computer and System Sciences 63, 512–530 (1998); Preliminary version in 39th Annual IEEE Symposium on Foundations of Computer Science, pp 653–662 (1998)

[PPSZ05]   Paturi, R., Pudlák, P., Saks, M.E., Zane, F.: An improved exponential-time algorithm for k-SAT. Journal of the ACM 52(3), 337–364 (2005); Preliminary version in 39th Annual IEEE Symposium on Foundations of Computer Science, pp. 628–637 (1998)

[PPZ99]   Paturi, R., Pudlák, P., Zane, F.: Satisfiability coding lemma. Chicago Journal of Theoretical Computer Science (December 1999); Preliminary version in 38th Annual Symposium on Foundations of Computer Science, pp. 566–574 (1997)

[Rol05]   Rolf, D.: Improved bound for the PPSZ/Schöning-algorithm for 3-SAT. Electronic Colloquium on Computational Complexity (ECCC) 12(159) (2005)

[Sch99]   Schöning, U.: A probabilistic algorithm for k-SAT and constraint satisfaction problems. In: FOCS, pp. 410–414 (1999)

[Sch05]   Schuler, R.: An algorithm for the satisfiability problem of formulas in conjunctive normal form. Journal of Algorithms 54(1), 40–44 (2005)

[Wil02]   Williams, R.: Algorithms for quantified boolean formulas. In: Proceedings of the 13th Annual ACM-SIAM Symposium on Discrete Algorithms, pp. 299–307 (2002)

Cluster Editing:
Kernelization Based on Edge Cuts⋆

Yixin Cao and Jianer Chen

Department of Computer Science and Engineering
Texas A&M University
{yixin,chen}@cse.tamu.edu

Abstract. Kernelization algorithms for the CLUSTER EDITING problem have been a popular topic in the recent research in parameterized computation. Thus far most kernelization algorithms for this problem are based on the concept of *critical cliques*. In this paper, we present new observations and new techniques for the study of kernelization algorithms for the CLUSTER EDITING problem. Our techniques are based on the study of the relationship between CLUSTER EDITING and graph edge-cuts. As an application, we present an $\mathcal{O}(n^2)$-time algorithm that constructs a $2k$ kernel for the *weighted* version of the CLUSTER EDITING problem. Our result meets the best kernel size for the unweighted version for the CLUSTER EDITING problem, and significantly improves the previous best kernel of quadratic size for the weighted version of the problem.

1 Introduction

Errors are ubiquitous in most experiments, and we have to find out the true information buried behind them, that is, to remove the inconsistences in data of experiment results. In most cases, we want to make the data consistent with the least amount of modifications, i.e., we assume the errors are not too much. This is an *everyday* problem in real life. Indeed, the problem has been studied by researchers in different areas [3,25]. A graph theoretical formulation of the problem is called the CLUSTER EDITING problem that seeks a collection of edge insertion/deletion operations of minimum cost that transforms a given graph into a union of disjoint cliques. The CLUSTER EDITING problem has applications in many areas, including machine learning [3], world wide web [12], data-minning [4], information retrieval [19], and computational biology [10]. The problem is also closely related to another interesting and important problem in algorithmic research, CLUSTERING AGGREGATION [1], which, given a set of clusterings on the same set of vertices, asks for a single clustering that agrees as much as possible with the input clusterings.

Let $G = (V, E)$ be an undirected graph, and let V^2 be the set of all unordered pairs of vertices in G (thus, for two vertices v and w, $\{v, w\}$ and $\{w, v\}$ will be regarded as the same pair). Let $\pi : V^2 \mapsto \mathbb{N} \cup \{+\infty\}$ be a *weight function*, where

⋆ Supported in part by the US NSF under the Grants CCF-0830455 and CCF-0917288.

V. Raman and S. Saurabh (Eds.): IPEC 2010, LNCS 6478, pp. 60–71, 2010.
© Springer-Verlag Berlin Heidelberg 2010

$\mathbb{N}$ is the set of positive integers. The *weight* of an edge $[v, w]$ in G is defined to be $\pi(v, w)$. If vertices v and w are not adjacent, and we add an edge between v and w, then we say that we *insert an edge* $[v, w]$ *of weight* $\pi(v, w)$.

The weighted CLUSTER EDITING problem is formally defined as follows:

> (Weighted) CLUSTER EDITING: Given (G, π, k), where $G = (V, E)$ is an undirected graph, $\pi : V^2 \mapsto \mathbb{N} \cup \{+\infty\}$ is a weight function, and k is an integer, is it possible to transform G into a union of disjoint cliques by edge deletions and/or edge insertions such that the weight sum of the inserted edges and deleted edges is bounded by k?

The problem is NP-complete even in its unweighted version [25]. Polynomial-time approximation algorithms for the problem have been studied. The best result is a randomized approximation algorithm of expected approximation ratio 3 by Ailon, Charikar, and Newman [1], which was later derandomized by van Zuylen and Williamson [27]. The problem has also been shown to be APX-hard by Charikar, Guruswami, and Wirth [8].

Recently, some researchers have turned their attention to exact solutions, and to the study of parameterized algorithms for the problem. A closely related problem is to study *kernelization algorithms* for the problem, which, on an instance (G, π, k) of CLUSTER EDITING, produces an "equivalent" instance (G', π, k') such that $k' \leq f(k)$[1] and that the *kernel size* (i.e., the number of vertices in the graph G') is small. For the unweighted version of the problem (i.e., assuming that for each pair v and w of vertices, $\pi(v, w) = 1$), Gramm et al. [17] presented the first parameterized algorithm running in time $\mathcal{O}(2.27^k + n^3)$ and a kernelization algorithm that produces a kernel of $\mathcal{O}(k^2)$ vertices. This result was immediately improved by a successive sequence of studies on kernelization algorithms that produce kernels of size $24k$ [15], of size $4k$ [18] and of size $2k$ [9]. The $24k$ kernel was obtained via *crown reduction*, while the later two results were both based on the concept of simple series module (*critical clique*), which is a restricted version of modular decomposition [11]. Basically, these algorithms iteratively construct the modular decomposition, find reducible simple series modules and apply reduction rules on them, until there are no any reducible modules found.

For the weighted version, to our best knowledge, the only non-trivial result on kernelization is the quadratic kernel developed by Böcker et al. [7].

The main result of this paper is the following theorem:

Theorem 1. *There is an $\mathcal{O}(n^2)$-time kernelization algorithm for the weighted* CLUSTER EDITING *problem that produces a kernel which contains at most $2k$ vertices.*

Compared to all previous results, Theorem 1 is better not only in kernel size and running time, but also more importantly in conceptual simplicity.

A more general version of weighted CLUSTER EDITING problem is defined with real weights, that is, the weight function π is replaced by $\pi' : V^2 \mapsto \mathbb{R}_{\geq 1} \cup \{+\infty\}$ where $\mathbb{R}_{\geq 1}$ is the set of all real numbers larger than or equal to 1, and

---

[1] $f(\cdot)$ is a computable function.

correspondingly k becomes a positive real number. Our result also works for this version, in the same running time, and with only a small relaxation in the consant of kernel size.

Our contribution. We report the first linear vertex kernel with very small constant, for the weighted version of the CLUSTER EDITING problem. Our contribution to this research includes:

1. the cutting lemmas (some of them are not used for our kernelization algorithm) are of potential use for future work on kernelizations and algorithms;
2. both the idea and the process are very simple with efficient implementations that run in time $\mathcal{O}(n^2)$. Indeed, we use only a single reduction rule, which works for both weighted and unweighted versions;
3. the reduction processes to obtain the above results are independent of k, and therefore are more general and applicable.

2   Cutting Lemmas

In this paper, graphs are always undirected and simple. A graph is a *complete graph* if each pair of vertices are connected by an edge. A *clique* in a graph G is a subgraph G' of G such that G' is a complete graph. By definition, a clique of h vertices contains $\binom{h}{2} = h(h-1)/2$ edges. If two vertices v and w are not adjacent, then we say that the edge $[v, w]$ is *missing*, and call the pair $\{v, w\}$ an *anti-edge*. The total number of anti-edges in a graph of n vertices is $n(n-1)/2 - |E(G)|$. The subgraph of the graph G induced by a vertex subset X is denoted by $G[X]$.

Let $G = (V, E)$ be a graph, and let $S \subseteq V^2$. Denote by $G \triangle S$ the graph obtained from G as follows: for each pair $\{v, w\}$ in S, if $[v, w]$ is an edge in G, then remove the edge $[v, w]$ in the graph, while if $\{v, w\}$ is an anti-edge, then insert the edge $[v, w]$ into the graph. A set $S \subseteq V^2$ is a *solution* to a graph $G = (V, E)$ if the graph $G \triangle S$ is a union of disjoint cliques.

For an instance (G, π, k) of CLUSTER EDITING, where $G = (V, E)$, the *weight* of a set $S \subseteq V^2$ is defined as $\pi(S) = \sum_{\{v, w\} \in S} \pi(v, w)$. Similarly, for a set E' of edges in G, the *weight* of E' is $\pi(E') = \sum_{[v, w] \in E'} \pi(v, w)$. Therefore, the instance (G, π, k) asks if there is a solution to G whose weight is bounded by k.

For a vertex v, denote by $N(v)$ the set of neighbors of v, and let $N[v] = N(v) \cup \{v\}$. For a vertex set X, $N[X] = \bigcup_{v \in X} N[v]$, and $N(X) = N[X] \backslash X$. For the vertex set X, define $\overline{X} = V \backslash X$. For two vertex subsets X and Y, denote by $E(X, Y)$ the set of edges that has one end in X and the other end in Y. For a vertex subset X, the edge set $E(X, \overline{X})$ is called the *cut of* X. The total cost of the cut of X is denoted by $\gamma(X) = \pi(E(X, \overline{X}))$. Obviously, $\gamma(X) = \gamma(\overline{X})$. For an instance (G, π, k) of the CLUSTER EDITING problem, denote by $\omega(G)$ the weight of an optimal (i.e., minimum weighted) solution to the graph G.

Behind all of the following lemmas is a very simple observation: in the objective graph $G \triangle S$ for any solution S to the graph G, each induced subgraph is also a union of disjoint cliques. Therefore, a solution S to the graph G restricted to an induced subgraph G' of G (i.e., the pairs of S in which both vertices are in G') is also a solution to the subgraph G'. This observation leads to the following *Cutting Lemma*.

Lemma 1. *Let $\mathcal{P} = \{V_1, V_2, \ldots, V_p\}$ be a vertex partition of a graph G, and let $E_{\mathcal{P}}$ be the set of edges whose two ends belong to two different parts in $\mathcal{P}$. Then $\sum_{i=1}^{p} \omega(G[V_i]) \leq \omega(G) \leq \pi(E_{\mathcal{P}}) + \sum_{i=1}^{p} \omega(G[V_i])$.*

Proof. Let S be an optimal solution to the graph G. For $1 \leq i \leq p$, let S_i be the subset of S such that each pair in S_i has both its vertices in V_i. As noted above, the set S_i is a solution to the graph $G[V_i]$, which imples $\omega(G[V_i]) \leq \pi(S_i)$. Thus,

$$\sum_{i=1}^{p} \omega(G[V_i]) \leq \sum_{i=1}^{p} \pi(S_i) \leq \pi(S) = \omega(G).$$

On the other hand, if we remove all edges in $E_{\mathcal{P}}$, and for each i, apply an optimal solution S_i' to the induced subgraph $G[V_i]$, we will obviously end up with a union of disjoint cliques. Therefore, these operations make a solution to the graph G whose weight is $\pi(E_{\mathcal{P}}) + \sum_{i=1}^{p} \pi(S_i') = \pi(E_{\mathcal{P}}) + \sum_{i=1}^{p} \omega(G[V_i])$. This gives immediately $\omega(G) \leq \pi(E_{\mathcal{P}}) + \sum_{i=1}^{p} \omega(G[V_i])$.   □

Lemma 1 directly implies the following corollaries. First, if there is no edge between two different parts in the vertex partition $\mathcal{P}$, then Lemma 1 gives

Corollary 1. *Let G be a graph with connected components $G_1, \ldots, G_p$, then $\omega(G) = \sum_{i=1}^{p} \omega(G_i)$, and every optimal solution to the graph G is a union of optimal solutions to the subgraphs $G_1, \ldots, G_p$.*

When $p = 2$, i.e., the vertex partition is $\mathcal{P} = \{X, \overline{X}\}$, the edge set $E_{\mathcal{P}}$ becomes the cut $E(X, \overline{X})$, and $\pi(E(X, \overline{X})) = \gamma(X)$. Lemma 1 gives

Corollary 2. *Let $X \subseteq V$ be a vertex set, then $\omega(G[X]) + \omega(G[\overline{X}]) \leq \omega(G) \leq \omega(G[X]) + \omega(G[\overline{X}]) + \gamma(X)$.*

Corollary 3. *Let G be a graph, and let S^* be an optimal solution to G. For any subset X of vertices in G, if we let $S^*(X, \overline{X})$ be the subset of pairs in which one vertex is in X and the other vertex is in $\overline{X}$, then $\pi(S^*(X, \overline{X})) \leq \gamma(X)$.*

Proof. The optimal solution S^* can be divided into three disjoint parts: the subset $S^*(X)$ of pairs in which both vertices are in X, the subset $S^*(\overline{X})$ of pairs in which both vertices are in $\overline{X}$, and the subset $S^*(X, \overline{X})$ of pairs in which one vertex is in X and the other vertex is in $\overline{X}$. By Corollary 2,

$$\omega(G) = \pi(S^*(X)) + \pi(S^*(\overline{X})) + \pi(S^*(X, \overline{X})) \leq \omega(G[X]) + \omega(G[\overline{X}]) + \gamma(X).$$

Since $\pi(S^*(X)) \geq \omega(G[X])$ and $\pi(S^*(\overline{X})) \geq \omega(G[\overline{X}])$, we get immediately $\pi(S^*(X, \overline{X})) \leq \gamma(X)$.   □

Corollary 3 can be informally described as "cut preferred" principle, which is fundamental for this problem. Similarly we have the following lemmas.

Lemma 2. *Let X be a subset of vertices in a graph G, and let S^* be any optimal solution to G. Let $S^*(V, \overline{X})$ be the set of pairs in S^* in which at least one vertex is in $\overline{X}$. Then $\omega(G) \geq \omega(G[X]) + \pi(S^*(V, \overline{X}))$.*

Proof. The optimal solution S^* is divided into two disjoint parts: the subset $S^*(X)$ of pairs in which both vertices are in X, and the subset $S^*(V, \overline{X})$ of pairs in which at least one vertex is in $\overline{X}$. The set $S^*(X)$ is a solution to the induced subgraph $G[X]$. Therefore, $\pi(S^*(X)) \geq \omega(G[X])$. This gives

$$\omega(G) = \pi(S^*) = \pi(S^*(X)) + \pi(S^*(V, \overline{X}) \geq \omega(G[X]) + \pi(S^*(V, \overline{X})),$$

which proves the lemma.   □

Lemma 3. *Let X be a subset of vertices in a graph G, and let B_X be the set of vertices in X that are adjacent to vertices in $\overline{X}$. Then for any optimal solution S^* to G, if we let $S^*(B_X)$ be the set of pairs in S^* in which both vertices are in B_X, then $\omega(G) + \pi(S^*(B_X)) \geq \omega(G[X]) + \omega(G[\overline{X} \cup B_X])$.*

Proof. Again, the optimal solution S^* can be divided into three disjoint parts: the subset $S^*(X)$ of pairs in which both vertices are in X, the subset $S^*(\overline{X})$ of pairs in which both vertices are in $\overline{X}$, and the subset $S^*(X, \overline{X})$ of pairs in which one vertex is in X and the other vertex is in $\overline{X}$. We also denote by $S^*(B_X, \overline{X})$ the subset of pairs in S^* in which one vertex is in B_X and the other vertex is in $\overline{X}$. Since $S^*(X)$ is a solution to the induced subgraph $G[X]$, we have

$$\begin{aligned}
\omega(G) + \pi(S^*(B_X)) &= \pi(S^*(X)) + \pi(S^*(\overline{X})) + \pi(S^*(X, \overline{X})) + \pi(S^*(B_X)) \\
&\geq \omega(G[X]) + \pi(S^*(\overline{X})) + \pi(S^*(X, \overline{X})) + \pi(S^*(B_X)) \\
&\geq \omega(G[X]) + \pi(S^*(\overline{X})) + \pi(S^*(B_X, \overline{X})) + \pi(S^*(B_X)).
\end{aligned}$$

The last inequality is because $B_X \subseteq X$, so $S^*(B_X, \overline{X}) \subseteq S^*(X, \overline{X})$. Since $S' = S^*(\overline{X}) \cup S^*(B_X, \overline{X}) \cup S^*(B_X)$ is the subset of pairs in S^* in which both vertices are in the induced subgraph $G[\overline{X} \cup B_X]$, S' is a solution to the induced subgraph $G[\overline{X} \cup B_X]$. This gives

$$\pi(S') = \pi(S^*(\overline{X})) + \pi(S^*(B_X, \overline{X})) + \pi(S^*(B_X)) \geq \omega(G[\overline{X} \cup B_X]),$$

which implies the lemma immediately.   □

The above results that reveal the relations between the structures of the CLUSTER EDITING problem and graph edge cuts not only form the basis for our kernelization results presented in the current paper, but also are of their own importance and interests.

3   The Kernelization Algorithm

Obviously, the number of different vertices included in a solution S of k vertex pairs to a graph G is upper bounded by $2k$. Thus, if we can also bound the number of vertices that are not included in S, we get a kernel. For such a vertex v, the clique containing v in $G \triangle S$ must be $G[N[v]]$. Inspired by this, our approach is to check the closed neighborhood $N[v]$ for each vertex v.

The observation is that if an induced subgraph (e.g. the closed neighborhood of a vertex) is very "dense inherently", while is also "loosely connected to outside", (*i.e.* there are very few edges in the cut of this subgraph), it might be cut off and solved separately. By the cutting lemmas, the size of a solution obtained as such should not be too far away from that of an optimal solution. Actually, we will figure out the conditions under which they are equal.

The subgraph we are considering is $N[v]$ for some vertex v. For the connection of $N[v]$ to outside, a good measurement is $\gamma(N[v])$. Thus, here we only need to define the density. A simple fact is that the fewer edges missing, the denser the subgraph is. Therefore, to measure the density of $N[v]$, we define the *deficiency* $\delta(v)$ of $N[v]$ as the total weight of anti-edges in $G[N[v]]$, which is formally given by $\delta(v) = \pi(\{\{x, y\} \mid x, y \in N(v), [x, y] \notin E\})$.

Suppose that $N[v]$ forms a single clique with no other vertices in the resulting graph $G \triangle S$. Then anti-edges of total weight $\delta(v)$ have to be added to make $N[v]$ a clique, and edges of total weight $\gamma(N[v])$ have to be deleted to make $N[v]$ disjoint. Based on this we define the *stable cost* of a vertex v as $\rho(v) = 2\delta(v) + \gamma(N[v])$, and we say $N[v]$ is *reducible* if $\rho(v) < |N[v]|$.

Lemma 4. *For any vertex v such that $N[v]$ is reducible, there is an optimal solution S^* to G such that the vertex set $N[v]$ is entirely contained in a single clique in the graph $G \triangle S^*$.*

Proof. Let S be an optimal solution to the graph G, and pick any vertex v such that $N[v]$ is reducible, i.e., $\rho(v) < |N[v]|$. Suppose that $N[v]$ is not entirely contained in a single clique in $G \triangle S$, i.e., $N[v] = X \cup Y$, where $X \neq \emptyset$ and $Y \neq \emptyset$, such that Y is entirely contained in a clique C_1 in $G \triangle S$ while $X \cap C_1 = \emptyset$ (note that we do not assume that X is in a single clique in $G \triangle S$).

Inserting all missing edges between vertices in $N[v]$ will transform the induced subgraph $G[N[v]]$ into a clique. Therefore, $\omega(G[N[v]]) \leq \delta(v)$. Combining this with Corollary 2, we get

$$\begin{aligned} \omega(G) &\leq \omega(G[N[v]]) + \omega(G[\overline{N[v]}]) + \gamma(N[v]) \\ &\leq \delta(v) + \omega(G[\overline{N[v]}]) + \gamma(N[v]) \\ &= \omega(G[\overline{N[v]}]) + \rho(v) - \delta(v). \end{aligned} \tag{1}$$

Let $S(V, N[v])$ be the set of pairs in the solution S in which at least one vertex is in $N[v]$, and let $S(X, Y)$ be the set of pairs in S in which one vertex is in X and the other vertex is in Y. Also, let $P(X, Y)$ be the set of all pairs (x, y) such that $x \in X$ and $y \in Y$. Obviously, $\pi(S(V, N[v])) \geq \pi(S(X, Y))$ because $X \subseteq V$ and $Y \subseteq N[v]$. Moreover, since the solution S places the sets X and Y in different cliques, S must delete all edges between X and Y. Therefore $S(X, Y)$ is exactly the set of edges in G in which one end is in X and the other end is in Y. Also, by the definition of $\delta(v)$ and because both X and Y are subsets of $N[v]$, the sum of the weights of all anti-edges between X and Y is bounded by $\delta(v)$. Thus, we have $\pi(S(X,Y)) + \delta(v) \geq \pi(P(X,Y))$. Now by Lemma 2,

$$
\begin{aligned}
\omega(G) &\geq \omega(G[\overline{N[v]}]) + \pi(S(V,N[v])) \\
&\geq \omega(G[\overline{N[v]}]) + \pi(S(X,Y)) \\
&\geq \omega(G[\overline{N[v]}]) + \pi(P(X,Y)) - \delta(v).
\end{aligned}
\tag{2}
$$

Combining (1) and (2), and noting that the weight of each vertex pair is at least 1, we get

$$
|X||Y| \leq \pi(P(X,Y)) \leq \rho(v) < |N[v]| = |X| + |Y|. \tag{3}
$$

This can hold true only when $|X| = 1$ or $|Y| = 1$. In both cases, we have $|X| \cdot |Y| = |X| + |Y| - 1$. Combining this with (3), and noting that all the quantities are integers, we must have

$$
\pi(P(X,Y)) = \rho(v),
$$

which, when combined with (1) and (2), gives

$$
\omega(G) = \omega(G[\overline{N[v]}]) + \rho(v) - \delta(v) = \omega(G[\overline{N[v]}]) + \gamma(N[v]) + \delta(v). \tag{4}
$$

Note that $\gamma(N[v]) + \delta(v)$ is the minimum cost to insert edges into and delete edges from the graph G to make $N[v]$ a disjoint clique. Therefore, Equality (4) shows that if we first apply edge insert/delete operations of minimum weight to make $N[v]$ a disjoint clique, then apply an optimal solution to the induced subgraph $G[\overline{N[v]}]$, then we have an optimal solution S^* to the graph G. This completes the proof of the lemma because the optimal solution S^* has the vertex set $N[v]$ entirely contained in a single clique in the graph $G \triangle S^*$. $\qquad \square$

Based on Lemma 4, we have the following reduction rule:

Step 1. *For a vertex v such that $N[v]$ is reducible, insert edges between anti-edges in $G[N[v]]$ to make $G[N[v]]$ a clique, and decrease k accordingly.*

After Step 1, the induced subgraph $G[N[v]]$ becomes a clique with $\delta(v) = 0$ and $\rho(v) = \gamma(N[v])$. Now we use the following rule to remove the vertices in $N(N[v])$ that are loosely connected to $N[v]$ (recall that $N(N[v])$ is the set of vertices that are not in $N[v]$ but adjacent to some vertices in $N(v)$, and that for two vertex subsets X and Y, $E(X,Y)$ denotes the set of edges that has one end in X and the other end in Y).

Step 2. *Let v be a vertex such that $N[v]$ is reducible on which Step 1 has been applied. For each vertex x in $N(N[v])$, if $\pi(E(x,N(v))) \leq |N[v]|/2$, then delete all edges in $E(x,N(v))$ and decrease k accordingly.*

We say that a reduction step R is *safe* if after edge operations of cost c_R by the step, we obtain a new graph G' such that the original graph G has a solution of weight bounded by k if and only if the new graph G' has a solution of weight bounded by $k - c_R$.

Lemma 5. *Step 2 is safe.*

Proof. By Lemma 4, there is an optimal solution S to the graph G such that $N[v]$ is entirely contained in a single clique C in the graph $G \triangle S$. We first prove, by contradiction, that the clique C containing $N[v]$ in the graph $G \triangle S$ has at most one vertex in $\overline{N[v]}$. Suppose that there are r vertices $u_1, \ldots, u_r$ in $\overline{N[v]}$ that are in C, where $r \geq 2$. For $1 \leq i \leq r$, denote by c_i the total weight of all edges between u_i and $N[v]$, and by c_i' the total weight of all pairs (both edges and anti-edges) between u_i and $N[v]$. Note that $c_i' \geq |N[v]|$ and $\sum_{i=1}^{r} c_i \leq \gamma(N[v])$. Then in the optimal solution S to G, the total weight of the edges inserted between $N[v]$ and $\overline{N[v]}$ is at least

$$\sum_{i=1}^{r}(c_i' - c_i) \geq \sum_{i=1}^{r}(|N[v]| - c_i) = r|N[v]| - \sum_{i=1}^{r} c_i$$
$$\geq r|N[v]| - \gamma(N[v]) \geq 2|N[v]| - \gamma(N[v])$$
$$> 2|N[v]| - |N[v]| = |N[v]| > \gamma(N[v]),$$

where we have used the fact $|N[v]| > \gamma(N[v])$ (this is because by the conditions of the step, $\rho(v) = 2\delta(v) + \gamma(N[v]) < |N[v]|)$. But this contradicts Corollary 3.

Therefore, there is at most one vertex x in $N(N[v])$ that is in the clique C containing $N[v]$ in the graph $G \triangle S$. Such a vertex x must satisfy the condition $\pi(E(x, N(v))) > |N[v]|/2$: otherwise deleting all edges in $E(x, N(v))$ would result in a solution that is at least as good as the one that inserts all missing edges between x and $N[v]$ and makes $N[v] \cup \{x\}$ a clique. Thus, for a vertex x in $N(N[v])$ with $\pi(E(x, N(v))) \leq |N[v]|/2$, we can always assume that x is not in the clique containing $N[v]$ in the graph $G \triangle S$. In consequence, deleting all edges in $E(x, N(v))$ for such a vertex x is safe. $\qquad \square$

The structure of $N[v]$ changes after the above steps. The result can be in two possible cases: (1) no vertex in $N(N[v])$ survives, and $N[v]$ becomes an isolated clique – then by Corollary 1, we can simply delete the clique; and (2) there is one vertex x remaining in $N(N[v])$ (note that there cannot be more than one vertices surviving – otherwise it would contradict the assumption $\gamma(N[v]) \leq \rho(v) < |N[v]|$). In case (2), the vertex set $N[v]$ can be divided into two parts $X = N[v] \cap N(x)$ and $Y = N[v] \backslash X$. From the proofs of the above lemmas, we are left with only two options: either disconnecting X from x with edge cost c_X, or connecting Y and x with edge cost c_Y. Obviously $c_X > c_Y$. Since both options can be regarded as connection or disconnection between the vertex set $N[v]$ and the vertex x, we can further reduce the graph using the following reduction step:

Step 3. *Let v be a vertex such that $N[v]$ is reducible on which Steps 1 and 2 have been applied. If there still exists a vertex x in $N(N[v])$, then merge $N[v]$ into a single vertex v', connect v' to x with weight $c_X - c_Y$, set weight of each anti-edge between v' and other vertex to $+\infty$, and decrease k by c_Y.*

The correctness of this step immediately follows from above argument.

Note that the conditions for all the above steps are only checked once. If they are satisfied, we apply all three steps one by one, or else we do nothing at all. So they are actually the parts of a single reduction rule presented as follows:

The Rule. Let v be a vertex satisfying $2\delta(v) + \gamma(N[v]) < |N[v]|$, then:

1. add edges to make $G[N[v]]$ a clique and decrease k accordingly;
2. for each vertex x in $N(N[v])$ with $\pi(E(x, N[v])) \leq |N[v]|/2$, remove all edges in $E(x, N[v])$ and decrease k accordingly;
3. if a vertex x in $N(N[v])$ survives, merge $N[v]$ into a single vertex (as described above) and decrease k accordingly.

Now the following lemma implies Theorem 1 directly.

Lemma 6. *If an instance of the weighted* CLUSTER EDITING *problem reduced by our reduction rule has more than $2k$ vertices, it has no solution of weight $\leq k$.*

Proof. We divide the cost of inserting/deleting a pair $\{u, v\}$ into two halves and assign them to u and v equally. Thereafter we count the costs on all vertices.

For any two vertices with distance 2, at most one of them is not shown in a solution S: otherwise they would have to belong to the same clique in $G \triangle S$ because of their common neighbors but the edge between them is missing. Thus, if we let $\{v_1, v_2, \ldots, v_r\}$ be the vertices not shown in S, then each two of their closed neighbors $\{N[v_1], N[v_2], \ldots, N[v_r]\}$ are either the same (when they are in the same simple series module) or mutually disjoint. The cost in each $N[v_i]$ is $\delta(v_i) + \gamma(N[v_i])/2 = \rho(v_i)/2$, which is at least $|N[v_i]|/2$, because by our reduction rule, in the reduced instance we have $\rho(v) \geq |N[v]|$ for each vertex v. Each of the vertices not in any of $N[v_i]$ is contained in at least one pair of S and therefore bears cost at least $1/2$. Summing them up, we get a lower bound for the total cost at least $|V|/2$. Thus, if the solution S has a weight bounded by k, then $k \geq |V|/2$, i.e., the graph has at most $2k$ vertices.   □

4   On Unweighted and Real-Weighted Versions

We now show how to adapt the algorithm in the previous section to support unweighted and real-weighted versions. Only slight modifications are required. Therefore, the proof of the correctness of them is omitted for the lack of space.

Unweighted version. The kernelization algorithm presented does not work for unweighted version. The trouble arises in Step 3, where merging $N[v]$ is not a valid operation in an unweighted graph. Fortunately, this can be easily circumvented, by replacing Step 3 by the following new rule:

Step 3 (U). *Let v be a vertex such that $N[v]$ is reducible on which Steps 1 and 2 have been applied. If there still exists a vertex x in $N(N[v])$, then replace $N[v]$ by a complete subgraph $K_{|X|-|Y|}$, and connect x to all vertices of this subgraph.*

The correctness of this new rule is similar to the arguments in last section, and it is easy to check the first two rules apply for the unweighted version. Moreover, the proof of Lemma 6 can be easily adapted with the new rule.

Real-weighted version. There are even more troubles when weights are allowed to be real numbers, instead of only positive integers. The first problem is that, without the integrality, (3) cannot imply (4). This is fixable by changing the definition of reducible closed neighborhood from $\rho(v) < |N(v)|$ to $\rho(v) \leq |N(v)| - 1$ (they are equivalent for integers), then (3) becomes

$$|X||Y| \leq \pi(P(X,Y)) \leq \rho(v) \leq |N[v]| - 1 = |X| + |Y| - 1. \tag{5}$$

Formulated on reducible closed neighborhood, Steps 1 and 2 remain the same.

The second problem is Step 3, in which we need to maintain the validity of weights. Recall that we demand all weights be at least 1 for weight functions. This, although trivially holds for integral weight functions, will be problematic for real weight functions. More specifically, in Step 3, the edge $[x, v']$ could be assigned a weight $c_X - c_Y < 1$ when c_X and c_Y differ by less than 1. This can be fixed with an extension of Step 3:

Step 3 (R). *Let v be a vertex such that $N[v]$ is reducible and that on which Steps 1 and 2 have been applied. If there still exists a vertex x in $N(N[v])$, then*

- *if $c_X - c_Y \geq 1$, merge $N[v]$ into a single vertex v', connect v' to x with weight $c_X - c_Y$, set weight of each anti-edge between v' and other vertex to $+\infty$, and decrease k by c_Y;*
- *if $c_X - c_Y < 1$, merge $N[v]$ into two vertices v' and v'', connect v' to x with weight 2, and v' to v'' with weight $2 - (c_X - c_Y)$, set weight of each anti-edge between v', v'' to other vertex to $+\infty$, and decrease k by $c_X - 2$.*

The new case is just to maintain the validity of the weight, and does not make a real difference from the original case. However, there does exist one subtlety we need to point out, that is, the second case might increase k slightly, and this happens when $c_X - 2 \leq 0$, then we are actually increase k by $2 - c_X$. We do not worry about this trouble due to both theoretical and practical reasons. Theoretically, the definition of kernelization does not forbid increasing k, and we refer readers who feel uncomfortable with this to the monographs [13,16,24]. Practically, 1) it will not really enlarge or complicate the graph, and therefore any reasonable algorithms will work as the same; 2) this case will not happen too much, otherwise the graph should be very similar to a star, and easy to solve; 3) even using the original value of k, our kernel size is bounded by $3k$.

The proof of Lemma 6 goes almost the same, with only the constant slightly enlarged. Due to the relaxation of the condition of reducible closed neighborhood from $\rho(v) < |N(v)|$ to $\rho(v) \leq |N(v)| - 1$, the number of vertices in the kernel for real-weighted version is bounded by $2.5k$.

5   Discussion

One very interesting observation is that for the unweighted version, by the definition of simple series modules, all of the following are exactly the same:

$$N[u] = N[M], \quad \delta(u) = \delta(M), \quad \text{and} \quad \gamma(N[u]) = \gamma(N[M]),$$

where M is the simple series module containing vertex u, and $\delta(M)$ is a natural generalization of definition $\delta(v)$. Thus it does not matter we use the module or any vertex in it, that is, every vertex is a full representative for the simple series module it lies in. Although there has been a long list of linear algorithms for finding modular decomposition for an undirected graph (see a comprehensive survey by de Montgolfier [23]), it is very time-comsuming because the big constant hidden behind the big-O [26], and considering that the modular decopmosition needs to be re-constructed after each iteration, this will be helpful. It is somehow surprising that the previous kernelization algorithms can be significantly simplified by avoiding modular decomposition. Being more suprising, this enables our approach to apply for the weighted version, because one major weakness of modular decomposition is its inability in handling weights.

One similar problem on inconsistant information is the FEEDBACK VERTEX SET ON TOURNAMENT (FAST) problem, which asks the reverse of minimum number of arcs to make a tournament transtive. Given the striking resemblances between CLUSTER EDITING and FAST, and a series of "one-stone-two-birds" approximation algorithms [1,27] which only take advantage of the commonalities between them, we are strongly attempted to compare the results of these two problems from the parameterized aspect.

For the kernelization, our result already matches the best kernel, $(2 + \epsilon)k$ for weighted FAST of Bessy et al. [5], which is obtained based on a complicated PTAS [21].

For the algorithms, Alon et al. [2] managed to generalize the famous color coding approach to give a subexponential FPT algorithm for FAST. This is the first subexponential FPT algorithm out of bidimensionality theory, which was a systematic way to obtain subexponential algorithms, and has been intensively studied. This is an exciting work, and opens a new direction for further work. Indeed, immediately after the appearance of [2], for unweighted version, Feige reported an improved algorithm [14] that is far simpler and uses pure combinatorial approach. Recently, Karpinski and Schudy reached the same result for weighted version [20]. Based on their striking resemblances, we conjecture that there is also a subexponential algorithm for the CLUSTER EDITING problem.

References

1. Ailon, N., Charikar, M., Newman, A.: Aggregating inconsistent information: ranking and clustering. J. ACM 55(5), Article 23, 1–27 (2008)
2. Alon, N., Lokshtanov, D., Saurabh, S.: Fast FAST. In: Albers, S., Marchetti-Spaccamela, A., Matias, Y., Nikoletseas, S., Thomas, W. (eds.) ICALP 2009. LNCS, vol. 5555, pp. 49–58. Springer, Heidelberg (2009)
3. Bansal, N., Blum, A., Chawla, S.: Correlation clustering. Machine Learning 56(1), 89–113 (2004)
4. Berkhin, P.: A survey of clustering data mining techniques. In: Grouping Multidimensional Data, pp. 25–71. Springer, Heidelberg (2006)
5. Bessy, S., Fomin, F.V., Gaspers, S., Paul, C., Perez, A., Saurabh, S., Thomassé, S.: Kernels for feedback arc set in tournaments. In: CoRR, abs/0907.2165 (2009)

6. Böcker, S., Briesemeister, S., Klau, G.W.: Exact algorithms for cluster editing: evaluation and experiments. Algorithmica (in press)
7. Böcker, S., Briesemeister, S., Bui, Q.B.A., Truss, A.: Going weighted: parameterized algorithms for cluster editing. Theoretical Computer Science 410, 5467–5480 (2009)
8. Charikar, M., Guruswami, V., Wirth, A.: Clustering with qualitative information. Journal of Computer and System Sciences 71(3), 360–383 (2005)
9. Chen, J., Meng, J.: A 2k kernel for the cluster editing problem. In: COCOON 2010. LNCS, vol. 6196, pp. 459–468. Springer, Heidelberg (2010)
10. Chen, Z.-Z., Jiang, T., Lin, G.: Computing phylogenetic roots with bounded degrees and errors. Siam J. Comp. 32(4), 864–879 (2003)
11. Cunningham, W.H., Edmonds, J.: A combinatorial decomposition theory. Canad. J. Math. 32(3), 734–765 (1980)
12. Dean, J., Henzinger, M.R.: Finding related pages in the World Wide Web. Computer Networks 31, 1467–1479 (1999)
13. Downey, R.G., Fellows, M.R.: Parameterized Complexity. Springer, Heidelberg (1999)
14. Feige, U.: Faster FAST. In: CoRR, abs/0911.5094 (2009)
15. Fellows, M.R., Langston, M.A., Rosamond, F.A., Shaw, P.: Efficient parameterized preprocessing for cluster editing. In: Csuhaj-Varjú, E., Ésik, Z. (eds.) FCT 2007. LNCS, vol. 4639, pp. 312–321. Springer, Heidelberg (2007)
16. Flum, J., Grohe, M.: Parameterized Complexity Theory. Springer, Heidelberg (2006)
17. Gramm, J., Guo, J., Hüffner, F., Niedermeier, R.: Graph-modeled data clustering: exact algorithms for clique generation. Theory of Computing Systems 38(4), 373–392 (2005)
18. Guo, J.: A more effective linear kernelization for cluster editing. Theor. Comput. Sci. 410(8-10), 718–726 (2009)
19. Hearst, M.A., Pedersen, J.O.: Reexamining the cluster hypothesis: scatter/gather on retrieval results. In: Proceedings of SIGIR, pp. 76–84 (1996)
20. Karpinski, M., Schudy, W.: Faster algorithms for feedback arc set tournament, Kemeny rank aggregation and betweenness tournament. In: CoRR, abs/1006.4396 (2010)
21. Kenyon-Mathieu, C., Schudy, W.: How to rank with few errors. In: ACM Symposium on Theory of Computing (STOC), pp. 95–103 (2007)
22. Möhring, R.H., Radermacher, F.J.: Substitution decomposition for discrete structures and connections with combinatorial optimization. Ann. Discrete Math. 19(95), 257–355 (1984)
23. de Montgolfier, F.: Décomposition modulaire des graphes. Théorie, extensions et algorithmes. Thése de doctorat, Université Montpellier II (2003)
24. Niedermeier, R.: Invitation to Fixed-Parameter Algorithms. Oxford University Press, Oxford (2006)
25. Shamir, R., Sharan, R., Tsur, D.: Cluster graph modification problems. Discrete Appl. Math. 144(1-2), 173–182 (2004)
26. Tedder, M., Corneil, D., Habib, M., Paul, C.: Simpler linear-time modular decomposition via recursive factorizing permutations. In: Aceto, L., Damgård, I., Goldberg, L.A., Halldórsson, M.M., Ingólfsdóttir, A., Walukiewicz, I. (eds.) ICALP 2008, Part I. LNCS, vol. 5125, pp. 634–645. Springer, Heidelberg (2008)
27. van Zuylen, A., Williamson, D.P.: Deterministic pivoting algorithms for constrained ranking and clustering problems. Mathematics of Operations Research 34(3), 594–620 (2009)

Computing the Deficiency of Housing Markets with Duplicate Houses⋆

Katarína Cechlárová[1] and Ildikó Schlotter[2]

[1] Institute of Mathematics, Faculty of Science, P.J. Šafárik University,
Jesenná 5, 040 01 Košice, Slovakia
katarina.cechlarova@upjs.sk
[2] Budapest University of Technology and Economics,
H-1521 Budapest, Hungary
ildi@cs.bme.hu

Abstract. The model of a housing market, introduced by Shapley and Scarf in 1974 [14], captures a fundamental situation in an economy where each agent owns exactly one unit of some indivisible good: a house. We focus on an extension of this model where duplicate houses may exist. As opposed to the classical setting, the existence of an economical equilibrium is no longer ensured in this case. Here, we study the *deficiency* of housing markets with duplicate houses, a notion measuring how close a market can get to an economic equilibrium. We investigate the complexity of computing the deficiency of a market, both in the classical sense and also in the context of parameterized complexity.

We show that computing the deficiency is NP-hard even under several severe restrictions placed on the housing market, and thus we consider different parameterizations of the problem. We prove W[1]-hardness for the case where the parameter is the value of the deficiency we aim for. By contrast, we provide an FPT algoritm for computing the deficiency of the market, if the parameter is the number of different house types.

Keywords: Housing market, Economic equilibrium, Parameterized complexity.

1   Introduction

The standard mathematical model of a housing market was introduced in the seminal paper of Shapley and Scarf [14], and has successfully been used in the analysis of real markets such as campus housing [15], assigning students to schools [1], and kidney transplantation [13]. In a housing market there is a set of agents, each one owns one unit of a unique indivisible good (house) and wants to exchange it for another, more preferred one; the preference relation of an agent is a linearly ordered list (possibly with ties) of a subset of goods.

⋆ This work was supported by the VEGA grants 1/0035/09 and 1/0325/10 (Cechlárová), by the Hungarian National Research Fund OTKA 67651 (Schlotter) and by the Slovak-Hungarian APVV grant SK-HU-003-08.

V. Raman and S. Saurabh (Eds.): IPEC 2010, LNCS 6478, pp. 72–83, 2010.

Shapley and Scarf proved that in such a market an economic equilibrium always exists. A constructive proof in the form of the Top Trading Cycles algorithm is attributed to Gale (see [14]).

However, if we drop the assumption that each agent's house is unique, it may happen that the economic equilibrium no longer exists, and it is even NP-complete to decide its existence, see Fekete, Skutella, and Woeginger [8]. Further studies revealed that the border between easy and hard cases is very narrow: if agents have strict preferences over house types then a polynomial algorithm to decide the existence of an equilibrium is possible, see Cechlárová and Fleiner [4]. Alas, the problem remains NP-complete even if each agent only distinguishes between three classes of house types (trichotomous model): better house types, the type of his own house, and unacceptable house types [4]. So it becomes interesting to study the so-called *deficiency* of the housing market, i.e. the minimum possible number of agents who cannot get a most preferred house in their budget set under some prices of the house types.

In the present paper we give several results concerning the computation of the deficiency of housing markets, also from the parameterized complexity viewpoint. First, we show that the deficiency problem is NP-hard even in the case when each agent prefers only one house type to his endowment, and the maximum number of houses of the same type is two. This result is the strongest possible one in the sense that each housing market without duplicate houses admits an equilibrium [14]. Then we show that the deficiency problem is W[1]-hard with the parameter α describing the desired value of the deficiency, even if each agent prefers at most two house types to his own house, and the preferences are strict. Notice that the parameterized complexity of the case when each agent prefers only one house type to his endowment remains open. On the other hand, assuming that the preferences are strict, we provide a brute force algorithm that decides whether the deficiency is at most α in polynomial time for each fixed constant α. This shows that the problem is contained in XP when parameterized by α. This is in a strict contrast with the trichotomous model where even the case $\alpha = 0$ is NP-hard [4]. Finally, we provide an FPT algorithm for computing the deficiency (that works irrespectively of the type of preferences) if the parameter is the number of different house types.

To put our results into a broader context, let us mention that for general markets with divisible goods the celebrated Arrow–Debreu Theorem [2] guarantees the existence of an equilibirum under some mild conditions on agents' preferences. By contrast, it is well-known that in case of indivisible goods an equilibrium may not exist. From many existing approaches trying to cope with this nonexistence, let us mention Deng, Papadimitriou, and Safra [6] who introduced a notion of an ε-approximate equilibrium as one where the market "clears approximately", and the utility of each agent is within ε from the maximum possible utility in his budget set. They concentrated on approximation possibilites, and as far as we know such questions have not been studied yet from the parameterized complexity viewpoint.

2 Preliminaries

The paper is organized as follows. First, we introduce the model under examination, and give a brief overview of the basic concepts of parameterized complexity. In Section 3 we present some hardness results, whilst Section 4 is devoted to the proposal of two algorithms concerned with the computation of deficiency.

2.1 Description of the Model

Let A be a set of N agents, H a set of M house types. The endowment function $\omega : A \to H$ assigns to each agent the type of house he originally owns. In the classical model of Shapley and Scarf [14], $M = N$ and ω is a bijection. If $N > M$ we say that the housing market has duplicate houses. Preferences of agent a are given in the form of a linear preference list $P(a)$. The house types appearing in the preference list of agent a are said to be acceptable, and we assume that $\omega(a)$ belongs to the least preferred acceptable house types for each $a \in A$. The notation $i \succeq_a j$ means that agent a prefers house type i to house type j. If $i \succeq_a j$ and simultaneously $j \succeq_a i$, we say that house types i and j are in a tie in a's preference list; if $i \succeq_a j$ and not $j \succeq_a i$, we write $i \succ_a j$ and say that agent a strictly prefers house type i to house type j. (If the agent is clear from the context, the subscript will be omitted.) The N-tuple of preferences $(P(a), a \in A)$ will be denoted by $\mathcal{P}$ and called the preference profile.

The housing market is the quadruple $\mathcal{M} = (A, H, \omega, \mathcal{P})$. We also define the submarket of $\mathcal{M}$ restricted to some agents of $S \subseteq A$ in the straightforward way. We say that $\mathcal{M}$ is a housing market with strict preferences if there are no ties in $\mathcal{P}$. The maximum house-multiplicity of a market $\mathcal{M}$, denoted by $\beta(\mathcal{M})$, is the maximum number of houses of the same type, i.e. $\beta(\mathcal{M}) = \max_{h \in H} |\{a \in A : \omega(a) = h\}|$. The maximum number of preferred house types in the market, denoted by $\gamma(\mathcal{M})$, is the maximum number of house types that any agent might strictly prefer to its own house, i.e. $\gamma(\mathcal{M}) = \max_{a \in A} |\{h \in H : h \succ_a \omega(a)\}|$. We say that the market $\mathcal{M}$ is simple, if $\gamma(\mathcal{M}) = 1$.

The set of types of houses owned by agents in $S \subseteq A$ is denoted by $\omega(S)$. For each agent $a \in A$ we denote by $f_T(a)$ the set of the most preferred house types from $T \subseteq H$. For a set of agents $S \subseteq A$ we let $f_T(S) = \bigcup_{b \in S} f_T(b)$. For one-element sets of the form $\{h\}$ we often write simply h in expressions like $\omega(S) = h$, $f_T(S) = h$, etc.

We say that a function $x : A \to H$ is an allocation if there exists a permutation π on A such that $x(a) = \omega(\pi(a))$ for each $a \in A$. Notation $x(S)$ for $S \subseteq A$ denotes the set $\bigcup_{a \in S} \{x(a)\}$. In the whole paper, we assume that allocations are individually rational, meaning that $x(a)$ is acceptable for each $a \in A$. Notice that for each allocation x, the set of agents can be partitioned into directed cycles (trading cycles) of the form $K = (a_0, a_1, \ldots, a_{\ell-1})$ in such a way that $x(a_i) = \omega(a_{i+1})$ for each $i = 0, 1, \ldots, \ell - 1$ (here and elsewhere, indices for agents on cycles are taken modulo ℓ). We say that agent a is trading in allocation x if $x(a) \neq \omega(a)$.

Given a price function $p : H \to \mathbb{R}$, the *budget set* of agent a according to p is the set of house types that a can afford, i.e. $\{h \in H : p(h) \le p(\omega(a))\}$. A pair (p, x), where $p : H \to \mathbb{R}$ is a price function and x is an allocation, is an *economic equilibrium* for market $\mathcal{M}$ if $x(a)$ is among the most preferred house types in the budget set of a.

It is known that if (p, x) is an economic equilibrium, then x is *balanced* with respect to p, i.e. $p(x(a)) = p(\omega(a))$ for each $a \in A$ (see [8,4]).

As a housing market with duplicate houses may admit no equilibrium, we are interested in price-allocation pairs that are "not far" from the equilibrium. One possible measure of this distance was introduced in [4] by the notion of *deficiency* of the housing market.

An agent is said to be *unsatisfied* with respect to (p, x) if $x(a)$ is not among the most preferred house types in his budget set according to p. We denote by $\mathcal{D}_{\mathcal{M}}(p, x)$ the set of unsatisfied agents in $\mathcal{M}$ w.r.t. (p, x); more formally

$$\mathcal{D}_{\mathcal{M}}(p, x) = \{a \in A : \exists h \in H \text{ such that } h \succ_a x(a) \text{ and } p(h) \le p(\omega(a))\}.$$

Given a price function p and an allocation x balanced w.r.t. p, we say that (p, x) is an α-*deficient equilibrium*, if $|\mathcal{D}_{\mathcal{M}}(p, x)| = \alpha$. Clearly, an economic equilibrium is a 0-deficient equilibrium. The deficiency of a housing market $\mathcal{M}$, denoted by $\mathcal{D}(\mathcal{M})$ is the minimum α such that $\mathcal{M}$ admits an α-deficient equilibrium. Given a housing market $\mathcal{M}$ and some $\alpha \in \mathbb{N}$, the task of the DEFICIENCY problem is to decide whether $\mathcal{D}(\mathcal{M}) \le \alpha$.

We shall deal with the computational complexity of DEFICIENCY. For computational purposes, we shall say that the *size of the market* is equal to the total length of all preference lists of the agents, denoted by L.

2.2 Parameterized Complexity

The aim of parameterized complexity theory is to study the computational complexity of NP-hard problems in a more detailed manner than in the classical setting. In this approach, we regard the running time of an algorithm as a function that depends not only on the size but also on some other crucial properties of the input. To this end, for each input of a given problem we define a so-called *parameter*, usually an integer, describing some important feature of the input.

Given a parameterized problem, we say that an algorithm is *fixed-parameter tractable* or *FPT*, if its running time on an input I with parameter k is at most $f(k)|I|^{O(1)}$ for some computable function f that only depends on k, and not on the size $|I|$ of the input. The intuitive motivation for this definition is that such an algorithm might be tractable even for large instances, if the parameter k is small. Hence, looking at some parameterized version of an NP-hard problem, an FPT algorithm may offer us a way to deal with a large class of typical instances.

The parameterized analysis of a problem might also reveal its W[1]-hardness, which is a strong argument showing that an FPT algorithm is unlikely to exist. Such a result can be proved by means of an FPT-reduction from an already

known W[1]-hard problem such as CLIQUE. Instead of giving the formal definitions, we refer to the books by Flum and Grohe [9] or by Niedermeier [12]. For
a comprehensive overview, see the monograph of Downey and Fellows [7].

Considering the DEFICIENCY problem, the most natural parameters, each
describing some key property of a market $\mathcal{M}$, are as follows: the number of
different houses types $|H| = M$, the maximum house-multiplicity $\beta(\mathcal{M})$, and the
maximum number of preferred house types $\gamma(\mathcal{M})$ in the market. The value α
describing the deficiency of the desired equilibrium can also be a meaningful
parameter, if we aim for a price-allocation pair that is "almost" an economic
equilibrium. The next sections investigate the influence of these parameters on
the computational complexity of the DEFICIENCY problem.

3   Hardness Results

We begin with a simple observation which will be used repeatedly later on.

Lemma 1. *Let $\mathcal{M} = (A, H, \omega, \mathcal{P})$ be a housing market, p a price function and x
a balanced allocation for p. Suppose $\omega(U) = u$ and $\omega(Z) = z$ for some sets $U, Z \subseteq
A$ of agents. Suppose also that $f_H(Z) = u$ and $f_T(U) = z$ where $T \subseteq H$
contains the budget sets of all agents in U. Then $p(u) \neq p(z)$ implies that at
least $\min\{|U|, |Z|\}$ agents in $U \cup Z$ are unsatisfied with respect to (p, x).*

Proof. If $p(u) \neq p(z)$ and the allocation is balanced, agents from the two sets
cannot trade with each other. Therefore, due to the assumptions, if $p(u) > p(z)$
then all the agents in U are unsatisfied; if $p(z) > p(u)$ then all the agents in Z
must be unsatisfied, and the assertion follows.   □

Theorem 1. *The DEFICIENCY problem is NP-complete even for simple markets $\mathcal{M}$ with $\beta(\mathcal{M}) = 2$.*

Proof. We provide a reduction from the DIRECTED FEEDBACK VERTEX SET.
We shall take its special version where the out-degree of each vertex is at most
2, which is also NP-complete, see Garey and Johnson [10], Problem GT7.

Given a directed graph $G = (V, E)$ with vertex set V and arc set E such that
the outdegree of each vertex is at most 2, and an integer k, we construct a simple
housing market $\mathcal{M}$ with $\beta(\mathcal{M}) = 2$ such that $\mathcal{D}(\mathcal{M}) \leq k$ if and only if G admits
a feedback vertex set of cardinality at most k.

First, there are two house types $\hat{v}, \hat{v}'$ for each vertex $v \in V$ and $k + 1$ house
types $\hat{e}_1, \ldots, \hat{e}_{k+1}$ for each arc $e \in E$. The agents and their preferences are given
in Table 1. Here and later on, we write $[n]$ for $\{1, 2, \ldots, n\}$. The last entry in the
list of each agent represents its endowment.

It is easy to see that $\mathcal{M}$ is simple, $\beta(\mathcal{M}) = 2$, the number of house types
in $\mathcal{M}$ is $2|V| + (k + 1)|E|$ and the number of agents $|V| + (2k + 3)|E|$. To make
the following arguments more straightforward, let us imagine $\mathcal{M}$ as a directed
multigraph $\bar{G}$, where vertices are house types, and an arc from vertex $h \in H$
to vertex $h' \in H$ corresponds to an agent a with $\omega(a) = h$ and $h' \succ_a h$. Now,

Table 1. Endowments and preferences of agents in the market

agent	preference list
one agent $\bar{v}$ for each $v \in V$	$\hat{v}' \succ \hat{v}$
one agent $\bar{e}$ for each $e = vu \in E$	$\hat{e}_1 \succ \hat{v}'$
two agents $\bar{e}_i$ for each $e = vu \in E; i \in [k]$	$\hat{e}_{i+1} \succ \hat{e}_i$
two agents $\bar{e}_{k+1}$ for each $e = vu \in E$	$\hat{u} \succ \hat{e}_{k+1}$

each directed cycle C in G has its counterpart $\bar{C}$ in $\bar{G}$, but each arc $e = vu$ on C corresponds to a "thick path" $\bar{v} \to \bar{u}$ containing $k + 1$ consecutive pairs of parallel arcs in $\bar{G}$ (agents $\bar{e}_i, i \in [k + 1]$). We shall also say that agents $\bar{e}, \bar{e}_i, i = 1, 2, \ldots, k + 1$ are associated with the arc $e = vu$.

Now suppose that G contains a feedback vertex set W with cardinality at most k. For each $v \in W$ we remove agent $\bar{v}$ (together with its endowed house of type $\hat{v}$) from $\mathcal{M}$. The obtained submarket is acyclic, so assigning prices to house types in this submarket according to a topological ordering, we get a price function and an allocation with no trading in $\mathcal{M}$, where the only possible unsatisfied agents are the agents $\{\bar{v} \mid v \in W\}$.

Conversely, suppose that $\mathcal{M}$ admits a k-deficient equilibrium (p, x). If x produced any trading, then each trading cycle would necessarily involve some thick path $\bar{v} \to \bar{u}$ and thus exactly one agent from each pair $\bar{e}_i, i \in [k + 1]$ on this thick path, making at least $k + 1$ agents unsatisfied. Hence, there is no trading in x. Now, take any cycle $C = (v_1, v_2, \ldots, v_r, v_1)$ in G. Since it is impossible that all the inequalities $p(\hat{v}_1) < p(\hat{v}_2), p(\hat{v}_2) < p(\hat{v}_3), \ldots, p(\hat{v}_r) < p(\hat{v}_1)$ along the vertices of C are fulfilled, at least one agent in $\bar{C}$ is unsatisfied. If this agent is $\bar{v}$ or belongs to the set of agents associated to an arc $e = vu$, we choose vertex v into a set W. It is easy to see that W is a feedback vertex set and $|W| \leq k$. □

Theorem 1 yields that DEFICIENCY remains NP-hard even if $\gamma(\mathcal{M}) = 1$ and $\beta(\mathcal{M}) = 2$ holds for the input market $\mathcal{M}$. This immediately implies that DEFICIENCY is not in the class XP w.r.t. the parameters $\beta(\mathcal{M})$, describing the maximum house-multiplicity, and $\gamma(\mathcal{M})$, denoting the maximum number of preferred house types. Next, we show that regarding α (the desired value of deficiency) as a parameter is not likely to yield an FPT algorithm, not even if $\gamma(\mathcal{M}) = 2$.

Theorem 2. *The* DEFICIENCY *problem for a market $\mathcal{M}$ with strict preferences and with $\gamma(\mathcal{M}) = 2$ is W[1]-hard with the parameter α.*

Proof. We are going to show a reduction from the W[1]-hard CLIQUE problem, parameterized by the size of the solution. Given a graph G and an integer k as the input of CLIQUE, we will construct a housing market $\mathcal{M} = (\mathcal{A}, H, \omega, \mathcal{P})$ with strict preferences and with $\gamma(\mathcal{M}) = 2$ in polynomial time such that $\mathcal{M}$ has deficiency at most $\alpha = k^2$ if and only if G has a clique of size k. Since α depends only on k, this construction yields an FPT-reduction, and we obtain that DEFICIENCY is W[1]-hard with the parameter α.

Table 2. The preference profile of the market $\mathcal{M}$

agent	preferences	"multiplicity"
$a \in A$	$\hat{c} \succ \hat{a}$	$|A| = n - k$
$b \in B$	$\hat{a} \succ \hat{d} \succ \hat{b}$	$|B| = 2m - k(k-1)$
$b \in B'$	$\hat{a} \succ \hat{b}$	$|B'| = t - (2m - k(k-1))$
$f \in F_1^c$	$\hat{c} \succ \hat{f}_2^c \succ \hat{f}_1^c$	$|F_1^c| = k$
$f \in F_2^c$	$\hat{f}_1^c \succ \hat{f}_2^c$	$|F_2^c| = k + 1$
$f \in F_1^d$	$\hat{d} \succ \hat{f}_2^d \succ \hat{f}_1^d$	$|F_1^d| = k(k-1)$
$f \in F_2^d$	$\hat{f}_1^d \succ \hat{f}_2^d$	$|F_2^d| = k(k-1) + 1$
$c_i \in C$	$\hat{a} \succ \hat{q}_i \succ \hat{c}$	$|C| = n$
$d \in D$	$\hat{b} \succ \hat{s}_i \succ \hat{d}$ if $d \in \{d_i^1, d_i^2\}$	$|D| = 2m$
$q_i \in Q$	$\hat{f}_1^c \succ \hat{d} \succ \hat{q}_i$	$|Q| = n$
$s_i^1 \in S$	$\hat{q}_x \succ \hat{f}_1^d \succ \hat{s}_i$ where $e_i = v_x v_y \in E$, $x < y$	$|\{s_i^1 \mid i \in [m]\}| = m$.
$s_i^2 \in S$	$\hat{q}_y \succ \hat{f}_1^d \succ \hat{s}_i$ where $e_i = v_x v_y \in E$, $x < y$	$|\{s_i^2 \mid i \in [m]\}| = m$.

Let $G = (V, E)$ with $V = \{v_1, v_2 \ldots, v_n\}$ and $E = \{e_1, e_2 \ldots, e_m\}$. We can clearly assume $n > k^2 + k$, as otherwise we could simply add the necessary number of isolated vertices to G, without changing the answer to the CLIQUE problem. Similarly, we can also assume $m > k^2$, as otherwise we can add the necessary number of independent edges (with newly introduced endvertices) to G.

The set of house types in $\mathcal{M}$ is $H = \{\hat{a}, \hat{b}, \hat{c}, \hat{d}, \hat{f}_1^c, \hat{f}_2^c, \hat{f}_1^d, \hat{f}_2^d\} \cup \hat{Q} \cup \hat{S}$, where $\hat{Q} = \{\hat{q}_i \mid i \in [n]\}$ and $\hat{S} = \{\hat{s}_i \mid i \in [m]\}$. Let $t = \max\{2m - k(k-1), n - k + \alpha + 1\}$. First, we define seven sets of agents, $A, B, B', F_1^c, F_2^c, F_1^d$ and F_2^d. The cardinality of these agent sets are shown in Table 2; note that there might be zero agents in the set B'. Any two agents will have the same preferences and endowments if they are contained in the same set among these seven sets. Additionally, we also define agents in $C \cup D \cup Q \cup S$, where $C = \{c_i \mid i \in [n]\}$, $Q = \{q_i \mid i \in [n]\}$, $D = \{d_i^1, d_i^2 \mid i \in [m]\}$, and $S = \{s_i^1, s_i^2 \mid i \in [m]\}$. The preference profile of the market is shown on Table 2. Again, the endowment of an agent is the last house type in its preference list.

First, suppose that $\mathcal{M}$ admits a balanced allocation x for some price function p such that (p, x) is α-deficient. Observe that $f_H(c) = \hat{a}$ for each $c \in C$, $f_H(a) = \hat{c}$ for each $a \in A$. By $|C| > |A| > \alpha$ and Lemma 1, we obtain that $p(\hat{a}) = p(\hat{c})$ must hold. Moreover, by $|C| = |A| + k$ we also know that there are at least k agents in C who cannot obtain a house of type $\hat{a}$, let $C^* \subseteq C$ be a set containing k such agents. Clearly, agents in C^* are unsatisfied. Moreover, if all agents in $C \setminus C^*$ are satisfied, then they must trade with the agents of A.

Second, note that $f_H(b) = \hat{a}$ for each $b \in B \cup B'$, so $|B \cup B'| > |A| + \alpha$ (which follows from the definition of t) implies that $p(\hat{a}) > p(\hat{b})$ must hold, as otherwise more than α agents in $B \cup B'$ could afford a house of type $\hat{a}$ but would not be able to buy one. Thus, the budget set of the agents $B \cup B'$ does not contain the house type $\hat{a}$. In particular, we get that no agent in B' is trading in x. Note also that $f_{H \setminus \{\hat{a}\}}(b) = \hat{d}$ and $f_H(d) = \hat{b}$ for each $b \in B$ and $d \in D$, so Lemma 1 and $|D| > |B| > \alpha$ yield that only $p(\hat{b}) = p(\hat{d})$ is possible. Taking into account that $|B| = |D| - k(k-1)$, we know that there must be at least $k(k-1)$ unsatisfied agents in D who are not assigned a house of type $\hat{b}$; let D^* denote this set of unsatisfied agents. Notice that if all the agents in $D \setminus D^*$ are satisfied, then they must be trading with the agents of B.

As $C^* \cup D^*$ contains α unsatisfied agents w.r.t. (p, x), and the deficiency of (p, x) is at most α, we get that no other agent can be unsatisfied. By the above arguments, this implies $x(A) = \hat{c}$, $x(C \setminus C^*) = \hat{a}$, $x(B) = \hat{d}$, and $x(D \setminus D^*) = \hat{b}$.

Next, we will show that $x(f) = \hat{c}$ for each $f \in F_1^c$ and $x(f) = \hat{d}$ for each $f \in F_1^d$. We will only prove the first claim in detail, as the other statement is symmetric. First, observe that $p(\hat{f}_2^c) \geq p(\hat{f}_1^c)$ is not possible, because by $f_H(F_2^c) = \hat{f}_1^c$ and $|F_2^c| > |F_1^c|$ such a case would imply at least one unsatisfied agent in F_2^c. Thus, we know $p(\hat{f}_2^c) < p(\hat{f}_1^c)$, which means that $\hat{f}_2^c$ is in the budget set of each agent in F_1^c. But since they do not buy such a house (as x is balanced), and they cannot be unsatisfied, we obtain that they must prefer their assigned house to $\hat{f}_2^c$. Thus, for each agent f in F_1^c we obtain $x(f) = \hat{c}$, proving the claim. The most important consequence of these facts is that every agent in $C^* \cup D^*$ must be trading according to x, as otherwise the agents in F_1^c and in F_1^d would not be able to get a house of type $\hat{c}$ or $\hat{d}$, respectively.

Recall that agents in C^* are unsatisfied, as they do not buy houses of type $\hat{a}$. But since they are trading, they must buy k houses from $\hat{Q}$; let $\hat{q}_{i_1}, \hat{q}_{i_2}, \ldots, \hat{q}_{i_k}$ be these houses. The agents $F_1^c, C^*, Q^* = \{q_{i_j} \mid j \in [k]\}$ trade with each other at price $p(\hat{c})$, yielding $x(F_1^c) = \hat{c}$, $x(C^*) = \omega(Q^*)$ and $x(Q^*) = \hat{f}_1^c$.

Similarly, the $k(k-1)$ agents in D^* must be trading, buying $k(k-1)$ houses of the set $\hat{S}$; let S^* denote the owners of these houses. Now, it should be clear that exactly $2m - k(k-1)$ houses of type $\hat{d}$ are assigned to the agents B, and the remaining $k(k-1)$ such houses are assigned to the agents F_1^d.

It should also be clear that the agents S^* are trading with agents F_1^d, so we obtain $x(F_1^d) = \omega(D^*) = \hat{d}$ and $x(D^*) = \omega(S^*)$. Thus, agents of $Q \setminus Q^*$ can neither be assigned a house of type $\hat{d}$ (as those are assigned to the agents $B \cup F_1^d*$), nor a house of type $\hat{f}_1^c$ (as those are assigned to agents in Q^*). As agents of $Q \setminus Q^*$ cannot be unsatisfied, we have that $p(\hat{q}_i) < p(\hat{d}) < p(\hat{f}_1^c)$ holds for each $q_i \in Q \setminus Q^*$, meaning that these agents do not trade according to x. (Recall that $p(\hat{d}) = p(\hat{b}) < p(\hat{a}) = p(\hat{c}) = p(\hat{f}_1^c)$.)

Now, if $x(d) = s_i$ for some agent $d \in D^*$ and $i \in [m]$, then we know that $p(\hat{s}_i) = p(\hat{d}) = p(\hat{f}_1^d)$. As neither of s_i^1 and s_i^2 can be unsatisfied, but neither of them can get a house from $\hat{Q}$, it follows that both of them must obtain a house of type $\hat{f}_1^d$. Therefore, the set S^* must contain pairs of agents owning the same type of house, i.e. $S^* = \{s_{j_i}^1, s_{j_i}^2 \mid i \in [k(k-1)]\}$.

Let us consider the agents s_j^1 and s_j^2 in S^*, and let v_x and v_y denote the two endpoints of the edge e_j, with $x < y$. Since s_j^1 prefers $\hat{q}_x$ to $x(s_j^1) = \hat{f}_1^d$, we must have $p(\hat{s}_j) < p(\hat{q}_x)$, since s_j^1 must not be unsatisfied. Similarly, s_j^2 prefers $\hat{q}_y$ to $x(s_j^2) = \hat{f}_1^d$, implying $p(\hat{s}_j) < p(\hat{q}_y)$. Taking into account that $p(\hat{s}_j) = p(\hat{d}) >$

$p(\hat{q}_i)$ for each $q_i \in Q \setminus Q^*$, we get that both q_x and q_y must be contained in Q^*. Hence, each edge in the set $E^* = \{e_j \mid s_j^1, s_j^2 \in S^*\}$ in G must have endpoints in the vertex set $V^* = \{v_i \mid q_i \in Q^*\}$. This means that the $\binom{k}{2}$ edges in E^* have altogether k endpoints, which can only happen if V^* induces a clique of size k in G. This finishes the soundness of the first direction of the reduction.

For the other direction, suppose that V^* is a clique in G of size k. We construct an α-deficient equilibrium (p, x) for $\mathcal{M}$ as follows. Let $I^* = \{i \mid v_i \in V^*\}$ and $J^* = \{j \mid e_j = v_x v_y, v_x \in V^*, v_y \in V^*\}$ denote the indices of the vertices and edges of this clique, respectively. We define $Q^* = \{q_i \mid i \in I^*\}$, $C^* = \{c_i \mid i \in I^*\}$, $S^* = \{s_j^1, s_j^2 \mid j \in J^*\}$, and $D^* = \{d_j^1, d_j^2 \mid j \in J^*\}$. Now, we are ready to define the price function p as follows.

$$p(\hat{a}) = p(\hat{c}) = p(\hat{f}_1^c) = p(\hat{q}_i) = 4 \text{ for each } q_i \in Q^*,$$
$$p(\hat{b}) = p(\hat{d}) = p(\hat{f}_1^d) = p(\hat{s}_i) = 3 \text{ for each } i \text{ where } s_i^1, s_i^2 \in S^*,$$
$$p(\hat{q}_i) = 2 \text{ for each } q_i \in Q \setminus Q^*,$$
$$p(h) = 1 \text{ for each remaining house type } h.$$

It is straightforward to verify that the above prices form an α-deficient equilibrium with the allocation x, defined below.

$$\begin{aligned}
&x(A) = \omega(C \setminus C^*), && x(C \setminus C^*) = \hat{a}, \\
&x(B) = \omega(D \setminus D^*), && x(D \setminus D^*) = \hat{b}, \\
&x(F_1^c) = \omega(C^*), && x(C^*) = \omega(Q^*), && x(Q^*) = \hat{f}_1^c, \\
&x(F_1^d) = \omega(D^*), && x(D^*) = \omega(S^*), && x(S^*) = \hat{f}_1^d, \\
&x(a) = \omega(a) \text{ for each remaining agent } a.
\end{aligned}$$

It is easy to see that $\mathcal{D}(p, x) = C^* \cup D^*$, implying that (p, x) is indeed α-deficient by $|C^* \cup D^*| = k + k(k - 1) = \alpha$. The only non-trivial observation we need during this verification is that $p(\hat{q}_x) > p(\hat{s}_i)$ and $p(\hat{q}_y) > p(\hat{s}_i)$ for any s_i, where v_x and v_y are the endpoints of e_i. These inequalities trivially hold if $s_i \notin S^*$. In the case $s_i \in S^*$ we know $v_x, v_y \in V^*$ (since e_i is an edge in the clique V^*), which yields $p(\hat{q}_x) = p(\hat{q}_y) = p(\hat{s}_i) + 1$.

Hence, the reduction is correct, proving the theorem. □

4 Algorithms for Computing the Deficiency

Theorem 2 implies that we cannot expect an algorithm with running time $f(\alpha)L^{O(1)}$ for some computable function f for deciding whether a given market has deficiency at most α. However, we present a simple brute force algorithm that solves the DEFICIENCY problem for strict preferences in $O(L^{\alpha+1})$ time, which is polynomial if α is a fixed constant. This means that DEFICIENCY is in XP with respect to the parameter α. Recall that due to the results of [4], no such algorithm is possible if ties are present in the preference lists, as even the case $\alpha = 0$ is NP-hard in the trichotomous model.

Theorem 3. *If the preferences are strict, then the DEFICIENCY problem can be solved in $O(L^{\alpha+1})$ time.*

Proof. Let $\mathcal{M} = (A, H, \omega, \mathcal{P})$ be the market given, and let α denote the deficiency what we aim for. Suppose (p, x) is an α-deficient equilibrium for $\mathcal{M}$, and let $\mathcal{D}_\mathcal{M}(p, x) = \{a_1, a_2, \ldots, a_\alpha\}$ be the set of unsatisfied agents. Let also $h_i = x(a_i)$ denote the house type obtained by the unsatisfied agent a_i for each $i \in [\alpha]$.

Now, we define a set of modified preference lists $\mathcal{P}[p, x]$ as follows: for each agent $a \in \mathcal{D}_\mathcal{M}(p, x)$ we delete every house type from its preference list, except for $x(a)$ and $\omega(a)$. We claim that (p, x) is an equilibrium allocation for the modified market $\mathcal{M}[p, x] = (A, H, \omega, \mathcal{P}[p, x])$. First, it is easy to see that x is balanced with respect to the price function p and for $\mathcal{M}[p, x]$, as neither the prices nor the allocation was changed. Thus, we only have to see that there are no unsatisfied agents in $\mathcal{M}[p, x]$ according to (p, x). By definition, in the market $\mathcal{M}[p, x]$ we know $x(a_i) = f_H(a_i)$ for each agent $a_i \in \mathcal{D}_\mathcal{M}(p, x)$. It should also be clear that for each other agent $b \notin \mathcal{D}_\mathcal{M}(p, x)$, we get that $x(b)$ is the first choice of b in its budget set according to p, since b was satisfied according to (p, x) in $\mathcal{M}$. Thus, b is also satisfied according to (p, x) in $\mathcal{M}[p, x]$. This means that (p, x) is indeed an equilibrium allocation for $\mathcal{M}[p, x]$.

For the other direction, it is also easy to verify that any equilibrium allocation (p', x') for $\mathcal{M}[p, x]$ results in an equilibrium for $\mathcal{M}$ with deficiency at most α, as only agents in $\mathcal{D}_\mathcal{M}(p, x)$ can be unsatisfied in $\mathcal{M}$ with respect to (p', x').

These observations directly indicate a simple brute force algorithm solving the DEFICIENCY problem. For any set $\{a_1, a_2, \ldots, a_\alpha\}$ of α agents, and for any α-tuple $h_1, h_2, \ldots, h_\alpha$ of house types such that h_i is in the preference list of a_i (for each $i \in [\alpha]$), find out whether there is an economic equilibrium for the modified market, constructed by deleting every house type except for h_i and $\omega(a_i)$ from the preference list of a_i, for each $i \in [\alpha]$. Finding an economic equilibrium for such a submarket can be carried out in $O(L)$ time using the algorithm provided by Cechlárová and Jelínková [5].

Note that we have L possibilities for choosing an arbitrary agent together with a house type from its preference list (as L is exactly the number of "feasible" agent-house pairs), so we have to apply the algorithm of [5] at most $\binom{L}{\alpha}$ times. Therefore, the running time of the whole algorithm is $O(L^{\alpha+1})$. The correctness of the algorithm follows directly from the above discussion.   □

Finally, we provide an FPT algorithm for the case where the parameter is the number of house types in the market.

Theorem 4. *There is a fixed-parameter tractable algorithm for computing the deficiency of a housing market with arbitrary preferences, where the parameter is the number M of house types in the market. The running time of the algorithm is $O(M^M \sqrt{N} L)$.*

Proof. Let $\mathcal{M} = (A, H, \omega, \mathcal{P})$ be a given housing market. If there is an α-deficient equilibrium (p, x) for $\mathcal{M}$ for some α, then we can modify the price function p to p' such that all prices are integers in $[M]$, and (p', x) forms an α-deficient equilibrium. Thus, we can restrict our attention to price functions from H to $[M]$.

The basic idea of the algorithm is the following: for each possible price function, we look for an allocation maximizing the number of satisfied agents. As a result, we get the minimum number of unsatisfied agents over all possible price functions. Note that we have to deal with exactly M^M price functions.

Given a price function $p : H \to [M]$ and an agent a, we denote by $T(a)$ the house types having the same price as $\omega(a)$, and by $B(a)$ the budget set of a.

Clearly, for any balanced allocation x w.r.t. p, we know $x(a) \in T(a)$. Thus, we can reduce the market by restricting the preference list of each agent a to the house types in $T(a)$; let $P'(a)$ denote the resulting list. The reduced market now defines a digraph G with vertex set A and arcs ab for agents $a, b \in A$ where b owns a house of type contained in $P'(a)$; note that each vertex has a loop attached to it. It is easy to see that any balanced allocation x indicates a cycle cover of G, and vice versa. (A cycle cover is a collection of vertex disjoint cycles covering each vertex.)

By definition, a is satisfied in some allocation x with respect to p, if $x(a) \in f_{B(a)}(a)$. We call an arc ab in G *important*, if $\omega(b)$ is contained in $f_{B(a)}(a)$. Hence, an agent a is satisfied in a balanced allocation if and only if the arc leaving a in the corresponding cycle cover is an important arc. By assigning weight 1 to each important arc in G and weight 0 to all other arcs, we get that any maximum weight cycle cover in G corresponds to an allocation with the maximum possible number of satisfied agents with respect to p.

To produce the reduced preference lists and construct the graph G, we need $O(L)$ operations. For finding the maximum weight cycle cover, a folklore method reducing this problem to finding a maximum weight perfect matching in a bipartite graph can be used (see e.g. [3]). Finding a maximum weight perfect matching in a bipartite graph with $|V|$ vertices, $|E|$ edges, and maximum edge weight 1 can be accomplished in $O(\sqrt{|V|}|E|)$ time [11]. With this method, our algorithm computes the minimum possible deficiency of a balanced allocation in time $O(\sqrt{N}L)$, given the fixed price function p. As the algorithm checks all possible price functions from H to $[M]$, the total running time is $O(M^M \sqrt{N}L)$.   □

5   Conclusion

We have dealt with the computation of the deficiency of housing markets. We showed that in general, if the housing market contains duplicate houses, this problem is hard even in the very restricted case where the maximum house-multiplicity in the market $\mathcal{M}$ is two $(\beta(\mathcal{M}) = 2)$ and each agent prefers only one house type to his own $(\gamma(\mathcal{M}) = 1)$.

To better understand the nature of the arising difficulties, we also looked at this problem within the context parameterized complexity. We proposed an FPT algorithm for computing the deficiency in the case where the parameter is the number of different house types. We also presented a simple algorithm that decides in $O(L^{\alpha+1})$ time if a housing market with strict preferences has deficiency at most α, where L is the length of the input. By contrast, we showed W[1]-hardness for the problem where the parameter is the value α describing the deficiency of the equilibrium we are looking for.

This W[1]-hardness result holds if $\gamma(\mathcal{M}) = 2$, leaving an interesting problem open: if each agent prefers only one house type to his endowment (i.e. $\gamma(\mathcal{M}) = 1$), is it possible to find an FPT algorithm with parameter α that decides whether the deficiency of the given market $\mathcal{M}$ is at most α? Looking at the digraph underlying such a market where vertices correspond to house types and arcs correspond to agents, and using the characterization of housing markets that admit an economic equilibrium given by Cechlárová and Fleiner [4], it is not hard to observe that this problem is in fact equivalent to the following natural graph modification problem: given a directed graph G, can we delete at most α edges from it such that each strongly connected component of the remaining graph is Eulerian?

References

1. Abdulkadiroğlu, A., Pathak, P.A., Roth, A.E.: Strategy-proofness versus efficiency in matching with indifferences: Redesigning the NYC high school match. American Economic Review 99(5), 1954–1978 (2009)
2. Arrow, K.J., Debreu, G.: Existence of an equilibrium for a competitive economy. Econometrica 22(3), 265–290 (1954)
3. Biró, P., Manlove, D.F., Rizzi, R.: Maximum weight cycle packing in optimal kidney exchange programs. Technical Report TR-2009-298, University of Glasgow, Department of Computing Science (2009)
4. Cechlárová, K., Fleiner, T.: Housing markets through graphs. Algorithmica 58, 19–33 (2010)
5. Cechlárová, K., Jelínková, E.: An efficient implementation of the equilibrium algorithm for housing markets with duplicate houses. Technical Report IM Preprint series A, no. 2/2010, P.J. Šafárik University, Faculty of Science, Institute of Mathematics (2010)
6. Deng, X., Papadimitriou, C., Safra, S.: On the complexity of equilibria. In: STOC 2002: Proceedings of the Thiry-Fourth Annual ACM Symposium on Theory of Computing, pp. 67–71. ACM, New York (2002)
7. Downey, R.G., Fellows, M.R.: Parameterized Complexity. Springer, Heidelberg (1999)
8. Fekete, S.P., Skutella, M., Woeginger, G.J.: The complexity of economic equilibria for house allocation markets. Inform. Process. Lett. 88(5), 219–223 (2003)
9. Flum, J., Grohe, M.: Parameterized Complexity Theory. Springer, Heidelberg (2006)
10. Garey, M.R., Johnson, D.S.: Computers and Intractability: A Guide to the Theory of NP-Completeness. W. H. Freeman & Co., New York (1979)
11. Kao, M.-Y., Lam, T.-W., Sung, W.-K., Ting, H.-F.: A decomposition theorem for maximum weight bipartite matchings with applications to evolutionary trees. In: Nešetřil, J. (ed.) ESA 1999. LNCS, vol. 1643, pp. 438–449. Springer, Heidelberg (1999)
12. Niedermeier, R.: Invitation to Fixed-Parameter Algorithms. Oxford University Press, Oxford (2006)
13. Roth, A.E., Sönmez, T., Ünver, M.U.: Kidney exchange. Quarterly J. of Econ. 119, 457–488 (2004)
14. Shapley, L., Scarf, H.: On cores and indivisibility. J. Math. Econ. 1, 23–37 (1974)
15. Sönmez, T., Ünver, M.U.: House allocation with existing tenants: an equivalence. Games and Economic Behavior 52, 153–185 (2005)

A New Lower Bound on the Maximum Number of Satisfied Clauses in Max-SAT and Its Algorithmic Application[*]

Robert Crowston, Gregory Gutin, Mark Jones, and Anders Yeo

Royal Holloway, University of London, United Kingdom
{robert,gutin,markj,anders}@cs.rhul.ac.uk

Abstract. For a formula F in conjunctive normal form (CNF), let $\mathrm{sat}(F)$ be the maximum number of clauses of F that can be satisfied by a truth assignment, and let m be the number of clauses in F. It is well-known that for every CNF formula F, $\mathrm{sat}(F) \geq m/2$ and the bound is tight when F consists of conflicting unit clauses (x) and $(\bar{x})$. Since each truth assignment satisfies exactly one clause in each pair of conflicting unit clauses, it is natural to reduce F to the unit-conflict free (UCF) form. If F is UCF, then Lieberherr and Specker (J. ACM 28(2):411-421, 1981) proved that $\mathrm{sat}(F) \geq \hat{\varphi}m$, where $\hat{\varphi} = (\sqrt{5} - 1)/2$.

We introduce another reduction that transforms a UCF CNF formula F into a UCF CNF formula F', which has a complete matching, i.e., there is an injective map from the variables to the clauses, such that each variable maps to a clause containing that variable or its negation. The formula F' is obtained from F by deleting some clauses and the variables contained only in the deleted clauses. We prove that $\mathrm{sat}(F) \geq \hat{\varphi}m + (1 - \hat{\varphi})(m - m') + n'(2 - 3\hat{\varphi})/2$, where n' and m' are the number of variables and clauses in F', respectively. This improves the Lieberherr-Specker lower bound on $\mathrm{sat}(F)$.

We show that our new bound has an algorithmic application by considering the following parameterized problem: given a UCF CNF formula F decide whether $\mathrm{sat}(F) \geq \hat{\varphi}m + k$, where k is the parameter. This problem was introduced by Mahajan and Raman (J. Algorithms 31(2):335–354, 1999) who conjectured that the problem is fixed-parameter tractable, i.e., it can be solved in time $f(k)(nm)^{O(1)}$ for some computable function f of k only. We use the new bound to show that the problem is indeed fixed-parameter tractable by describing a polynomial-time algorithm that transforms any problem instance into an equivalent one with at most $\lfloor (7 + 3\sqrt{5})k \rfloor$ variables.

1 Introduction

Let $F = (V, C)$ be a CNF formula, with a set V of variables and a multiset C of clauses, $m = |C|$, and $\mathrm{sat}(F)$ the maximum number of clauses that can be satisfied by a truth assignment. With a random assignment of truth values to the variables, the probability of a clause being satisfied is at least $1/2$. Thus, $\mathrm{sat}(F) \geq m/2$ for any F. This bound

---

[*] Research of Gutin, Jones and Yeo was supported in part by an EPSRC grant. Research of Gutin was also supported in part by the IST Programme of the European Community, under the PASCAL 2 Network of Excellence.

V. Raman and S. Saurabh (Eds.): IPEC 2010, LNCS 6478, pp. 84–94, 2010.

is tight when F consists of pairs of *conflicting unit clauses* (x) and $(\bar{x})$. Since each truth assignment satisfies exactly one clause in each pair of conflicting unit clauses, it is natural to reduce F to the *unit-conflict free (UCF)* form by deleting all pairs of conflicting clauses. If F is UCF, then Lieberherr and Specker [8] proved that sat$(F) \geq \hat{\varphi}m$, where $\hat{\varphi} = (\sqrt{5} - 1)/2$ (golden ratio inverse), and that for any $\epsilon > 0$ there are UCF CNF formulae F for which sat$(F) < m(\hat{\varphi}+\epsilon)$. Yannakakis [13] gave a short probabilistic proof that sat$(F) \geq \hat{\varphi}m$ by showing that if the probability of every variable appearing in a unit clause being assigned TRUE is $\hat{\varphi}$ (here we assume that for all such variables x the unit clauses are of the form (x)) and the probability of every other variable being assigned TRUE is $1/2$, then the expected number of satisfied clauses is $\hat{\varphi}m$.

In this paper, we introduce another reduction. We say that a UCF CNF formula $F = (V, C)$ has a *complete matching* if there is an injective map from the variables to the clauses, such that each variable maps to a clause containing that variable or its negation. We show that if a UCF CNF formula $F = (V, C)$ has no complete matching, then there is a subset C^* of clauses that can be found in polynomial time, such that $F' = (V \setminus V^*, C \setminus C^*)$ has a complete matching, where V^* is the set of variables of V appearing only in the clauses C^* (in positive or negative form). In addition, we show that all clauses of C^* can be satisfied by assigning appropriate truth values to variables in V^*. We also prove that for any UCF CNF $F'' = (V'', C'')$ with a complete matching, sat$(F'') \geq \hat{\varphi}|C''| + (2 - 3\hat{\varphi})|V''|/2$. These results imply that for a UCF CNF formula $F = (V, C)$, we have sat$(F) \geq \hat{\varphi}m + (1 - \hat{\varphi})(m - m') + n'(2 - 3\hat{\varphi})/2$, where $m = |C|$, $m' = |C \setminus C^*|$ and $n' = |V \setminus V^*|$. The last inequality improves the Lieberherr-Specker lower bound on sat(F).

Mahajan and Raman [10] were the first to recognize both practical and theoretical importance of parameterizing maximization problems above tight lower bounds. (We give some basic terminology on parameterized algorithms and complexity in the next section.) They considered MAX-SAT parameterized above the tight lower bound $m/2$. The problem is to decide whether we can satisfy at least $m/2 + k$ clauses, where k is the parameter. Mahajan and Raman proved that this parameterization of MAX-SAT is fixed-parameter tractable by obtaining a problem kernel with $O(k)$ variables.

Since $\hat{\varphi}m$ rather than $m/2$ is an asymptotically tight lower bound for UCF CNF formulae, Mahajan and Raman [10] introduced the following parameterization of MAX-SAT: given a UCF CNF formula F, decide whether sat$(F) \geq \hat{\varphi}m + k$, where k is the parameter. Mahajan and Raman [10] conjectured that this parameterized problem is fixed-parameter tractable. To solve the conjecture in the affirmative, we show the existence of a proper $O(k)$-variable kernel for MAX-SAT parameterized above $\hat{\varphi}m$ which follows from our improvement of the Lieberherr-Specker lower bound. Here we try to optimize the number of variables rather than the number of clauses in the kernel as the number of variables is more important than the number of clauses from the computational point of view.

The rest of this paper is organized as follows. In Section 2, we give further terminology and notation. Section 3 proves the improvement of the Lieberherr-Specker lower bound on sat(F) assuming correctness of the following lemma: if $F = (V, C)$ is a compact CNF formula, then sat$(F) \geq \hat{\varphi}|C| + (2 - 3\hat{\varphi})|V|/2$ (we give definition of a compact CNF formula in the next section). This non-trivial lemma is proved in Section 4. In

Section 5 we solve the conjecture of Mahajan and Raman [10] in the affirmative. We conclude the paper with discussions and open problems.

2   Terminology and Notation

A *parameterized problem* is a subset $L \subseteq \Sigma^* \times \mathbb{N}$ over a finite alphabet Σ. L is *fixed-parameter tractable* if the membership of an instance (I, k) in $\Sigma^* \times \mathbb{N}$ can be decided in time $f(k)|I|^{O(1)}$ where f is a computable function of the *parameter k* only [3,4,11]. Given a parameterized problem L, a *kernelization of L* is a polynomial-time algorithm that maps an instance (x, k) to an instance (x', k') (the *kernel*) such that (i) $(x, k) \in L$ if and only if $(x', k') \in L$, (ii) $k' \leq h(k)$, and (iii) $|x'| \leq g(k)$ for some functions h and g. It is well-known [3,4,11] that a decidable parameterized problem L is fixed-parameter tractable if and only if it has a kernel. By replacing Condition (ii) in the definition of a kernel by $k' \leq k$, we obtain a definition of a *proper kernel* (sometimes, it is called a *strong kernel*); cf. [1,2].

We let $F = (V, C)$ denote a CNF formula with a set of variables V and a multiset of clauses C. It is normally assumed that each clause may appear multiple times in C. For the sake of convenience, we assume that each clause appears at most once, but allow each clause to have an integer *weight*. (Thus, instead of saying a clause c appears t times, we will say that c has weight t). If at any point a particular clause c appears more than once in C, we replace all occurrences of c with a single occurrence of the same total weight. We use $w(c)$ to denote the weight of a clause c. For any clause $c \notin C$ we set $w(c) = 0$. If $C' \subseteq C$ is a subset of clauses, then $w(C')$ denotes the sum of the weights of the clauses in C'.

A *solution* to a CNF formula F is a truth assignment to the variables of the formula. The *weight* of the solution is the sum of the weights of all clauses satisfied by the solution. The maximum weight of a solution for F is denoted by $sat(F)$.

We call a CNF formula $F = (V, C)$ *compact* if the following conditions hold:

1. All clauses in F have the form (x) or $(\bar{x} \vee \bar{y})$ for some $x, y \in V$.
2. For every variable $x \in V$, the clause (x) is in C.

In a graph, for some vertex v, $N(v)$ denotes the set of vertices adjacent to v. For a set of vertices S, $N(S)$ denotes the set of vertices which are adjacent to a vertex in S. A bipartite graph G with vertex classes U_1, U_2 is said to have a *complete matching* from U_1 to U_2 if it is possible to choose a unique vertex u_2 from U_2 for every vertex u_1 in U_1, such that there is an edge between u_1 and u_2. Thus, a CNF formula F has a complete matching if and only if the *bipartite graph* B_F has a complete matching from V to C, where the partite sets of B_F are V and C, and the edge vc is in B_F if and only if the variable v or its negation $\bar{v}$ appears in the clause c.

3   New Lower Bound for sat(F)

We start by recalling the Augmenting Path Algorithm (cf. [14]), which is used to augment matchings in bipartite graphs. Let G be a bipartite graph with partite sets U_1 and

U_2, let M be a matching in G and let $u \in U_1$ be a vertex uncovered by M. The Augmenting Path Algorithm run from u proceeds as follows. Initially, no vertex in G is marked or explored apart from u which is marked, and set $T := \emptyset$. Choose a marked but unexplored vertex $x \in U_1$, declare it explored and consider each $y \in N(x) \setminus T$ as follows: add y to T and if y is uncovered by M then augment M, otherwise mark the neighbor z of y in M.

Let $F = (V, C)$ be a UCF CNF formula. Borrowing notation from graph theory, for $S \subseteq V$ we let $N(S)$ be the set of clauses which contain a variable (positive or negative instance) from S. Consider the following reduction rule.

Reduction Rule 1. *If $F = (V, C)$ does not have a complete matching, do the following. Find a maximum size matching M in B_F and a vertex $u \in V$ not covered by M, and run the Augmenting Path Algorithm from u. Let S be the vertices of V marked by the algorithm apart from u. Delete, from F, all the clauses in $N(S)$ and all the variables only appearing in $N(S)$ (in positive or negative form).*

This rule is of interest due to the following:

Theorem 1. *Let $F' = (V', C')$ be the formula derived from a UCF CNF formula $F = (V, C)$ by an application of Rule 1, and let S and $N(S)$ be defined as in Rule 1. Then $\mathrm{sat}(F') = \mathrm{sat}(F) - w(N(S))$.*

Proof. Let M' be the subset of M covering S. By the existence of M', we have $|N(S)| \geq |S|$. We claim that $|N(S)| = |S|$. Suppose it is not true. Then there is a vertex $c \in N(S)$ uncovered by M', and either c is uncovered by M in which case an M-augmenting path between u and c will augment M, which is impossible, or c is covered by M and its neighbor in M is unmarked, which is in contradiction with the Augmenting Path Algorithm. Thus, the matching M' provides a bijection between S and $N(S)$ that allows us to satisfy all clauses of $N(S)$ by assigning appropriate values to variables in S and this does not affect any clause outside of $N(S)$. Therefore, we conclude that $\mathrm{sat}(F') = \mathrm{sat}(F) - w(N(S))$. $\qquad\square$

Our improvement of the Lieberherr-Specker lower bound on $\mathrm{sat}(F)$ for a UCF CNF formula F will follow immediately from Theorems 1 and 2. It is much harder to prove Theorem 2 than Theorem 1, and our proof of Theorem 2 is based on the following quite non-trivial lemma that will be proved in the next section.

Lemma 1. *If $F = (V, C)$ is a compact CNF formula, then there exists a solution to F with weight at least*

$$\hat{\varphi} w(C) + \frac{|V|(2 - 3\hat{\varphi})}{2},$$

where $\hat{\varphi} = (\sqrt{5} - 1)/2$, and such a solution can be found in polynomial time.

The next proof builds on some of the basic ideas in [8].

Theorem 2. *If $F = (V, C)$ is a UCF CNF formula with a complete matching, then there exists a solution to F with weight at least*

$$\hat{\varphi} w(C) + \frac{|V|(2 - 3\hat{\varphi})}{2},$$

where $\hat{\varphi} = (\sqrt{5} - 1)/2$ and such a solution can be found in polynomial time.

Proof. We will describe a polynomial-time transformation from F to a compact CNF formula F', such that $|V'| = |V|$ and $w(C') = w(C)$, and any solution to F' can be turned into a solution to F of greater or equal weight. The theorem then follows from Lemma 1.

Consider a complete matching from V to C (in B_F), and for each $x \in V$ let c_x be the unique clause associated with x in this matching. For each variable x, if the unit clause (x) or $(\bar{x})$ appears in C, leave c_x as it is for now. Otherwise, remove all variables except x from c_x. We now have that for every x, exactly one of (x), $(\bar{x})$ appears in C.

If $(\bar{x})$ is in C, replace every occurrence of the literal $\bar{x}$ in the clauses of C with x, and replace every occurrence of x with $\bar{x}$. We now have that Condition 2 in the definition of a compact formula is satisfied. For any clause c which contains more than one variable and at least one positive literal, remove all variables except one that occurs as a positive. For any clause which contains only negative literals, remove all but two variables. We now have that Condition 1 is satisfied. This completes the transformation.

In the transformation, no clauses or variables were completely removed, so $|V'| = |V|$ and $w(C') = w(C)$. Observe that the transformation takes polynomial time, and that any solution to the compact formula F' can be turned into a solution to F of greater or equal weight. Indeed, for some solution to F', flip the assignment to x if and only if we replaced occurrences of x with $\bar{x}$ in the transformation. This gives a solution to F in which every clause will be satisfied if its corresponding clause in F' is satisfied. □

Our main result follows immediately from Theorems 1 and 2 (since $(w(C) - w(C')) + \hat{\varphi}w(C') + \frac{|V'|(2-3\hat{\varphi})}{2} = \hat{\varphi}w(C) + (1 - \hat{\varphi})(w(C) - w(C')) + \frac{|V'|(2-3\hat{\varphi})}{2})$.

Theorem 3. *Let $F = (V, C)$ be a UCF CNF formula and let $F' = (V', C')$ be obtained from F by repeatedly applying Rule 1 as long as possible. Then* $\mathrm{sat}(F) \geq \hat{\varphi}w(C) + (1 - \hat{\varphi})(w(C) - w(C')) + \frac{|V'|(2-3\hat{\varphi})}{2}$.

4   Proof of Lemma 1

In this section, we use the fact that $\hat{\varphi} = (\sqrt{5} - 1)/2$ is the positive root of the polynomial $\hat{\varphi}^2 + \hat{\varphi} - 1$. We call a clause $(\bar{x} \vee \bar{y})$ *good* if for every literal $\bar{z}$, the set of clauses containing $\bar{z}$ is not equal to $\{(\bar{x} \vee \bar{z}), (\bar{y} \vee \bar{z})\}$. We define $w_v(x)$ to be the total weight of all clauses containing the literal x, and $w_v(\bar{x})$ the total weight of all clauses containing the literal $\bar{x}$. (Note that $w_v(\bar{x})$ is different from $w(\bar{x})$, which is the weight of the particular clause $(\bar{x})$.) Let $\epsilon(x) = w_v(x) - \hat{\varphi}w_v(\bar{x})$. Let $\gamma = (2 - 3\hat{\varphi})/2 = (1 - \hat{\varphi})^2/2$ and let $\Delta(F) = \mathrm{sat}(F) - \hat{\varphi}w(C)$.

To prove Lemma 1, we will use an algorithm, Algorithm A, described below. We will show that, for any compact CNF formula $F = (V, C)$, Algorithm A finds a solution with weight at least $\hat{\varphi}w(C) + \gamma|V|$. Step 3 of the algorithm removes any clauses which are satisfied or falsified by the given assignment of truth values to the variables. The purpose of Step 4 is to make sure the new formula is compact.

In order to show that the algorithm finds a solution with weight at least $\hat{\varphi}w(C) + \frac{|V|(2-3\hat{\varphi})}{2}$, we need the following two lemmas.

> **ALGORITHM A**
>
> While $|V| > 0$, repeat the following steps:
>
> 1. For each $x \in V$, calculate $w_v(x)$ and $w_v(\bar{x})$.
> 2. We mark some of the variables as TRUE or FALSE, according to the following cases:
>
>   **Case A:** *There exists $x \in V$ with $w_v(x) \geq w_v(\bar{x})$.* Pick one such x and assign it TRUE.
>
>   **Case B:** *Case A is false, and there exists $x \in V$ with $(1 - \hat{\varphi})\epsilon(x) \geq \gamma$.* Pick one such x and assign it TRUE.
>
>   **Case C:** *Cases A and B are false, and there exists a good clause.* Pick such a good clause $(\bar{x} \vee \bar{y})$, with (without loss of generality) $\epsilon(x) \geq \epsilon(y)$, and assign y FALSE and x TRUE.
>
>   **Case D:** *Cases A, B and C are false.* Pick any clause $(\bar{x} \vee \bar{y})$ and pick z such that both clauses $(\bar{x} \vee \bar{z})$ and $(\bar{y} \vee \bar{z})$ exist. Consider the six clauses $(x), (y), (z), (\bar{x} \vee \bar{y}), (\bar{x} \vee \bar{z}), (\bar{y} \vee \bar{z})$, and all 2^3 combinations of assignments to the variables x, y, z, picking the assignment satisfying the maximum possible weight from the six clauses.
>
> 3. Perform the following simplification: For any variable x assigned FALSE, remove any clause containing $\bar{x}$, remove the unit clause (x), and remove x from V. For any variable x assigned TRUE, remove the unit clause (x), remove $\bar{x}$ from any clause containing $\bar{x}$ and remove x from V.
> 4. For each y remaining, if there is a clause of the form $(\bar{y})$, do the following: If the weight of this clause is greater than $w_v(y)$, then replace all clauses containing the variable y (that is, literals y or $\bar{y}$) with one clause (y) of weight $w_v(\bar{y}) - w_v(y)$. Otherwise remove $(\bar{y})$ from C and change the weight of (y) to $w(y) - w(\bar{y})$.

Lemma 2. *For a formula F, if we assign a variable x TRUE, and run Steps 3 and 4 of the algorithm, the resulting formula $F^* = (V^*, C^*)$ satisfies the following:*

$$\Delta(F) \geq \Delta(F^*) + (1 - \hat{\varphi})\epsilon(x).$$

Furthermore, we have $|V^*| = |V| - 1$, unless there exists $y \in V^*$ such that (y) and $(\bar{x} \vee \bar{y})$ are the only clauses containing y and they have the same weight. In this case, y is removed from V^*.

Proof. Observe that at Step 3, the clause (x) (of weight $w_v(x)$) is removed, clauses of the form $(\bar{x} \vee \bar{y})$ (total weight $w_v(\bar{x})$) become $(\bar{y})$, and the variable x is removed from V.

At Step 4, observe that for each y such that $(\bar{y})$ is now a clause, $w(C)$ is decreased by $2w_y$ and sat(F) is decreased by w_y, where $w_y = \min\{w(y), w(\bar{y})\}$. Let $q = \sum_y w_y$, and observe that $q \leq w_v(\bar{x})$. A variable y will only be removed at this stage if the clause $(\bar{x} \vee \bar{y})$ was originally in C. We therefore have

1. sat$(F^*) \leq$ sat$(F) - w_v(x) - q$
2. $w(C^*) = w(C) - w_v(x) - 2q$

Using the above, we get

$$
\begin{aligned}
\Delta(F) &= \mathrm{sat}(F) - \hat{\varphi} \cdot w(C) \\
&\geq (w_v(x) + \mathrm{sat}(F^*) + q) - \hat{\varphi}(w(C^*) + 2q + w_v(x)) \\
&= \Delta(F^*) + (1 - \hat{\varphi})w_v(x) - (2\hat{\varphi} - 1)q \\
&\geq \Delta(F^*) + (1 - \hat{\varphi})(\epsilon(x) + \hat{\varphi} \cdot w_v(\bar{x})) - (2\hat{\varphi} - 1)w_v(\bar{x}) \\
&= \Delta(F^*) + (1 - \hat{\varphi} - \hat{\varphi}^2)w_v(\bar{x}) + (1 - \hat{\varphi})\epsilon(x) \\
&= \Delta(F^*) + (1 - \hat{\varphi})\epsilon(x). \qquad \square
\end{aligned}
$$

Lemma 3. *For a formula F, if we assign a variable x FALSE, and run Steps 3 and 4 of the algorithm, the resulting formula $F^{**} = (V^{**}, C^{**})$ has $|V^{**}| = |V| - 1$ and satisfies the following: $\Delta(F) \geq \Delta(F^{**}) - \hat{\varphi}\epsilon(x)$.*

Proof. Observe that at Step 3, every clause containing the variable x is removed, and no other clauses will be removed at Steps 3 and 4. Since the clause (y) appears for every other variable y, this implies that $|V^{**}| = |V| - 1$. We also have the following: $\mathrm{sat}(F^{**}) \leq \mathrm{sat}(F) - w_v(\bar{x})$ and $w(C^{**}) = w(C) - w_v(\bar{x}) - w_v(x)$. Thus,

$$
\begin{aligned}
\Delta(F) &= \mathrm{sat}(F) - \hat{\varphi}w(C) \\
&\geq (w_v(\bar{x}) + \mathrm{sat}(F^{**})) - \hat{\varphi}(w(C^{**}) + w_v(\bar{x}) + w_v(x)) \\
&= \Delta(F^{**}) + (1 - \hat{\varphi})w_v(\bar{x}) - \hat{\varphi}w_v(x) \\
&= \Delta(F^{**}) + (1 - \hat{\varphi})w_v(\bar{x}) - \hat{\varphi}(\epsilon(x) + \hat{\varphi} \cdot w_v(\bar{x})) \\
&= \Delta(F^{**}) + (1 - \hat{\varphi} - \hat{\varphi}^2)w_v(\bar{x}) - \hat{\varphi}\epsilon(x) \\
&= \Delta(F^{**}) - \hat{\varphi}\epsilon(x). \qquad \square
\end{aligned}
$$

Now we are ready to prove Lemma 1.

Proof of Lemma 1: We will show that Algorithm A finds a solution with weight at least $\hat{\varphi}w(C) + \frac{|V|(2-3\hat{\varphi})}{2}$. Note that the inequality in the lemma can be reformulated as $\Delta(F) \geq \gamma|V|$.

Let F and $\hat{\varphi}$ be defined as in the statement of the lemma. Note that at each iteration of the algorithm, at least one variable is removed. Therefore, we will show the lemma by induction on $|V|$. If $|V| = 0$ then we are done trivially and if $|V| = 1$ then we are done as $\mathrm{sat}(F) = w(C) \geq \hat{\varphi}w(C) + \gamma$ (as $w(C) \geq 1$). So assume that $|V| \geq 2$.

For the induction step, let $F' = (V', C')$ be the formula resulting from F after running Steps 1-4 of the algorithm, and assume that $\Delta(F') \geq \gamma|V'|$. We show that $\Delta(F) \geq \gamma|V|$, by analyzing each possible case in Step 2 separately.

Case A: $w_v(x) \geq w_v(\bar{x})$ *for some $x \in V$.* In this case we let x be TRUE, which by Lemma 2 implies the following:

$$
\begin{aligned}
\Delta(F) &\geq \Delta(F') + (1 - \hat{\varphi})\epsilon(x) \\
&= \Delta(F') + (1 - \hat{\varphi})(w_v(x) - \hat{\varphi}w_v(\bar{x})) \\
&\geq \Delta(F') + (1 - \hat{\varphi})(w_v(x) - \hat{\varphi}w_v(x)) \\
&= \Delta(F') + (1 - \hat{\varphi})^2 w_v(x) \\
&= \Delta(F') + 2\gamma w_v(x).
\end{aligned}
$$

If $y \in V \setminus V'$, then either $y = x$ or $(\bar{x} \vee \bar{y}) \in C$. Therefore $|V| - |V'| \leq w_v(\bar{x}) + 1 \leq w_v(x) + 1$. As $w_v(x) \geq 1$ we note that $2\gamma w_v(x) \geq \gamma(w_v(x) + 1)$. This implies the following, by induction, which completes the proof of Case A.

$$\Delta(F) \geq \Delta(F') + \gamma(w_v(x) + 1)$$
$$\geq \gamma|V'| + \gamma(w_v(x) + 1) \geq \gamma|V|.$$

Case B: *Case A is false, and* $(1 - \hat{\varphi})\epsilon(x) \geq \gamma$ *for some* $x \in V$.

Again we let x be TRUE. Since $w_v(y) < w_v(\bar{y})$ for all $y \in V$, we have $|V| = |V'| + 1$. Analogously to Case A, using Lemma 2, we get the following:

$$\Delta(F) \geq \Delta(F') + (1 - \hat{\varphi})\epsilon(x)$$
$$\geq \gamma|V'| + \gamma = \gamma|V|.$$

For Cases C and D, we generate a graph G from the set of clauses. The vertex set of G is the variables in V (i.e. $V(G) = V$) and there is an edge between x and y if and only if the clause $(\bar{x} \vee \bar{y})$ exists in C. A *good* edge in G is an edge $uv \in E(G)$ such that no vertex $z \in V$ has $N(z) = \{u, v\}$ (that is, an edge is good if and only if the corresponding clause is good).

Case C: *Cases A and B are false, and there exists a good clause* $(\bar{x} \vee \bar{y})$. Without loss of generality assume that $\epsilon(x) \geq \epsilon(y)$. We will first let y be FALSE and then we will let x be TRUE. By letting y be FALSE we get the following by Lemma 3, where F^{**} is defined in Lemma 3: $\Delta(F) \geq \Delta(F^{**}) - \hat{\varphi}\epsilon(x)$.

Note that the clause $(\bar{x} \vee \bar{y})$ has been removed so $w_v^{**}(\bar{x}) = w_v(\bar{x}) - w(\bar{x} \vee \bar{y})$ and $w_v^{**}(x) = w_v(x)$ (where $w_v^{**}(.)$ denote the weights in F^{**}). Therefore using Lemma 2 on F^{**} instead of F we get the following, where the formula F^* in Lemma 2 is denoted by F' below and $w^0 = w(\bar{x} \vee \bar{y})$:

$$\Delta(F^{**}) \geq \Delta(F') + (1 - \hat{\varphi})(w_v(x) - \hat{\varphi}(w_v(\bar{x}) - w^0)).$$

First we show that $|V'| = |V^{**}| - 1 = |V| - 2$. Assume that $z \in V \setminus (V' \cup \{x, y\})$ and note that $N(z) \subseteq \{x, y\}$. Clearly $|N(z)| = 1$ as xy is a good edge. If $N(z) = \{y\}$ then $(z) \in C'$, so we must have $N(z) = \{x\}$. However the only way $z \notin V'$ is if $w_v(z) = w_v(\bar{z})$, a contradiction as Case A is false. Therefore, $|V'| = |V| - 2$, and the following holds by the induction hypothesis.

$$\Delta(F) \geq \Delta(F^{**}) - \hat{\varphi}\epsilon(x)$$
$$\geq \Delta(F') + (1 - \hat{\varphi})(w_v(x) - \hat{\varphi}(w_v(\bar{x}) - w^0)) - \hat{\varphi}\epsilon(x)$$
$$\geq \gamma|V'| + (1 - \hat{\varphi})(\epsilon(x) + \hat{\varphi}w^0) - \hat{\varphi}\epsilon(x)$$
$$= \gamma|V| - 2\gamma + (1 - \hat{\varphi})\hat{\varphi}w^0 - (2\hat{\varphi} - 1)\epsilon(x).$$

We would be done if we can show that $2\gamma \leq (1-\hat{\varphi})\hat{\varphi}w^0 - (2\hat{\varphi}-1)\epsilon(x)$. As $w^0 \geq 1$ and we know that, since Case B does not hold, $(1-\hat{\varphi})\epsilon(x) < \gamma$, we would be done if we can show that $2\gamma \leq (1 - \hat{\varphi})\hat{\varphi} - (2\hat{\varphi} - 1)\gamma/(1 - \hat{\varphi})$. This is equivalent to $\gamma = (1 - \hat{\varphi})^2/2 \leq \hat{\varphi}(1 - \hat{\varphi})^2$, which is true, completing the proof of Case C.

Case D: *Cases A, B and C are false.* Then G has no good edge.

Assume xy is some edge in G and $z \in V$ such that $N(z) = \{x, y\}$. As xz is not a good edge there exists a $v \in V$, such that $N(v) = \{x, z\}$. However v is adjacent to z and, thus,

$v \in N(z) = \{x, y\}$, which implies that $v = y$. This shows that $N(y) = \{x, z\}$. Analogously we can show that $N(x) = \{y, z\}$. Therefore, the only clauses in C that contain a variable from $\{x, y, z\}$ form the following set: $S = \{(x), (y), (z), (\bar{x} \vee \bar{y}), (\bar{x} \vee \bar{z}), (\bar{y} \vee \bar{z})\}$.

Let F' be the formula obtained by deleting the variables x, y and z and all clauses containing them. Now consider the three assignments of truth values to x, y, z such that only one of the three variables is assigned FALSE. Observe that the total weight of clauses satisfied by these three assignments equals

$$w_v(\bar{x}) + w_v(\bar{y}) + w_v(\bar{z}) + 2(w(x) + w(y) + w(z)) = 2W,$$

where W is the total weight of the clauses in S. Thus, one of the three assignments satisfies the weight of at least $2W/3$ among the clauses in S. Observe also that $w(C) - w(C') \geq 6$, and, thus, the following holds.

$$\begin{aligned}
\Delta(F) &\geq 2(w(C) - w(C'))/3 + \text{sat}(F') - \hat{\varphi}(w(C') + w(C) - w(C')) \\
&\geq \gamma|V'| + 2(w(C) - w(C'))/3 - \hat{\varphi}(w(C) - w(C')) \\
&= \gamma|V| - 3\gamma + (2 - 3\hat{\varphi})(w(C) - w(C'))/3 \\
&\geq \gamma|V| - 3\gamma + 2(2 - 3\hat{\varphi}) > \gamma|V|.
\end{aligned}$$

This completes the proof of the correctness of Algorithm A. It remains to show that Algorithm A takes polynomial time.

Each iteration of the algorithm takes $O(nm)$ time. The algorithm stops when V is empty, and at each iteration some variables are removed from V. Therefore, the algorithm goes through at most n iterations and, in total, it takes $O(n^2 m)$ time. This completes the proof of Lemma 1. □

Note that the bound $(2 - 3\hat{\varphi})/2$ in Lemma 1 cannot be improved due to the following example. Let l be any positive integer and let $F = (V, C)$ be defined such that $V = \{x_1, x_2, \ldots, x_l, y_1, y_2, \ldots, y_l\}$ and C contain the constraints $(x_1), (x_2), \ldots, (x_l), (y_1),$ $(y_2), \ldots, (y_l)$ and $(\bar{x}_1 \vee \bar{y}_1), (\bar{x}_2 \vee \bar{y}_2), \ldots, (\bar{x}_l \vee \bar{y}_l)$. Let the weight of every constraint be one and note that for every i we can only satisfy two of the three constraints $(x_i), (y_i)$ and $(\bar{x}_i \vee \bar{y}_i)$. Therefore $\text{sat}(F) = 2l$ and the following holds:

$$\hat{\varphi}w(C) + \frac{|V|(2 - 3\hat{\varphi})}{2} = 3l\hat{\varphi} + \frac{2l(2 - 3\hat{\varphi})}{2} = l(3\hat{\varphi} + 2 - 3\hat{\varphi}) = 2l = \text{sat}(F).$$

5   Parameterized Complexity Result

Recall that Mahajan and Raman [10] conjectured that the following parameterization of MAX-SAT is fixed-parameter tractable: given a UCF CNF formula F, decide whether $\text{sat}(F) \geq \hat{\varphi}m + k$, where k is the parameter.

Theorem 4. *The parameterized problem has a proper kernel with at most $\lfloor (7 + 3\sqrt{5})k \rfloor$ variables.*

Proof. Consider an instance $F = (V, C)$ of the problem. If F has no complete matching, then apply Rule 1 to F as long as possible. As a result, we obtain either an empty formula or a UCF CNF formula $F' = (V', C')$ with a complete matching. In the first case, by Theorem 3, $\text{sat}(F) = w(C)$, and the kernel is trivial. In the second case, we want to choose a parameter k' such that (F, k) is a YES-instance of the problem if and only if (F', k') is a YES-instance of the problem. It is enough for k' to satisfy $k - \text{sat}(F) + \lfloor \hat{\varphi} w(C) \rfloor = k' - \text{sat}(F') + \lfloor \hat{\varphi} w(C') \rfloor$. By Theorem 1, $\text{sat}(F') = \text{sat}(F) - w(C) + w(C')$. Therefore, we can set $k' = k - w(C) + w(C') + \lfloor \hat{\varphi} w(C) \rfloor - \lfloor \hat{\varphi} w(C') \rfloor$. Since $w(C) - w(C') \geq \lceil \hat{\varphi}(w(C) - w(C')) \rceil \geq \lfloor \hat{\varphi} w(C) \rfloor - \lfloor \hat{\varphi} w(C') \rfloor$, we have $k' \leq k$.

By Theorem 2, if $k' \leq \frac{|V'|(2-3\hat{\varphi})}{2}$, then F is a YES-instance of the problem. Otherwise, $|V'| < \frac{2k}{2-3\hat{\varphi}} = (7 + 3\sqrt{5})k$. Note that F' is not necessarily a kernel as $w(C')$ is not necessarily bounded by a function of k. However, if $w(C') \geq 2^{2k/(2-3\hat{\varphi})}$ then we can solve the instance (F', k') in time $O(w(C')^2)$ and, thus, we may assume that $w(C') < 2^{2k/(2-3\hat{\varphi})}$, in which case, F' is the required kernel.   □

6 Discussion

A CNF formula I is *t-satisfiable* if any subset of t clauses of I can be satisfied simultaneously. In particular, a CNF formula is unit-conflict free if and only if it is 2-satisfiable. Let r_t be the largest real such that in any t-satisfiable CNF formula at least r_t-th fraction of its clauses can be satisfied simultaneously. Note that $r_1 = 1/2$ and $r_2 = (\sqrt{5} - 1)/2$. Lieberherr and Specker [9] and, later, Yannakakis [13] proved that $r_3 \geq 2/3$. Käppeli and Scheder [6] proved that $r_3 \leq 2/3$ and, thus, $r_3 = 2/3$. Král [7] established the value of r_4: $r_4 = 3/(5 + [(3\sqrt{69} - 11)/2]^{1/3} - [3\sqrt{69} + 11)/2]^{1/3}) \approx 0.6992$.

For general t, Huang and Lieberherr [5] showed that $\lim_{t \to \infty} r_t \leq 3/4$ and Trevisan [12] proved that $\lim_{t \to \infty} r_t = 3/4$ (a different proof of this result is later given by Král [7]).

It would be interesting to establish parameterized complexity of the following parameterized problem: given a 3-satisfiable CNF formula $F = (V, C)$, decide whether $\text{sat}(F) \geq 2|C|/3 + k$, where k is the parameter. This question can be extended to $t > 3$.

References

1. Abu-Khzam, F.N., Fernau, H.: Kernels: Annotated, proper and induced. In: Bodlaender, H.L., Langston, M.A. (eds.) IWPEC 2006. LNCS, vol. 4169, pp. 264–275. Springer, Heidelberg (2006)
2. Chen, Y., Flum, J., Müller, M.: Lower bounds for kernelizations and other preprocessing procedures. In: Proc. CiE 2009, vol. 5635, pp. 118–128 (2009)
3. Downey, R.G., Fellows, M.R.: Parameterized Complexity. Springer, Heidelberg (1999)
4. Flum, J., Grohe, M.: Parameterized Complexity Theory. Springer, Heidelberg (2006)
5. Huang, M.A., Lieberherr, K.J.: Implications of forbidden structures for extremal algorithmic problems. Theoret. Comput. Sci. 40, 195–210 (1985)
6. Käppeli, C., Scheder, D.: Partial satisfaction of k-satisfiable formulas. Electronic Notes in Discrete Math. 29, 497–501 (2007)
7. Král, D.: Locally satisfiable formulas. In: Proc. SODA 2004, pp. 330–339 (2004)
8. Lieberherr, K.J., Specker, E.: Complexity of partial satisfaction. J. ACM 28(2), 411–421 (1981)

 9. Lieberherr, K.J., Specker, E.: Complexity of partial satisfaction, II. Tech. Report 293 of Dept. of EECS, Princeton Univ. (1982)
10. Mahajan, M., Raman, V.: Parameterizing above guaranteed values: MaxSat and MaxCut. J. Algorithms 31(2), 335–354 (1999)
11. Niedermeier, R.: Invitation to Fixed-Parameter Algorithms. Oxford University Press, Oxford (2006)
12. Trevisan, L.: On local versus global satisfiability. SIAM J. Discret. Math. 17(4), 541–547 (2004)
13. Yannakakis, M.: On the approximation of maximum satisfiability. J. Algorithms 17, 475–502 (1994)
14. West, D.B.: Introduction to Graph Theory, 2nd edn. Prentice-Hall, Englewood Cliffs (2001)

An Improved FPT Algorithm and Quadratic Kernel for Pathwidth One Vertex Deletion*

Marek Cygan, Marcin Pilipczuk,
Michał Pilipczuk, and Jakub Onufry Wojtaszczyk

Faculty of Mathematics, Computer Science and Mechanics,
University of Warsaw, Poland
{cygan@,malcin@,michal.pilipczuk@students.,onufry@}mimuw.edu.pl

Abstract. The PATHWIDTH ONE VERTEX DELETION (POVD) problem, given an undirected graph G and an integer k, asks whether one can delete at most k vertices from G so that the remaining graph has pathwidth at most 1. Recently Philip et al. [14] initiated the study of the parameterized complexity of POVD and have shown a quartic kernel and a $7^k n^{O(1)}$ algorithm. In this paper we improve these results by showing a quadratic kernel and a $4.65^k n^{O(1)}$ algorithm.

1 Introduction

In the parameterized complexity setting, an input instance comes with an integer parameter k — formally, a parameterized problem Q is a subset of $\Sigma^* \times \mathbb{N}$ for some finite alphabet Σ. We say that the problem is *fixed parameter tractable* (*FPT*) if there exists an algorithm solving any instance (x, k) in time $f(k)\text{poly}(|x|)$ for some (usually exponential) computable function f. It is known that a decidable problem is FPT iff it is kernelizable: a kernelization algorithm for a problem Q is an algorithm that takes an instance (x, k) and in time polynomial in $|x| + k$ produces an equivalent instance (x', k') (i.e., $(x, k) \in Q$ iff $(x', k') \in Q$) such that $|x'| + k' \leq g(k)$ for some computable function g. The function g is the *size of the kernel* and if it is polynomial, we say that Q admits a polynomial kernel. Kernelization techniques can be viewed as polynomial time preprocessing routines for tackling NP-hard problems. Parameterized complexity provides a formal framework for the analysis of such algorithms [7,8,11].

The notions of *pathwidth* and *treewidth*, introduced by Robertson and Seymour [15,16], measure how much a given graph resembles a path and a tree, respectively. Both play an important role in the proof of the Graph Minor Theorem, and there exists a large family of fixed parameter algorithms (usually based on the dynamic programming principle) for problems parameterized by treewidth.

One of the most extensively studied problems in the parameterized complexity community, FEEDBACK VERTEX SET, asks for a set of vertices of size at most k such that their deletion results in a graph of treewidth at most 1, i.e., a forest. Currently the fastest known algorithm for the FEEDBACK VERTEX SET problem runs in $3.83^k n^{O(1)}$ time [4] and a kernel of size $O(k^2)$ is due to Thomassé [17]. Very recently, Philip et al.

---

* This work was partially supported by the Polish Ministry of Science grants N206 491038 and N206 491238.

V. Raman and S. Saurabh (Eds.): IPEC 2010, LNCS 6478, pp. 95–106, 2010.

[14] initiated a study of the parameterized complexity of a closely related problem —
PATHWIDTH ONE VERTEX DELETION.

PATHWIDTH ONE VERTEX DELETION (POVD)   **Parameter:** k
Input: An undirected graph $G = (V, E)$ and a positive integer k
Question: Does there exist a set $T \subseteq V$ such that $|T| \leq k$ and $G[V \setminus T]$ is a graph of
pathwidth at most one?

We omit the formal definitions of pathwidth and treewidth (an interested reader is
invited to read, e.g., [6]), as the following simple characterisation will be sufficient for
our purposes:

Definition 1. *A graph is a caterpillar iff it is a tree and after removing all vertices
of degree one it becomes a path (possibly empty or consisting of a single vertex). A
caterpillar forest is a graph whose every connected component is a caterpillar.*

Lemma 1 ([2]). *A graph has pathwidth at most one if and only if it is a caterpillar
forest.*

The class of caterpillars and caterpillar forests has been studied as a further simplifica-
tions of trees. A number of problems which have been proved to be difficult even in the
class of trees have efficient solutions for caterpillars. Examples include BANDWIDTH
[3,9,13], PROPER INTERVAL COLORED GRAPH and PROPER COLORED LAYOUT [1].

Philip et al. [14] have shown a simple branching algorithm that solves POVD in
$7^k n^{O(1)}$ time and a kernel consisting of $O(k^4)$ vertices. In this paper we improve these
bounds: we show a $4.65^k n^{O(1)}$ FPT algorithm and a kernel consisting of $O(k^2)$ vertices
and edges. Let us note that our kernel is a *weak* kernel in the following sense: the kernelization algorithm takes a POVD instance (G, k) and outputs an equivalent instance
(G', k') such that $|G'|, k' \leq O(k^2)$ (by $|G'|$ we mean $|V(G')| + |E(G')|$). However, it
may happen that the new parameter k' is indeed of size $\Omega(k^2)$. As it may happen that
the running time of an algorithm solving the kernelized instance is strongly dependent
on k (for instance if one wants to run an FPT algorithm, or mix it with another algorithm
in some win–win approach) this may be very undesirable. This motivates a search for
strong kernels, ie. kernels in which $|G'| \leq g(k)$ and $k' \leq k$. As far as *strong* kernels
are considered, we obtain a cubic kernel ($|G'| \leq O(k^3)$) for POVD. A strong quadratic
kernel (that is $|G'| \leq O(k^2)$, $k' \leq k$) may be obtained for a version of the problem
which allows multiple edges:

MULTIGRAPH PATHWIDTH ONE VERTEX DELETION (MPOVD)   **Parameter:** k
Input: An undirected multigraph $G = (V, E)$ and a positive integer k
Question: Does there exist a set $T \subseteq V$ such that $|T| \leq k$ and $G[V \setminus T]$ is a simple
graph of pathwidth at most one?

Note that in MPOVD we require that the resulting graph is simple, i.e., the solution
is required to hit all multiple edges.

Finally, let us note that the first of our kernelization results almost matches known kernel lower bounds for POVD. It follows from a general result of Dell and van Melkebeek [5] that there cannot exist a kernel for POVD that can be encoded in $O(k^\gamma)$ bits for any $\gamma < 2$ unless coNP $\subseteq$ NP/poly, as it may be easily seen that the class of graphs of pathwidth at most one is nontrivial and hereditary (the appropriate definitions are given in [5]).

Notation. For a graph or multigraph G we denote by $V(G)$ the set of vertices of the graph. The degree of a vertex $v \in V(G)$ is the number of edge ends incident to the vertex (so a loop contributes 2 to the degree). K_3 is used to denote the 3–element cycle, while C_4 denotes the four–element cycle. By $N(v)$ we denote the set of neighbours of the vertex v. For $S \subseteq V$ by $N(S)$ we denote $\left(\bigcup_{s \in S} N(s)\right) \setminus S$.

Organization. In Section 2 we introduce some notation and give several introductory observations and results. In Section 3 we give a simple $4.65^k n^{O(1)}$ branching algorithm that solves POVD. In Section 4 we describe the kernelization algorithms. Finally, Section 5 contains some concluding remarks and open problems. Due to space limitations some proofs are sketched only and will appear in the full version of the paper.

2 Preliminaries

First let us recall the characterization of graphs of pathwidth one proved in [14].

Lemma 2 ([14]). *A simple graph G has pathwidth at most one iff it does not contain cycles and subgraphs isomorphic to the graph T_2 depicted in Figure 1.*

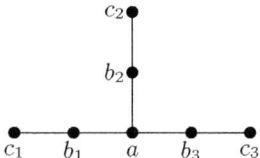

Fig. 1. A forbidden subgraph T_2 in graphs of pathwidth at most one. The notation for the vertices is used in Lemma 6 and in the proof of Theorem 3.

The following corollary of Lemma 2 is proved in [14] in a slightly different formulation. We include the proof for the sake of completeness.

Lemma 3 ([14]). *If a simple graph G does not contain T_2, K_3 or C_4 as subgraphs, the PATHWIDTH ONE VERTEX DELETION problem on G can be solved in polynomial time.*

Proof. We begin by introducing the following simple definition:

Definition 2. *If w is a vertex of degree 1, and v is its sole neighbour, we call w a pendant of v.*

Let G be a graph that does not contain T_2, K_3 and C_4 as subgraphs. Let H be any connected component of G. If H is a tree, it does not contain any cycles as subgraphs, and due to Lemma 2 it already has pathwidth at most one. Otherwise, let C be a shortest cycle in H. As we excluded K_3 and C_4 as subgraphs, C has length at least 5. We aim to prove that all vertices of H either lie on C or are pendants of vertices lying on C.

Consider any vertex v on C, and any w adjacent to v, but not lying on C. If w has another neighbour $u \in V(C)$, then the cycle formed by w and the shorter of the arcs connecting u and v on C would be shorter than C, a contradiction (as $|V(C)| \geq 5$, the length of the longer arc is at least 3). On the other hand, if w has a neighbour $u \notin V(C)$, then v, w, u and arcs of length two in both directions from v on C form a T_2 subgraph of H (the two arcs are disjoint as $|V(C)| \geq 5$), again a contradiction. Thus any vertex adjacent to C is a pendant of some vertex of C, and as H is connected, there are no other vertices in H.

Additionally we know that there are no internal edges (chords) in C, since we could get a shorter cycle using such a chord. Now note that any solution of POVD needs to delete at least one vertex of C, while the deletion of any vertex of C makes H a caterpillar. Thus, the minimum number of deleted vertices in the graph G is exactly the number of connected components of G that are not trees.   □

Now we state a simple observation which enables us to add pendant vertices to the input graph.

Lemma 4. *If G is a caterpillar forest where a vertex $v \in V(G)$ is of degree at least two, then after adding a pendant vertex w adjacent to v the graph remains a caterpillar forest.*

Proof. After adding w the set of non-pendant vertices does not change. Hence the set of non-pendant vertices induces a set of paths. This implies that the graph remains a caterpillar forest.   □

Lemma 5. *If (G, k) is an instance of MPOVD and u is a vertex with at least $k + 2$ neighbours in G then (G, k) is equivalent to (G', k), where G' is constructed from G by adding w — a pendant neighbour of u.*

Proof. Obviously, the intersection of every solution to MPOVD in (G', k) with $V(G)$ is a solution to MPOVD in (G, k).

Now take a solution $T \subset V(G)$ to MPOVD in (G, k). If $u \in T$, then T is also a solution to MPOVD in (G', k). Otherwise there are at least two neighbours of u not in T, thus u is of degree at least 2 in the caterpillar forest $G[V(G) \setminus T]$. But then $G[V(G') \setminus T]$ is also a caterpillar forest by Lemma 4, and so T is a solution to MPOVD in (G', k) also in this case.   □

Let us now recall two results from matching theory, which will be extensively used in the construction of the kernel:

Theorem 1 (Ore's formula [12], see also [10], Theorem 1.3.1). *Let H be a nonempty bipartite graph with a bipartition (X, Y). Then the maximum cardinality of a matching in H is equal to $|X| - \max_{A \subseteq X} def(A)$, where $def(A) = |A| - |N(A)|$. If $A \subseteq X$ is* *a set which maximizes the value of* $def(A)$, *then every maximal matching in* H *matches* $|A| - def(A)$ *vertices from* A *and all vertices from* $X \setminus A$.

Moreover, a set A maximizing the value of $def(A)$ can be computed in time polynomial in the size of H.

Theorem 2 (α–**Expansion Lemma, Theorem 2.3 in [17]**). *Let* H *be a nonempty bipartite graph with a bipartition* (X, Y) *with* $|Y| \geq \alpha |X|$, *where* α *is a positive integer, and such that every vertex of* Y *has at least one neighbour in* X. *Then there exist nonempty subsets* $X' \subseteq X$, $Y' \subseteq Y$ *and a function* $\phi : X' \to \binom{Y'}{\alpha}$ *(the set of subsets of* Y' *of cardinality* α*) such that*

- $N(Y') = X'$
- *the sets* $\phi(x)$ *are disjoint for* $x \in X'$.

The elements of $\phi(x)$ *will be denoted by* $y_1^x, y_2^x, \ldots, y_\alpha^x$, *and will be jointly referred to as the* private *neighbours of* x.

Moreover, such X', Y' and ϕ can be computed in time polynomial in the size of H.

3   The $4.65^k n^{O(1)}$ Algorithm for POVD

In this section we develop a $4.65^k n^{O(1)}$ branching FPT algorithm for POVD. First, let us recall the approach used by Philip et al. [14]. By Lemma 3 we know that if we hit all subgraphs that are isomorphic to K_3, C_4 or T_2, the problem is solvable in polynomial time. Thus, if a graph contains one of the forbidden subgraphs, we branch and guess which of its vertices should be deleted in the solution. There are at most 7 choices (for T_2), thus we obtain a $7^k n^{O(1)}$ solution described by Philip et al. [14].

Our algorithm uses the same approach, but we use the following combinatorial observation to enhance branching rules for T_2.

Lemma 6. *Let* G *be a graph and let* H *be its subgraph isomorphic to* T_2. *Denote the vertices of* H *as on Figure 1. Assume that the degree of vertex* b_1 *in the whole graph* G *is two. If* (G, k) *is a YES instance of* POVD *and there exists a solution* A *that satisfies* $A \cap \{a, b_1, b_2, b_3\} = \{b_1\}$ *then there exists a solution* A', *such that* $|A'| \leq |A|$ *and* $A' \cap \{a, b_1, b_2, b_3\} = \emptyset$.

Proof. Assume that we have a solution A to POVD that satisfies $b_1 \in A$, $b_2, b_3, a \notin A$. We argue that $A' := A \cup \{c_1\} \setminus \{b_1\}$ is also a valid solution to POVD. As $|A'| \leq |A|$ and $A' \cap \{a, b_1, b_2, b_3\} = \emptyset$, this proves the Lemma.

If $G[V \setminus A]$ is a caterpillar forest then obviously $G[V \setminus (A \cup \{c_1\})]$ also is. However, since $\{a, b_2, b_3\} \cap A = \emptyset$, we know that the vertex a is of degree at least two in $G[V \setminus A]$. Hence by Lemma 4 we can add a pendant vertex adjacent to a and the graph remains a caterpillar forest. In this way we obtain a graph which is isomorphic to $G[V \setminus A']$, thus A' is a valid solution to POVD instance (G, k). □

We are now ready to present branching rules that prove the main theorem of this section.

Theorem 3. *There exists a $4.65^k n^{O(1)}$ FPT algorithm for* PATHWIDTH ONE VERTEX DELETION.

Proof. As in the algorithm of Philip et al. [14], the algorithm hits all subgraphs isomorphic to K_3, C_4 or T_2 and then solves the remaining instance in polynomial time using Lemma 3. At each step, the algorithm first looks for subgraphs isomorphic to K_3 or C_4 and if it finds one, it guesses which vertex of the forbidden subgraph should be included in the solution. We have at most 4 branches and each branch decreases k by one, thus this branching rule fits into the claimed time bound.

In the rest of the analysis we assume that the girth of the graph G is at least 5 and there exists a subgraph H isomorphic to T_2. Denote vertices of H as in Figure 1. We are going to guess which vertices of H are included in the solution, however we use Lemma 6 to limit the number of choices or to delete more than one vertex in some branches. In order to obtain the claimed time bound in each branching point the recurrence should not be worse than $T(k) \leq 4T(k-1) + 3T(k-2)$, which leads to a $4.65^k n^{O(1)}$ running time since $4 \cdot 4.65^{-1} + 3 \cdot 4.65^{-2} \leq 1$. Let us distinguish four cases, depending on how many vertices b_i are of degree 2 in G.

Case 0. All vertices b_i have degree at least 3. For $i = 1, 2, 3$ denote by c_i' any neighbour of b_i different than c_i and a. Since G has girth at least 5, vertices c_i' are pairwise different and different from vertices in $V(H)$. Let us branch on the following seven options. In the first four branches, either a, b_1, b_2 or b_3 is included in the solution. We have 4 branches, each decreases k by one. In the last three branches c_i and c_i' are included in the solution for some $i = 1, 2, 3$, hence in each branch k is decreased by two. To prove the correctness of this branching rule note that if we have a solution to POVD that is disjoint with $\{a, b_1, b_2, b_3\}$ and does not delete both c_1 and c_1' and both c_2 and c_2', the six remaining vertices form a subgraph isomorphic to T_2 with both c_3 and c_3'. Thus both c_3 and c_3' need to be included in the solution.

Case 1. Exactly one (say b_1) vertex b_i has degree 2. For $i = 2, 3$ denote by c_i' any neighbour of b_i different than c_i and a. As before, $c_2' \neq c_3'$ and $c_2', c_3' \notin V(H)$. Let us branch on the following six options. In the first four branches, either a, b_2, b_3 or c_1 is included in the solution. In the fifth branch we include c_2 and c_2' in the solution and in the sixth branch we include c_3 and c_3'. We have four branches that decrease k by one and two that decrease k by two. Let us now check correctness. If a, b_2 and b_3 are not included in an optimal solution, we may also assume that b_1 is not included either, due to Lemma 6. If an optimal solution is disjoint with $\{a, b_1, b_2, b_3, c_1\}$ and does not contain both c_2 and c_2', it needs to contain both c_3 and c_3', as otherwise it misses a subgraph isomorphic to T_2.

Case 2. Exactly two (say b_2 and b_3) vertices b_i have degree 2. Let c_1' be any neighbour of b_1 different than c_1 and a; as before, $c_1' \notin V(H)$. Let us branch on the following six options. In the first four branches, one of a, b_1, c_2 and c_3 is included in the solution. In the fifth branch, b_2 and b_3 are included in the solution. In the sixth branch, c_1 and c_1' are included in the solution. We have four branches that decrease k by one and two that decrease k by two. Let us now check correctness. If a and b_1 are not included in an optimal solution A, we may assume that either $b_2, b_3 \in A$ or $b_2, b_3 \notin A$ by Lemma 6. In the first case, we fit into the fifth branch. In the second case, if c_2 and c_3 are not included in the solution, both c_1 and c_1' are — otherwise a subgraph isomorphic to T_2 is left.

Case 3. All vertices b_i are of degree 2 in G. We branch into seven options. In the first four cases, one of a, c_1, c_2 and c_3 is included in the solution. In the last three cases, one of the subsets of $\{b_1, b_2, b_3\}$ of size two is included in the solution. We have four branches that decrease k by one and three that decrease k by two, thus we fit into the time bound. To check correctness note that if neither a nor any of the vertices c_i is included in the solution, we need to include at least one vertex b_i. But, due to Lemma 6, there exists a solution that contains at least two of them. □

4 $O(k^2)$ **Kernel for** POVD

4.1   **A Quadratic Kernel for** MPOVD

First, we focus on a kernelization algorithm for MULTIGRAPH PATHWIDTH ONE VER-TEX DELETION that transforms an instance (G, k) of MPOVD into an equivalent instance (G', k'), where $k' \leq k$ and $|G'| = O(k^2)$. We provide a set of reduction rules, each of which transforms a MPOVD instance (G, k) into another instance (G', k') in time polynomial in $|G|+k$, where $|G'| < |G|$ and $k' \leq k$. For each of the reduction rules we will check *correctness*, ie. that the output instance (G', k') is a MPOVD instance equivalent to (G, k). The kernelization algorithm tries to apply the lowest numbered applicable rule. If none of the rules is applicable, we claim that the size of the multigraph is already bounded by $O(k^2)$. Let (G, k) denote the MPOVD instance we are dealing with.

Definition 3. *We say that vertices u, v are connected by a* multiedge *if the number of edges between u, v is at least two. In such a case we say that vertices u, v are incident to a multiedge.*

The number of multiedges can be reduced quite easily.

Reduction 1. *If there is a loop at a vertex v, delete v and decrease k by one. If vertices u and v are connected by $\gamma \geq 3$ edges, delete $\gamma - 2$ of them, leaving only two.*

Reduction 2. *For a vertex u in G, if it is connected by multiedges to at least $k+1$ other vertices, then delete u and decrement k by one. The resulting instance is $(G[V(G) \setminus \{u\}], k-1)$.*

Proof (of correctness). Observe that if u is connected by multiedges to at least $k + 1$ other vertices then each solution not containing u would need to contain the (at least $k+1$) other endpoints of these edges, a contradiction. Thus any solution has to contain u. □

We follow up with a group of reductions already provided by Philip et al. in [14].

Reduction 3 ([14]). *If a vertex u in G has two or more pendant neighbours, then delete all but one of these pendant neighbours to obtain G'. The resulting instance is (G', k).*

Reduction 4 ([14]). *For a vertex u in G, if there is a matching M of size $k + 3$ in $G[V(G) \setminus \{u\}]$, where each edge in M has at least one end in $N(u)$, then delete u and decrement k by one. The resulting instance is $(G[V(G) \setminus \{u\}], k - 1)$.*

Reduction 5 (Rule 5 from [14] rephrased). *Let $v_0 - v_1 - v_2 - \ldots - v_p - v_{p+1}$ be a path in G, such that for each vertex v_i $(1 \leq i \leq p)$ its neighbours other than v_{i-1}, v_{i+1} are pendant, and v_i is not incident to any multiedges. If $p \geq 5$ then contract the edge (v_2, v_3) in G to obtain the graph G'. The resulting instance is (G', k).*

The correctness of the above rules has already been proved in [14]. It is easy to observe that the rules are also correct in the multigraph model we are working with.

Let us now take some integer $\alpha \geq 5$ (we apply the following for $\alpha = 5$ in this section and $\alpha = \max\{k + 3, 5\}$ in the next one). We are now going to bound the degrees of vertices in G. The following lemma is crucial for the next reduction.

Lemma 7. *Let $\alpha \geq 5$ and consider a MPOVD instance (G, k). Assume that there is a vertex $u \in V(G)$ of degree at least $(5k + 7) + \alpha(k + 2)$ and Reductions 1–5 cannot be applied. Then one can in polynomial time find disjoint sets of vertices $X', Y' \subseteq V(G) \setminus \{u\}$ and a function $\phi : X' \to \binom{Y'}{\alpha}$ satisfying the following properties:*

1. *every vertex in Y' is connected by a single edge to u,*
2. *Y' is an independent set in G,*
3. *$N(Y') = X' \cup \{u\}$, and each vertex from Y' has a neighbour in X',*
4. *the sets $\phi(x)$ are disjoint for any two different $x \in X'$ (as before we denote the elements of $\phi(x)$ by $y_1^x, y_2^x, \ldots, y_\alpha^x$ in Y' and call them the private neighbours of x).*

Proof. The proof is an application of Theorems 1 and 2 as in the paper by Thomassé [17]. We omit the details due to space limitations. □

Reduction 6. *Let u be a vertex of degree at least $(5k+7)+\alpha(k+2)$. Let X' and Y' be the nonempty sets found using Lemma 7. Obtain G' by deleting all the edges between u and Y' and adding double edges between u and every vertex from X'. Also, if u had no pendant neighbour, add one. The resulting instance is (G', k).*

Proof (of correctness). First, note during the reduction we delete at least $\alpha|X'|$ edges and add at most $2|X'| + 1$ edges and one vertex, thus (as $\alpha \geq 5$) strictly reducing the graph size. Now let us prove the correctness of this reduction.

We have to show that the resulting instance (G', k) is equivalent to the input instance (G, k). Observe that the degree of u in G is at least $(5k + 7) + \alpha(k + 2)$, so using Lemma 5 one can add a pendant vertex to u without changing the answer to the POVD problem. Thus, without loss of generality we may assume that the pendant vertex has already been attached to u and therefore there are no additional vertices in G'.

Denote $V = V(G) = V(G')$ and $l = |X'|$. Suppose that there is a set $A \subseteq V$ such that $|A| \leq k$ and $G[V \setminus A]$ has pathwidth at most one. As $\alpha \geq 2$ then one can find l cycles of length 4 that meet only at u: for each $x \in X'$ we form a cycle consisting of x, y_1^x, u and y_2^x. Observe that if $u \in A$ then $G[V \setminus A] = G'[V \setminus A]$ and A is also a solution to MPOVD in (G', k). Now suppose that $u \notin A$. In this situation, the set A has to contain at least l vertices from $X' \cup Y'$ in order to have at least one vertex on each of the indicated cycles. Obtain $A' = (A \setminus Y') \cup X'$ by replacing these vertices with simply the set X' (of cardinality l). Observe that $|A'| \leq k$ and A' is also a solution to (G, k) instance of MPOVD — as deleting X' results in Y' being a set of pendant vertices of u, irrelevant from the point of view of the MPOVD problem, as there exists another pendant vertex of u (see Reduction 3). Now observe that A' is also a solution to MPOVD for (G', k), as $G'[V \setminus A']$ is the graph obtained from $G[V \setminus A']$ by removing all the edges from u to Y'.

Now suppose that B is a solution to MPOVD in (G', k). Recall that both in G' and G the vertex u has a pendant neighbour w. If $w \in B$ we can modify B by taking $B := B \cup \{u\} \setminus \{w\}$, which is also a solution, thus we can assume $w \notin B$. As before, if $u \in B$ then $G'[V \setminus B] = G[V \setminus B]$ and B is also a solution to MPOVD in (G, k). Now suppose that $u \notin B$. As u is connected to all the vertices from X' by double edges, $X' \subseteq B$. Therefore in $G[V \setminus B]$ the set Y' is a set of pendant neighbours of u, irrelevant to the answer to MPOVD as there exists a pendant neighbour of u in the original graph G. $G'[V \setminus B]$ is simply $G[V \setminus B]$ where vertices from Y' are isolated vertices instead of pendant neighbours of u, so B is a solution to MPOVD in (G, k) as well. □

Let us denote $\beta = (5k + 7) + \alpha(k + 2)$. Using Reduction 6 we already bounded the maximal degree by β. Let us introduce two more reductions in order the make the graph more dense and thus bound its size.

Reduction 7. *Let u be a vertex which is not contained in any cycle or T_2. Obtain G' by deleting u. The resulting instance is (G', k).*

Proof (of correctness). Consider any induced subgraph H of G containing u. Then u is not contained in any cycle or T_2 in H. Thus $G[V(H) \setminus \{u\}]$ is a caterpillar forest if and only if H is a caterpillar forest. This implies that a set $A \subset V(G)$ is a solution to MPOVD in (G, k) iff $A \setminus \{u\}$ is a solution. Thus u is not contained in any minimal solution, and a set A is a solution to MPOVD in (G, k) iff it is a solution in (G', k). □

The next reduction can be viewed as greedy deletion of vertices in treelike parts of the graph.

Reduction 8. *Let $W \subseteq V(G)$ be a set of vertices such that $G[W]$ is a tree (without loops or multiple edges) containing at least one T_2 and there are no edges between $W \setminus \{w\}$ and $V(G) \setminus W$ for some $w \in W$. Root W in w and in every T_2 contained in W mark the vertex nearest to w. Let c be the vertex farthest from w among the marked ones. Delete c to obtain G' and decrement k by one. The resulting instance is $(G', k-1)$.*

Proof (of correctness). Let F be a T_2 subgraph of $G[W]$, due to which c has been marked. From the extremality of c, the deletion of c from G cuts out a set of caterpillars. Observe that at least one of the vertices from $V(F)$ has to be deleted and that the deletion of every other vertex can be substituted by the deletion of c. Thus one can safely assume that c is in the solution. □

Note that the situation described in Reduction 8 can be recognized in polynomial time — for every vertex w one deletes w and chooses as W the union of w and these connected components derived from the connected component in which w was, which are trees.

Now we claim that if none of the Reductions 1–8 can be applied and there is a solution to MPOVD in (G, k) then the size of G is bounded by $O(k^2)$.

Lemma 8. *Let (G, k) be a YES instance of MPOVD. If Reductions 1–8 are not applicable then $|V(G)| \leq (182\beta + 70)k$ and $|E(G)| \leq (210\beta + 79)k$.*

Proof. Due to space limitations we onle sketch the proof. Three steps of the proof are as follows. Firstly from the graph G we recursively remove vertices of degree one obtaining a graph H. Because of Reductions 3,7,8 we infer $|G| \leq c_1|H|$ for come constant c_1.

In the second step we transform the graph H into a multigraph H' with possible loops by contracting edges with at least one endpoint of degree 2. Using Reduction 5 we bound $|H| \leq c_2|H'|$ for some constant c_2.

Finally using standard techniques and Reduction 6 we bound the size of $|H'|$ by a function of k and the lemma follows.   □

This Lemma justifies the final reduction.

Reduction 9. *If $|E(G)| > (210\beta + 79)k$ or $|V(G)| > (182\beta + 70)k$, the resulting instance is $(K_3, 0)$, which is a trivial NO instance of MPOVD.*

Taking $\alpha = 5$ (and thus $\beta = 10k + 17$) we obtain the kernelization algorithm for MPOVD.

Theorem 4. *There exists a kernelization algorithm for MULTIGRAPH PATHWIDTH ONE VERTEX DELETION that from an instance (G, k) produces an equivalent instance (G', k') satisfying $k' \leq k$, $|E(G')| \leq 2100k^2 + 3649k$ and $|V(G')| \leq 1820k^2 + 3164k$, i.e., $|G'| = O(k^2)$.*

4.2 Kernels for POVD

First, let us establish an equivalence between POVD and MPOVD instances. Obviously, any instance of POVD can be treated as an instance of MPOVD. In the other direction, the following lemma shows that each multiedge can be replaced by a small gadget at the cost of a small increase of the parameter.

Lemma 9. *Let (G, k) be a MPOVD instance. One can compute in $|G|^{O(1)}$ time complexity an equivalent instance (G', k') of POVD, where $|G'| = O(|G|)$ and $k' \leq k + |E(G)|$.*

Proof. Firstly if there exists a vertex v with a loop, we remove this vertex and decrease k by one since this vertex has to be contained in any solution. Next for each pair of vertices $v_1, v_2 \in V(G)$ connected by a multiedge, replace all edges $v_1 v_2$ with the gadget shown on Figure 2 and increase the parameter by one. The new instance (G', k') clearly satisfies $|G'| = O(|G|)$ and $k' \leq k + |E(G)|$. Let us now check the correctness of this construction. We need to check only the correctness of a single replacement. Let us assume that v_1 and v_2 and connected by a multiedge in the instance (G, k) and the instance $(G', k+1)$ is created by replacing the multiedge $v_1 v_2$ with the gadget. We now argue that those instances are equivalent.

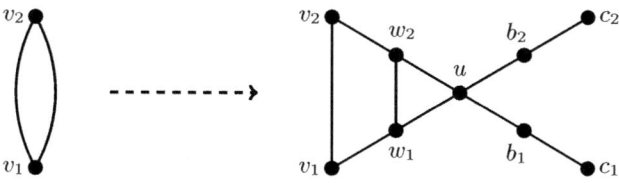

Fig. 2. A gadget that replaces a multiedge $v_1 v_2$

First assume that (G, k) is a YES instance and let $A \subseteq V(G)$ be a valid solution. If $\{v_1, v_2\} \subseteq A$, then $A \cup \{u\}$ is a valid solution to $(G', k + 1)$. Otherwise, as $A \cap \{v_1, v_2\} \neq \emptyset$, w.l.o.g. assume that $v_1 \in A$ and $v_2 \notin A$. Then $A \cup \{w_2\}$ is a valid solution to $(G', k + 1)$.

Now assume that $(G', k + 1)$ is a YES instance and let $A' \subseteq V(G')$ be a valid solution. Denote $X = \{c_1, c_2, b_1, b_2, u, w_1, w_2, v_1, v_2\}$. Note that $X \setminus \{w_1, v_1\}$ and $X \setminus \{w_2, v_2\}$ form two subgraphs isomorphic to T_2, $\{v_1, v_2, w_1, w_2\}$ is a 4-cycle in G' and those three forbidden subgraphs have an empty intersection. Thus $|A' \cap X| \geq 2$. If $|A' \cap X| \geq 3$, then $A'' := A' \setminus X \cup \{v_1, v_2, u\}$ is also a valid solution to $(G', k + 1)$ and $A'' \cap V(G)$ is a valid solution to (G, k). Otherwise, if $|A' \cap X| = 2$ note that $A' \cap X \neq \{v_1, v_2\}$, because otherwise a 3-cycle u, w_1, w_2 is left untouched. W.l.o.g assume that $v_1 \notin A'$. Then $A'' := A' \setminus X \cup \{v_2, w_1\}$ is a valid solution to $(G', k + 1)$ and $A'' \cap V(G)$ is a valid solution to (G, k). $\square$

We are now ready to conclude with the kernelization results for POVD.

Theorem 5. *There exists a kernelization algorithm for* PATHWIDTH ONE VERTEX DELETION *that from an instance (G, k) produces an equivalent instance (G', k') satisfying $|G'|, k' = O(k^2)$.*

Proof. First treat the instance (G, k) as a MPOVD instance and obtain a MPOVD kernel with $O(k^2)$ vertices and edges. Then use Lemma 9 to obtain a POVD instance (G', k') with $O(k^2)$ vertices and edges and with $k' = O(k^2)$. $\square$

Theorem 6. *There exists a kernelization algorithm for* PATHWIDTH ONE VERTEX DELETION *that from an instance (G, k) produces an equivalent instance (G', k') satisfying $k' \leq k$ and $|G'| = O(k^3)$.*

Proof. We use the same reductions as for the MPOVD kernel in Section 4.1. Notice that Reduction 6 is the only one which introduces multiedges. Thus, we need to modify it so that it will use only single edges. We achieve this goal by transforming each double edge into a cycle K_3. We omit the details due to space limitations. $\square$

5   Conclusions and Open Problems

In this paper we have improved the FPT algorithm and kernelization upper bounds for PATHWIDTH ONE VERTEX DELETION. The kernelization bounds are almost tight comparing to lower bounds derived from the general result of Dell and van Melkebeek [5]. However, we do not know of any exact (not FPT) algorithm solving POVD that runs significantly faster than the trivial one (the trivial one, that tries to hit all subgraphs isomorphic to T_2, K_3 and C_4 runs in $127^{n/7} \cdot n^{O(1)} = O(1.9978^n)$ time). One approach to obtain a faster exact algorithm would be to develop an FPT algorithm that runs in time $c^k n^{O(1)}$ for some constant $c < 4$ and use the well-known win-win approach. This, however, seems really nontrivial, as it requires to do something clever with C_4 subgraphs instead of simply branching over them.

References

1. Àlvarez, C., Serna, M.J.: The proper interval colored graph problem for caterpillar trees (extended abstract). Electronic Notes in Discrete Mathematics 17, 23–28 (2004)
2. Arnborg, S., Proskurowski, A., Seese, D.: Monadic Second Order Logic, tree automata and forbidden minors. In: Schönfeld, W., Börger, E., Kleine Büning, H., Richter, M.M. (eds.) CSL 1990. LNCS, vol. 533, pp. 1–16. Springer, Heidelberg (1991)
3. Assmann, S.F., Peck, G.W., Syso, M.M., Zak, J.: The bandwidth of caterpillars with hairs of length 1 and 2. SIAM Journal on Algebraic and Discrete Methods 2(4), 387–393 (1981)
4. Cao, Y., Chen, J., Liu, Y.: On feedback vertex set new measure and new structures. In: Proc. of SWAT 2010, pp. 93–104 (2010)
5. Dell, H., van Melkebeek, D.: Satisfiability allows no nontrivial sparsification unless the polynomial-time hierarchy collapses. In: Proc. of STOC 2010, pp. 251–260 (2010)
6. Diestel, R.: Graph Theory. Springer, Heidelberg (2005)
7. Downey, R.G., Fellows, M.R.: Parameterized Complexity. Springer, Heidelberg (1999)
8. Flum, J., Grohe, M.: Parameterized Complexity Theory, 1st edn. Texts in Theoretical Computer Science. An EATCS Series. Springer, Heidelberg (March 2006)
9. Lin, M., Lin, Z., Xu, J.: Graph bandwidth of weighted caterpillars. Theor. Comput. Sci. 363(3), 266–277 (2006)
10. Lovász, L., Plummer, M.D.: Matching Theory. AMS Chelsea Publishing (2009)
11. Niedermeier, R.: Invitation to Fixed Parameter Algorithms. Oxford Lecture Series in Mathematics and Its Applications. Oxford University Press, USA (March 2006)
12. Ore, O.: Graphs and matching theorems. Duke Math. J. 22(4), 625–639 (1955)
13. Papadimitriou, C.H.: The NP-completeness of the bandwidth minimization problem. Computing 16(3), 263–270 (1976)
14. Philip, G., Raman, V., Villanger, Y.: A quartic kernel for pathwidth-one vertex deletion. In: Proc. of WG 2010 (to appear, 2010)
15. Robertson, N., Seymour, P.D.: Graph minors. I. Excluding a forest. J. Comb. Theory, Ser. B 35(1), 39–61 (1983)
16. Robertson, N., Seymour, P.D.: Graph minors. II. Algorithmic aspects of tree-width. J. Algorithms 7(3), 309–322 (1986)
17. Thomassé, S.: A quadratic kernel for feedback vertex set. In: Proc. of SODA 2009, pp. 115–119 (2009)

Multivariate Complexity Analysis of
Swap Bribery

Britta Dorn[1] and Ildikó Schlotter[2,*]

[1] Wilhelm-Schickard-Institut für Informatik, Universität Tübingen
Sand 13, 72076 Tübingen, Germany
bdorn@informatik.uni-tuebingen.de
[2] Budapest University of Technology and Economics
H-1521 Budapest, Hungary
ildi@cs.bme.hu

Abstract. We consider the computational complexity of a problem modeling *bribery* in the context of voting systems. In the scenario of SWAP BRIBERY, each voter assigns a certain price for swapping the positions of two consecutive candidates in his preference ranking. The question is whether it is possible, without exceeding a given budget, to bribe the voters in a way that the preferred candidate wins in the election.

We initiate a parameterized and multivariate complexity analysis of SWAP BRIBERY, focusing on the case of k-approval. We investigate how different cost functions affect the computational complexity of the problem. We identify a special case of k-approval for which the problem can be solved in polynomial time, whereas we prove NP-hardness for a slightly more general scenario. We obtain fixed-parameter tractability as well as W[1]-hardness results for certain natural parameters.

1 Introduction

In the context of voting systems, the question of how to manipulate the votes in some way in order to make a preferred candidate win the election is a very interesting question. One possibility is *bribery*, which can be described as spending money on changing the voters' preferences over the candidates in such a way that a preferred candidate wins, while respecting a given budget. There are various situations that fit into this scenario: The act of bribing the voters in order to make them change their preferences, or paying money in order to get into the position of being able to change the submitted votes, but also the setting of systematically spending money in an election campaign in order to convince the voters to change their opinion on the ranking of candidates.

The study of bribery in the context of voting systems was initiated by Faliszewski, Hemaspaandra, and Hemaspaandra in 2006 [11]. Since then, various models have been analyzed. In the original version, each voter may have a different but fixed price which is independent of the changes made to the bribed

---

* Supported by the Hungarian National Research Fund (OTKA 67651).

V. Raman and S. Saurabh (Eds.): IPEC 2010, LNCS 6478, pp. 107–122, 2010.
© Springer-Verlag Berlin Heidelberg 2010

Table 1. Overview of known and new results for SWAP BRIBERY for k-approval. The results obtained in this paper are printed in bold. Here, m and n denote the number of candidates and votes, respectively, and β is the budget. For the parameterized complexity results, the parameters are indicated in brackets. If not stated otherwise, the value of k is fixed.

	Result	Reference
$k = 1$ or $k = m - 1$	P	[9]
$1 \leq k \leq m$, m or n constant	P	[9]
$1 \leq k \leq m$, all costs $= 1$	**P**	Thm. 1
$k = 2$	NP-complete	[2]
$3 \leq k \leq m - 2$, costs in $\{0, 1, 2\}$	NP-complete	[9]
$2 \leq k \leq m - 2$, costs in $\{0, 1\}$ and $\beta = 0$	**NP-complete**	[2], Prop. 2
$2 \leq k \leq m - 2$ is part of the input, costs in $\{0, 1\}$ and $\beta = 0$, n constant	**NP-complete**	[3], Prop. 2
$2 \leq k \leq m - 2$, costs in $\{1, 1 + \varepsilon\}$, $\varepsilon > 0$	**NP-complete, W[1]-hard** (β)	Thm. 3
$1 \leq k \leq m$	**FPT** (m)	Thm. 4
$1 \leq k \leq m$ is part of the input	**FPT** (β, n) **by kernelization**	Thm. 5
$1 \leq k \leq m$	**FPT** (β, n, k) **by kernelization**	Thm. 5

vote. The scenario of nonuniform bribery introduced by Faliszewski [10] and the case of microbribery studied by Faliszewski, Hemaspaandra, Hemaspaandra, and Rothe in [12] allow for prices that depend on the amount of change the voter is asked for by the briber.

In addition, the SWAP BRIBERY problem as introduced by Elkind, Faliszewski, and Slinko [9] takes into consideration the ranking aspect of the votes: In this model, each voter may assign different prices for swapping two consecutive candidates in his preference ordering. This approach is natural, since it captures the notion of small changes and comprises the preferences of the voters. Elkind et al. [9] prove complexity results for this problem for several election systems such as Borda, Copeland, Maximin, and approval voting. In particular, they provide a detailed case study for k-approval. In this voting system, every voter can specify a group of k preferred candidates which are assigned one point each, whereas the remaining candidates obtain no points. The candidates which obtain the highest sum of points over all votes are the winners of the election. Two prominent special cases of k-approval are plurality, (where $k = 1$, i.e., every voter can vote for exactly one candidate) and veto (where $k = m - 1$ for m candidates, i.e., every voter assigns one point to all but one disliked candidate). Table 1 shows a summary of research considering SWAP BRIBERY for k-approval, including both previously known and newly achieved results.

This paper contributes to the further investigation of the case study of k-approval that was initiated in [9], this time from a parameterized point of view.

This approach seems to be appealing in the context of voting systems, where NP-hardness is a desired property for various problems, like MANIPULATION (where certain voters, the manipulators, know the preferences of the remaining voters and try to adjust their own preferences in such a way that a preferred candidate wins), LOBBYING (here, a lobby affects certain voters on their decision for several issues in an election), CONTROL (where the chair of the election tries to make a certain candidate win (or lose) by deleting or adding either candidates or votes), or, as in our case, SWAP BRIBERY. However, NP-hardness does not necessarily constitute a guarantee against such dishonest behavior. As Conitzer et al. [7] point out for the MANIPULATION problem, an NP-hardness result in these settings would lose relevance if an efficient fixed-parameter algorithm with respect to an appropriate parameter was found. Parameterized complexity can hence provide a more robust notion of hardness. The investigation of problems from voting theory under this aspect has started, see for example [1,3,4,6,18].

We show NP-hardness as well as fixed-parameter intractability of SWAP BRIBERY for certain very restricted cases of k-approval if the parameter is the budget, whereas we identify a natural special case of the problem which can be solved in polynomial time. By contrast, we obtain fixed-parameter tractability with respect to the parameter 'number of candidates' for k-approval and a large class of other voting systems, and a polynomial kernel for k-approval if we consider certain combined parameters.

The paper is organized as follows. After introducing notation in Section 2, we investigate the complexity of SWAP BRIBERY depending on the cost function in Section 3, where we show the connection to the POSSIBLE WINNER problem, identify a polynomial-time solvable case of k-approval and a hardness result. In Section 4, we consider the parameter 'number of candidates' and obtain an FPT result for SWAP BRIBERY for a large class of voting systems. We also consider the combination of parameters 'number of votes' and 'size of the budget'. We conclude with a discussion of open problems and further directions that might be interesting for future investigations.

2   Preliminaries

Elections. An $\mathcal{E}$-*election* is a pair $E = (C, V)$, where $C = \{c_1, \ldots, c_m\}$ denotes the set of *candidates*, $V = \{v_1, \ldots, v_n\}$ is the set of *votes* or *voters*, and $\mathcal{E}$ is the *election system* which is a function mapping (C, V) to a set $W \subseteq C$ called the *winners* of the election. We will express our results for the *winner case* where several winners are possible, but our results can be adapted to the *unique winner case* where W consists of a single candidate only.

In our context, each vote is a strict linear order over the set C, and we denote by $\mathrm{rank}(c, v)$ the position of candidate $c \in C$ in a vote $v \in V$.

For an overview of different election systems, we refer to [5]. We will mainly focus on election systems that are characterized by a given *scoring rule*, expressed as a vector $(s_1, s_2, \ldots, s_m)$. Given such a scoring rule, the *score* of a candidate c in a vote v, denoted by $\mathrm{score}(c, v)$, is $s_{\mathrm{rank}(c,v)}$. The score of a candidate c in a set of votes V is score$(c, V) = \sum_{v \in V}$ score(c, v), and the winners of the election are the candidates that receive the highest score in the given votes.

The election system we are particularly interested in is k-approval, which is defined by the scoring vector $(1, \ldots, 1, 0, \ldots, 0)$, starting with k ones. In the case of $k = 1$, this is the *plurality* rule, whereas $(m - 1)$-approval is also known as *veto*. Given a vote v, we will say that a candidate c with $1 \leq \text{rank}(c, v) \leq k$ takes a *one-position* in v, whereas a candidate c' with $k + 1 \leq \text{rank}(c', v) \leq m$ takes a *zero-position* in v.

Swap Bribery, Possible Winner, Manipulation. Given V and C, a *swap* in some vote $v \in V$ is a triple (v, c_1, c_2) where $\{c_1, c_2\} \subseteq C, c_1 \neq c_2$. Given a vote v, we say that a swap $\gamma = (v, c_1, c_2)$ is *admissible in* v, if $\text{rank}(c_1, v) = \text{rank}(c_2, v) - 1$. Applying this swap means exchanging the positions of c_1 and c_2 in the vote v, we denote by v^γ the vote obtained this way. Given a vote v, a set Γ of swaps is *admissible in* v, if the swaps in Γ can be applied in v in a sequential manner, one after the other, in some order. Note that the obtained vote, denoted by v^Γ, is independent from the order in which the swaps of Γ are applied. We also extend this notation for applying swaps in several votes, in the straightforward way.

In a SWAP BRIBERY instance, we are given V, C, and $\mathcal{E}$ forming an election, a preferred candidate $p \in C$, a cost function $c : C \times C \times V \to \mathbb{N}$ mapping each possible swap to a non-negative integer, and a budget $\beta \in \mathbb{N}$. The task is to determine a set of admissible swaps Γ whose total cost is at most β, such that p is a winner in the $\mathcal{E}$-election (C, V^Γ). Such a set of swaps is called a *solution* of the SWAP BRIBERY instance. The underlying decision problem is the following.

SWAP BRIBERY
Given: An $\mathcal{E}$-election $E = (C, V)$, a preferred candidate $p \in C$, a cost function c mapping each possible swap to a non-negative integer, and a budget $\beta \in \mathbb{N}$.
Question: Is there a set of swaps Γ whose total cost is at most β such that p is a winner in the $\mathcal{E}$-election (C, V^Γ)?

We will also show the connection between SWAP BRIBERY and the POSSIBLE WINNER problem. In this setting, we have an election where some of the votes may be *partial* orders over C instead of complete linear ones. The question is whether it is possible to extend the partial votes to complete linear orders in such a way that a preferred candidate wins the election. For a more formal definition, we refer to the article by Konczak and Lang [16] who introduced this problem. The corresponding decision problem is defined as follows.

POSSIBLE WINNER
Given: A set of candidates C, a set of partial votes $V' = (v'_1, \ldots, v'_n)$ over C, an election system $\mathcal{E}$, and a preferred candidate $p \in C$.
Question: Is there an extension $V = (v_1, \ldots, v_n)$ of V' such that each v_i extends v'_i to a complete linear order, and p is a winner in the $\mathcal{E}$-election (C, V)?

A special case of POSSIBLE WINNER is MANIPULATION (see e.g. [7,15]). Here, the given set of partial orders consists of two subsets; one subset contains complete preference orders and the other one completely unspecified votes.

Parameterized complexity, Multivariate complexity. We assume the reader to be familiar with the concepts of parameterized complexity, parameterized reductions and kernelization [8,14,19]. Multivariate complexity is the natural sequel of the parameterized approach when expanding to multidimensional parameter spaces, see the recent survey by Niedermeier [20]. For the hardness reduction in Theorem 3, we will use the following W[1]-hard problem [13]:

> MULTICOLORED CLIQUE
> **Given:** An undirected graph $G = (V_1 \cup V_2 \cup \cdots \cup V_k, E)$ with $V_i \cap V_j = \emptyset$ for $1 \leq i < j \leq k$ where the vertices of V_i induce an independent set for $1 \leq i \leq k$.
> **Question:** Is there a complete subgraph (clique) of G of size k?

3   Complexity Depending on the Cost Function

In this section, we focus our attention on SWAP BRIBERY for k-approval. We start with the case where all costs are equal to 1, for which we obtain polynomial-time solvability.

Theorem 1. SWAP BRIBERY *for k-approval is polynomial-time solvable, if all costs are 1.*

Theorem 1 provides an algorithm which checks for every possible s, if there is a solution in which the preferred candidate wins with score s. This can be carried out by solving a minimum cost maximum flow problem. Due to lack of space, we omit the proof of Theorem 1; see the full version for it.

   Note that Theorem 1 also implies a polynomial-time approximation algorithm for SWAP BRIBERY for k-approval with approximation ratio δ, if all costs are in $\{1, \delta\}$ for some $\delta \geq 1$.

   Proposition 2 shows the connection between SWAP BRIBERY and POSSIBLE WINNER. This result is an easy consequence of a reduction given by Elkind et al. [9]. For the proof of the other direction, see again the full paper.

Proposition 2. *The special case of SWAP BRIBERY where the costs are in $\{0, \delta\}$ for some $\delta > 0$ and the budget is zero is equivalent to the POSSIBLE WINNER problem.*

As a corollary, SWAP BRIBERY with costs in $\{0, \delta\}$, $\delta > 0$ and budget zero is NP-complete for almost all election systems based on scoring rules [2]. For many voting systems such as k-approval, Borda, and Bucklin, it is NP-complete even for a fixed number of votes [3].

   We now turn to the case with two different positive costs, addressing 2-approval.

Theorem 3. *Suppose that $\varepsilon > 0$.*
(1) SWAP BRIBERY *for 2-approval with costs in $\{1, 1 + \varepsilon\}$ is NP-complete.*
(2) SWAP BRIBERY *for 2-approval with costs in $\{1, 1 + \varepsilon\}$ is W[1]-hard, if the parameter is the budget β, or equivalently, the maximum number of swaps allowed.*

Proof. We present a reduction from the MULTICOLORED CLIQUE problem. Let $\mathcal{G} = (V, E)$ with the k-partition $V = V_1 \cup V_2 \cup \cdots \cup V_k$ be the given instance of MULTICOLORED CLIQUE. Here and later, we write $[k]$ for $\{1, 2, \ldots, k\}$. For each $i \in [k]$, $x \in V_i$, and $j \in [k] \setminus \{i\}$ we let $E_x^j = \{xy \mid y \in V_j, xy \in E\}$. We construct an instance $I_{\mathcal{G}}$ of SWAP BRIBERY as follows.

The set $\mathcal{C}$ of candidates will be the union of the sets $A, B, C, \widetilde{C}, F, H, \widetilde{H}, M, \widetilde{M}, D, G, T$, and $\{p, r\}$ where

$A = \{a^{i,j} \mid i, j \in [k]\}$,
$B = \{b_v^j \mid j \in [k], v \in V\}$, and $C, \widetilde{C}, F, H$ are defined analogously,
$\widetilde{H} = \{h_v^j \mid j \in [k], v \in \bigcup_{i < j} V_i\}$,
$M = \{m^{i,j} \mid 1 \le i \le j \le k\}$,
$\widetilde{M} = \{\widetilde{m}^{i,j} \mid 1 \le i < j \le k\}$.

Our preferred candidate is p. The sets $D = \{d_1, d_2, \ldots\}$, $G = \{g_1, g_2, \ldots\}$, and $T = \{t_1, t_2, \ldots\}$ will contain *dummies*, *guards*, and *transporters*, respectively. Our budget will be $\beta = k^3 + 10k^2$. Regarding the indices i and j, we suppose $i, j \in [k]$ if not stated otherwise.

The set of votes will be $W = W_G \cup W_I \cup W_S \cup W_C$. Votes in W_G will define guards (explained later), votes in W_I will set the initial scores, votes in W_S will represent the selection of k vertices, and finally, votes in W_C will be responsible for checking that the selected vertices are pairwise neighboring. We construct W such that the following will hold for some (even) integer K, determined later:

$\mathrm{score}(r, W) = 0$,
$\mathrm{score}(a, W) = K + 1$ for each $a \in A$,
$\mathrm{score}(q, W) = K$ for each $q \in \mathcal{C} \setminus (A \cup D \cup \{r\})$,
$\mathrm{score}(d, W) \le 1$ for each $d \in D$.

We define the cost function c such that each swap has cost 1 or $1 + \varepsilon$. We will define each cost to be 1 if not explicitly stated otherwise. Since each cost is at least 1, none of the candidates ranked after the position $\beta + 2$ in a vote v can receive non-zero points in v without violating the budget. Thus, we can represent votes by listing only their first $\beta + 2$ positions. A candidate does not *appear* in some vote, if he is not contained in these positions.

Dummies, guards, truncation, and transporters. First, let us clarify the concept of dummy candidates: we will ensure that no dummy can receive more than one point in total, by letting each $d \in D$ appear in exactly one vote. Since we will use at most one dummy in each vote, this can be ensured easily by using at most $|W|$ dummies in total. We will use the sign $*$ to denote dummies in votes.

Now, we define $\beta + 2$ guards using the votes W_G. We let W_G contain votes of the form $w_G(h)$ for each $h \in [\beta + 2]$, each such vote having multiplicity $K/2$ in W_G. We let $w_G(h) = (g_h, g_{h+1}, g_{h+2}, \ldots, g_{\beta+2}, g_1, g_2, \ldots g_{h-1})$. Note that $\text{score}(g, W_G) = K$ for each $g \in G$, and the total score obtained by the guards in W_G cannot decrease. As we will make sure that p cannot receive more than K points without exceeding the budget, this yields that in any possible solution, each guard must have score exactly K.

Using guards, we can *truncate* votes at any position $h > 2$ by putting arbitrarily chosen guards at the positions $h, h+1, \ldots, \beta + 2$. This way we ensure that only candidates on the first $h - 1$ positions can receive a point in this vote. We will denote truncation at position h by using a sign $\dagger$ at that position.

Sometimes we will need votes which ensure that some candidate q_1 can "transfer" one point to some candidate q_2 using a cost of c from the budget ($c \in \mathbb{N}^+$, $q_1, q_2 \in \mathcal{C} \setminus (D \cup D \cup G)$). In such cases, we construct c votes by using exactly $c - 1$ *transporter* candidates, say $t_1, t_2, \ldots, t_{c-1}$, none of which appears in any other vote in $W \setminus W_I$. The constructed votes are as follows: for each $h \in [c-2]$ we add a vote $(*, t_h, t_{h+1}, \dagger)$, and we also add the votes $(*, q_1, t_1, \dagger)$ and $(*, t_{c-1}, q_2, \dagger)$. We let the cost of any swap here be 1, and we denote the obtained set of votes by $q_1 \leadsto^c q_2$. (Note that $q_1 \leadsto^1 q_2$ only consists of the vote $(*, q_1, q_2, \dagger)$.)

Observe that the votes $q_1 \leadsto^c q_2$ ensure that q_1 can transfer one point to q_2 at cost c. Later, we will make sure $\text{score}(t, W) = K$ for each transporter $t \in T$. Thus, no transporter can increase its score in a solution, and q_1 only loses a point in these votes if q_2 gets one.

Setting initial scores. Using dummies and guards, we define W_I to adjust the initial scores of the relevant candidates as follows. We put the following votes into W_I:

$(p, *, \dagger)$ with multiplicity K,
$(a^{i,j}, *, \dagger)$ with multiplicity $K + 1 - |V_j|$ for each $i, j \in [k]$,
$(h_x^i, *, \dagger)$ with multiplicity $K - |E_x^i|$ for each $i \in [k], x \in \bigcup_{i<j} V_j$,
$(\widetilde{h}_x^i, *, \dagger)$ with multiplicity $K - |E_x^i|$ for each $i \in [k], x \in \bigcup_{i>j} V_j$,
$(m^{i,j}, *, \dagger)$ with multiplicity $K - 2$ for each $i < j$,
$(q, *, \dagger)$ with multiplicity $K - 1$ for each remaining $q \notin D \cup G \cup \{r\}$.

The preferred candidate p will not appear in any other vote, implying $\text{score}(p, W) = K$

Selecting vertices. The set W_S consists of the following votes:

$a^{i,j} \leadsto^1 b_x^i$ for each $i, j \in [k]$ and $x \in V_j$,
$b_x^1 \leadsto^2 \widetilde{c}_x^1$ for each $x \in V$,
$w_S(i, x) = (b_x^i, c_x^{i-1}, \widetilde{c}_x^i, f_x^{i-1}, \dagger)$ for each $2 \le i \le k, x \in V$,
$c_x^k \leadsto^2 f_x^k$ for each $x \in V$,
$\widetilde{c}_x^i \leadsto^1 c_x^i$ for each $i \in [k], x \in V$, and
$f_x^i \leadsto^{2(k-i)+1} h_x^i$ for each $i \in [k], x \in V$.

Swapping candidate b_x^i with c_x^{i-1}, and swapping candidate $\widetilde{c}_x^i$ with f_x^{i-1} in $w_S(i, x)$ for some $2 \le i \le k$, $x \in V$ will have cost $1 + \varepsilon$.

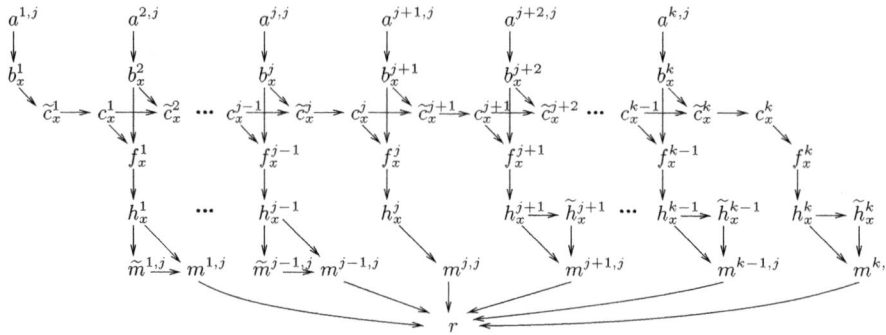

Fig. 1. Part of the instance $I_{\mathcal{G}}$ in the proof of Theorem 3, assuming $x \in V_j$ in the figure. An arc goes from q_1 to q_2 if q_1 can transfer a point to q_2 using one or several swaps.

Checking incidency. The set W_C will contain the votes

$h_x^i \leadsto^1 \tilde{h}_x^i$	for each $i \in [k]$, $x \in \bigcup_{i>j} V_j$,
$w_C(i,j,y,x) = (h_x^i, \tilde{h}_y^j, \tilde{m}^{i,j}, m^{i,j}, \dagger)$	for each $i < j$, $x \in V_j$, $y \in V_i$, $xy \in E$,
$\tilde{m}^{i,j} \leadsto^1 m^{i,j}$	for each $i < j$,
$h_x^i \leadsto^3 m^{i,i}$	for each $i \in [k]$, $x \in V_i$, and
$m^{i,j} \leadsto^1 r$ with multiplicity 2	for each $i < j$,
$m^{i,i} \leadsto^1 r$	for each $i \in [k]$.

Again, swapping candidate h_x^i with $\tilde{h}_y^j$, and also candidate $\tilde{m}^{i,j}$ with $m^{i,j}$ in a vote of the form $w_C(i,j,y,x)$ will have cost $1 + \varepsilon$.

It remains to define K properly. To this end, we let $K \geq 2$ be the minimum even integer not smaller than the integers in the set $\{|E_x^j| \mid j \in [k], x \notin V_j\} \cup \{|V_i| \mid i \in [k]\} \cup \{k^2\}$. This finishes the construction. It is straightforward to verify that the initial scores of the candidates are as claimed above. The constructed instance is illustrated in Fig. 1.

Construction time. Note $|W_G| = (\beta + 2)K/2$, $|W_I| = O(Kk^2 + Kk|V|) = O(Kk|V|)$, $|W_S| = O(k^2|V|)$, and $|W_C| = O(k|V| + |E|)$. Hence, the number of votes is polynomial in the size of the input graph $\mathcal{G}$. This also implies that the number of candidates is polynomial as well, and the whole construction takes polynomial time. Note also that β is only a function of k, hence this yields an FPT reduction as well.

If for some vote v, exactly one candidate q_1 gains a point and exactly one candidate q_2 loses a point as a result of the swaps in Γ, then we say that q_2 *sends* one point to q_1, or equivalently, q_1 *receives* one point from q_2 in v according to Γ. Also, if Γ consists of swaps that transform a vote $(a, b, c, d, \dagger)$ into a vote $(c, d, a, b, \dagger)$, then we say that a sends one point to c, and b sends one point to d. A point is *transferred* from q_1 to q_2 in Γ, if it is sent from q_1 to q_2 possibly through some other candidates.

Our aim is to show the following: $\mathcal{G}$ has a k-clique if and only if the constructed instance $I_\mathcal{G}$ is a yes-instance of SWAP BRIBERY. This will prove both (1) and (2).

Direction $\Longrightarrow$. Suppose that $\mathcal{G}$ has a clique consisting of the vertices $x_1, x_2, \ldots, x_k$ with $x_i \in V_i$. We are going to define a set Γ of swaps transforming W into W^Γ with total cost β such that p wins in W^Γ according to 2-approval.

First, we define the swaps applied by Γ in W_S:

- Swap $a^{i,j}$ with $b^i_{x_j}$ for each i and j in $a^{i,j} \rightsquigarrow^1 b^i_{x_j}$. Cost: k^2.
- Transfer one point from $b^1_{x_j}$ to $\tilde{c}^1_{x_j}$ for each $j \in [k]$ in $b^1_{x_j} \rightsquigarrow^2 \tilde{c}^1_{x_j}$. Cost: $2k$.
- Apply four swaps in each vote $w_S(i, x_j) = (b^i_{x_j}, c^{i-1}_{x_j}, \tilde{c}^i_{x_j}, f^{i-1}_{x_j}, \dagger)$ transforming it to $(\tilde{c}^i_{x_j}, f^{i-1}_{x_j}, b^i_{x_j}, c^{i-1}_{x_j}, \dagger)$, sending one point from $b^i_{x_j}$ to $\tilde{c}^i_{x_j}$ and simultaneously, also one point from $c^{i-1}_{x_j}$ to $f^{i-1}_{x_j}$. Cost: $4k(k-1)$.
- Swap $\tilde{c}^i_{x_j}$ with $c^i_{x_j}$ for each $i, j \in [k]$ in $\tilde{c}^i_{x_j} \rightsquigarrow^1 c^i_{x_j}$. Cost: k^2.
- Transfer one point from $c^k_{x_j}$ to $f^k_{x_j}$ for each $j \in [k]$ in $c^k_{x_j} \rightsquigarrow^2 f^k_{x_j}$. Cost: $2k$.
- Transfer one point from $f^i_{x_j}$ to $h^i_{x_j}$ for each $i, j \in [k]$ in $f^i_{x_j} \rightsquigarrow^{2(k-i)+1} h^i_{x_j}$. Cost: k^3.

The above swaps transfer one point from $a^{i,j}$ to $h^i_{x_j}$ via the candidates $b^i_{x_j}, \tilde{c}^i_{x_j}, c^i_{x_j}$, and $f^i_{x_j}$ for each i and j. These swaps of Γ, applied in the votes W_S, have total cost $k^3 + 6k^2$.

Now, we define the swaps applied by Γ in the votes W_C.

- Swap $h^i_{x_j}$ with $\tilde{h}^i_{x_j}$ for each $j < i$ in $h^i_{x_j} \rightsquigarrow^1 \tilde{h}^i_{x_j}$. Cost: $k(k-1)/2$.
- Apply four swaps in each vote $w_C(i, j, x_i, x_j) = (h^i_{x_j}, \tilde{h}^j_{x_i}, \tilde{m}^{i,j}, m^{i,j}, \dagger)$ transforming it to $(\tilde{m}^{i,j}, m^{i,j}, h^i_{x_j}, \tilde{h}^j_{x_i}, \dagger)$, sending one point from $h^i_{x_j}$ to $\tilde{m}^{i,j}$ and, simultaneously, also one point from $\tilde{h}^j_{x_i}$ to $m^{i,j}$. Note that $w_C(i, j, x_i, x_j)$ is indeed defined for each i and j, since x_i and x_j are neighboring. Cost: $2k(k-1)$.
- Swap $\tilde{m}^{i,j}$ with $m^{i,j}$ for each $i < j$ in $\tilde{m}^{i,j} \rightsquigarrow^1 m^{i,j}$. Cost: $k(k-1)/2$.
- Transfer one point from $h^i_{x_i}$ to $m^{i,i}$ for each $i \in [k]$ in $h^i_{x_i} \rightsquigarrow^3 m^{i,i}$. Cost: $3k$.
- Swap $m^{i,j}$ with r in both of the votes $m^{i,j} \rightsquigarrow^1 r$ for each $i < j$. Cost: $k(k-1)$.
- Swap $m^{i,i}$ with r for each $i \subset [k]$ in $m^{i,i} \rightsquigarrow^1 r$. Cost: k.

Candidate r receives k^2 points after all these swaps in Γ. Easy computations show that the above swaps have cost $4k^2$, so the total cost of Γ is $\beta = k^3 + 10k^2$. Clearly,

$\quad$ score$(p, W^\Gamma) = K$,
$\quad$ score$(r, W^\Gamma) = k^2 \leq K$,
$\quad$ score$(a, W^\Gamma) = K$ for each $a \in A$, and
$\quad$ score$(q, W^\Gamma) =$ score$(q, W) \leq K$ for all the remaining candidates q.

This means that p is a winner in W^Γ according to 2-approval. Hence, Γ is indeed a solution for $I_\mathcal{G}$, proving the first direction of the reduction.

Direction $\Longleftarrow$. Suppose that $I_\mathcal{G}$ is solvable, and there is a set Γ of swaps transforming W into W^Γ with total cost at most β such that p wins in W^Γ

according to 2-approval. We also assume w.l.o.g. that Γ is a solution having minimum cost.

As argued above, score$(p, W^\Gamma) \leq K$ and score$(g, W^\Gamma) \geq K$ for each $g \in G$ follow directly from the construction. Thus, only score$(p, W^\Gamma) = $ score$(g, W^\Gamma) = K$ for each $g \in G$ is possible. Hence, for any $i, j \in [k]$, by score$(a^{i,j}, W) = K+1$ we get that $a^{i,j}$ must lose at least one point during the swaps in Γ. As no dummy can have more points in W^Γ than in W (by their positions), and each candidate in $\mathcal{C} \setminus (A \cup D \cup \{r\})$ has K points in W, the k^2 points lost by the candidates in A can only be transferred by Γ to the candidate r.

By the optimality of Γ, this means that $a^{i,j}$ sends a point to b_x^i in Γ for some unique $x \in V_j$; we define $\sigma(i, j) = x$ in this case. First, we show $\sigma(1, j) = \sigma(2, j) = \cdots = \sigma(k, j)$ for each $j \in [k]$, and then we prove that the vertices $\sigma(1, 1), \ldots, \sigma(k, k)$ form a k-clique in $\mathcal{G}$.

Let B^* be the set of candidates in B that receive a point from some candidate in A according to Γ; $|B^*| = k^2$ follows from the minimality of Γ. Observing the votes in $W_S \cup W_C$, we can see that some $b_x^i \in B^*$ can only transfer one point to r by transferring it to $h_x^{i'}$ via $f_x^{i'}$ for some i' using swaps in the votes W_S, and then transferring the point from $h_x^{i'}$ to r using swaps in the votes W_C. Basically, there are three ways to transfer a point from b_x^i to $h_x^{i'}$:

(A) b_x^i sends one point to f_x^{i-1} in $w_S(i, x)$ at a cost of $3 + 2\varepsilon$, and then f_x^{i-1} transfers one point to h_x^{i-1}. This can be carried out applying exactly $3 + 2(k - i + 1) + 1 = 6 + 2(k - i)$ swaps, having total costs $6 + 2(k - i) + 2\varepsilon$.

(B) b_x^i sends one point to $\widetilde{c}_x^i$ in $w_S(i, x)$, $\widetilde{c}_x^i$ sends one point to c_x^i, c_x^i sends one point to f_x^i, and then the point gets transferred to h_x^i. Again, the number of used swaps is exactly $5 + 2(k - i) + 1 = 6 + 2(k - i)$, and the total cost is at least $6 + 2(k - i)$.

(C) b_x^i sends one point to $\widetilde{c}_x^i$ in $w_S(i, x)$, and then the point is transferred to a candidate $f_x^{i'}$ for some $i' > i$ via the candidates $c_x^i, \widetilde{c}_x^{i+1}, c_x^{i+1}, \ldots, c_x^{i'}$. Again, the number of used swaps is exactly $5 + 2(k - i) + 1 = 6 + 2(k - i)$, and the total cost is at least $6 + 2(k - i)$.

Summing up these costs for each $b_x^i \in B^*$, and taking into account the cost of sending the k^2 points from the candidates of A to B^*, we get that the swaps of Γ applied in the votes W_S must have total cost at least $k^2 + k \left(\sum_{j=1}^k 6 + 2(k - i) \right) = k^3 + 6k^2$. Equality can only hold if each $b_x^i \in B^*$ transfers one point to $h_x^{i'}$ for some $i' \geq i$, i.e. either case B or C happens.

Let H^* be the set of those k^2 candidates in H that receive a point transferred from a candidate in B^*, and let us consider now the swaps of Γ applied in the votes W_C that transfer one point from a candidate $h_x^i \in H^*$ to r. Let j be the index such that $x \in V_j$. First, note that h_x^i must transfer one point to $m^{i,j}$ (if $i \leq j$) or to $m^{j,i}$ (if $i > j$). Moreover, independently of whether $i < j$, $i = j$, or $i > j$ holds, this can only be done using exactly 3 swaps, thanks to the role of the candidates in $\widetilde{H}$ and in $\widetilde{M}$. To see this, note that only the below possibilities are possible:

– If $i < j$, then h_x^i sends one point in $w_C(i, j, y, x)$ for some $y \in V_i$ either to $\widetilde{m}^{i,j}$ via two swaps, or to $m^{i,j}$ via three swaps. In the former case, $\widetilde{m}^{i,j}$ must further transfer the point to $m^{i,j}$, which is the third swap needed.

– If $i > j$, then h_x^i first sends one point to $\widetilde{h}_x^i$, and then $\widetilde{h}_x^i$ sends this point either to $\widetilde{m}^{j,i}$ via one swap, or to $m^{j,i}$ via two swaps applied in the vote $w_C(j, i, x, y)$ for some $y \in V_i$. In the former case, $\widetilde{m}^{j,i}$ transfers the point to $m^{j,i}$ via an additional swap. Note that in any of these cases, Γ applies 3 swaps (maybe having cost $3 + \varepsilon$ or $3 + 2\varepsilon$).

– If $i = j$, then h_x^i sends one point to $m^{i,i}$ through 3 swaps.

Thus, transferring a point from h_x^i to r needs 4 swaps in total, and hence the number of swaps applied by Γ in the votes W_C is at least $4k^2$. Now, by $\beta = k^3 + 10k^2$ we know that equality must hold everywhere in the previous reasonings. Therefore, as argued above, each b_x^i must transfer a point to $h_x^{i'}$ for some $i' \geq i$, i.e., only cases B and C might happen from the above listed possibilities. Now, we are going to argue that only case B can occur.

Let us consider the multiset I_B containing k^2 pairs of indices, obtained by putting (i, j) into I_B for each $b_x^i \in B^*$ with $x \in V_j$. It is easy to see that $I_B = \{(i, j) \mid 1 \leq i, j \leq k\}$. Similarly, we also define the multiset I_H containing k^2 pairs of indices, obtained by putting (i, j) into I_H for each $h_x^i \in H^*$ with $x \in V_j$. By the previous paragraph, I_H can be obtained from I_B by taking some pair (i, j) from I_B and replacing them with corresponding pairs (i', j) where $i' > i$. Let the *measure* of a multiset of pairs I be $\mu(I) = \sum_{(i,j) \in I} i + j$. Then, $\mu(I_H) \geq \mu(I_B) = k^2(k + 1)$.

By the above arguments, if for some $i < j$ the pair (i, j) is contained with multiplicity m_1 in I_H, and (j, i) is contained with multiplicity m_2 in I_H, then the candidate $m^{i,j}$ has to send $m_1 + m_2$ points to r. Similarly, if (i, i) is contained in I_H with multiplicity m, then $m^{i,i}$ has to send m points to r. Thus, $\mu(I_H)$ equals the value obtained by summing up $i + j$ for each $m^{i,j}$ and for each point transferred from $m^{i,j}$ to r. However, each $m^{i,j}$ (where $i < j$) can only send two points to r, and each $m^{i,i}$ can only send one point to r, implying $\mu(I_H) \leq \sum_{i \in [k]}(i + i) + 2\sum_{1 \leq i < j \leq k}(i + j) = k^2(k + 1) = \mu(I_B)$. Hence, the measures of I_B and I_H must be equal, from which $I_H = I_B$ follows. Thus, only case B can happen.

Therefore, Γ must send one point from b_x^i to $\widetilde{c}_x^i$ at a cost of 2, and apply three more swaps of cost 3 to transfer one point from $\widetilde{c}_x^i$ to f_x^i. But in the case $i \geq 2$, this can only be done avoiding any swap of cost $1 + \varepsilon$ in the vote $w_S(i, x)$, if f_x^{i-1} simultaneously receives one point from c_x^{i-1} in $w_S(i, x)$ as well, which implies $b_x^{i-1} \in B^*$. Applying this argument iteratively, this shows that $b_x^i \in B^*$ implies $\{b_x^h \mid h < i\} \subseteq B^*$. Hence, B^* is the union of k sets of the form $\{h_x^1, h_x^2, \ldots, h_x^k\}$, implying $\sigma(1, j) = \sigma(2, j) = \cdots = \sigma(k, j)$ for each $j \in [k]$.

Finally, consider the swaps that transfer one point from $h_x^i \in H^*$ to $m^{i,j}$ in W_C where $x \in V_j$ and $i < j$. We know that if $x \in V_j$, then this must be done by applying some swaps in the vote $w_C(i, j, y, x)$ for some $y \in V_i$ such that $xy \in E$. But because of our budget, each such swap must have cost 1 and not $1 + \varepsilon$, which can only happen if Γ transforms $w_C(i, j, y, x) = (h_x^i, \widetilde{h}_y^j, \widetilde{m}^{i,j}, m^{i,j}, \dagger)$

into $(\widetilde{m}^{i,j}, m^{i,j}, h_x^i, \widetilde{h}_y^j, \dagger)$. But this implies that h_y^j must also be in B^*, implying $y = \sigma(j, i)$. Therefore we obtain that $\sigma(i, j)$ and $\sigma(j, i)$ must be vertices connected by an edge in $\mathcal{G}$. This proves the existence of a k-clique in $\mathcal{G}$, proving the theorem.   □

Looking into the proof of Theorem 3, we can see that the results hold even if the costs are uniform in the sense that swapping two given candidates has the same price in any vote, and the maximum number of swaps allowed in a vote is at most 4.

By applying minor modifications to the given reduction, Theorem 3 can be generalized to hold for the case when we want p to be the unique winner; we only have to set $\text{score}(p, W) = K + 1$ in the construction. Also, Theorem 3 remains true for the case of k-approval for some $3 \leq k \leq |C| - 2$ instead of 2-approval: to prove this, it suffices to insert $k - 2$ dummies into the first $k - 2$ positions of each vote.

4   Other Parameterizations

In this section, we will consider different kinds of parameterizations. First, we will look at the parameter 'number of candidates'. For this case, the following observation is helpful.

Let $S_m = \{\pi_1, \pi_2, \ldots, \pi_{m!}\}$ be the set of permutations of size m. We say that an election system is *described by linear inequalities*, if for a given set $C = \{c_1, c_2, \ldots, c_m\}$ of candidates it can be characterized by $f(m)$ sets $A_1, A_2, \ldots A_{f(m)}$ (for some computable function f) of linear inequalities over $m!$ variables $x_1, x_2, \ldots, x_{m!}$ in the following sense: if n_i denotes the number of those votes in a given election E that order C according to π_i, then the first candidate c_1 is a winner of the election if and only if for at least one index i, the setting $x_j = n_j$ for each j satisfies all inequalities in A_i.

It is easy to see that many election systems can be described by linear inequalities: any system based on scoring rules, Copeland$^\alpha$ ($0 \leq \alpha \leq 1$), Maximin, Bucklin, Ranked pairs. For example, k-approval is described by the following set A_1 of linear inequalities:

$$A_1: \quad \sum\nolimits_{i:\text{rank}(c_1, v_i) \leq k} x_i \geq \sum\nolimits_{i:\text{rank}(c_j, v_i) \leq k} x_i \quad \text{for each } 2 \leq j \leq m.$$

Theorem 4. SWAP BRIBERY *is FPT if the parameter is the number of candidates, for any election system described by linear inequalities.*

Proof. Let $C = \{c_1, c_2, \ldots, c_m\}$ be the set of candidates, where c_1 is the preferred one, and let $A_1, A_2, \ldots A_{f(m)}$ be the sets of linear inequalities over variables $x_1, \ldots, x_{m!}$ describing the given election system $\mathcal{E}$. For some $\pi_i \in S_m$, let v_i denote the vote that ranks C according to π_i. We describe the set V of votes by writing n_i for the multiplicity of the vote v_i in V.

Our algorithm solves $f(m)$ integer linear programs with variables $T = \{t_{i,j} \mid i \neq j, 1 \leq i, j \leq m!\}$. We will use $t_{i,j}$ to denote the number of votes v_i that we transform into votes v_j; we will require $t_{i,j} \geq 0$ for each $i \neq j$. Let V^T denote the set of votes obtained by transforming the votes in V according to the variables $t_{i,j}$ for each $i \neq j$. Such a transformation from V is feasible if $\sum_{j \neq i} t_{i,j} \leq n_i$ holds for each $i \in [m!]$ (inequality $\mathcal{A}$). By [9], we can compute the price $c_{i,j}$ of transforming the vote v_i into v_j in $O(m^3)$ time. Transforming V into V^T can be done with total cost at most β, if $\sum_{i,j \in [m!]} t_{i,j} c_{i,j} \leq \beta$ (inequality $\mathcal{B}$).

We can express the multiplicity x_i' of the vote v_i in V^T as $x_i' = n_i + \sum_{j \neq i} t_{j,i} - \sum_{i \neq j} t_{i,j}$. For some $i \in [f(m)]$, let A_i' denote the set of linear inequalities over the variables in T that are obtained from the linear inequalities in A_i by substituting x_i with the above given expression for x_i'. Using the description of $\mathcal{E}$ with the given linear inequalitites, we know that the preferred candidate c_1 wins in the $\mathcal{E}$-election (C, V^T) for some values of the variables $t_{i,j}$ if and only if these values satisfy the inequalities of A_i' for at least one $i \in [f(m)]$. Thus, our algorithm solves SWAP BRIBERY by finding a non-negative assignment for the variables in T that satisfies both the inequalities $\mathcal{A}$, $\mathcal{B}$, and all inequalities in A_i' for some i.

Solving such a system of linear inequalities can be done in linear FPT time, if the parameter is the number of variables [17]. By $|T| = (m! - 1)m!$ the theorem follows.   □

Similarly, we can also show fixed-parameter tractability for other problems if the parameter is the number of candidates, e.g. for POSSIBLE WINNER (this was already obtained by Betzler et al. for several voting systems, [3]), MANIPULATION (both for weighted and unweighted voters), several variants of CONTROL (this result was obtained for Llull and Copeland voting by Faliszewski et al., [12]), or LOBBYING [6] (here, the parameter would be the number of issues in the election). Since our topic is SWAP BRIBERY, we omit the details.

Finally, we consider a combined parameter and obtain fixed-parameter tractability.

Theorem 5. *If the minimum cost is 1, then SWAP BRIBERY for k-approval (where k is part of the input) with combined parameter $(|V|, \beta)$ admits a kernel with $O(|V|^2 \beta)$ votes and $O(|V|^2 \beta^2)$ candidates. Here, V is the set of votes and β is the budget.*

Proof. Let V, C, $p \in C$, and β denote the set of votes, the set of candidates, the preferred candidate, and the budget given, respectively. The idea of the kernelization algorithm is that not all candidates are interesting for the problem: only candidates that can be moved within the budget β from a zero-position to a one-position or vice versa are relevant.

Let Γ be a set of swaps with total cost at most β. Clearly, as the minimum possible cost of a swap is 1, we know that there are only 2β candidates c in a vote $v \in V$ for which $\operatorname{score}(c, v) \neq \operatorname{score}(c, v^\Gamma)$ is possible, namely, such a c has to fulfill $k - \beta + 1 \leq \operatorname{rank}(c, v) \leq k + \beta$. Thus, there are at most $2\beta|V|$ candidates for which $\operatorname{score}(c, V) \neq \operatorname{score}(c, V^\Gamma)$ is possible; let us denote the set of these candidates by $\widetilde{C}$. Let c^* be a candidate in $C \setminus \widetilde{C}$ whose score is the maximum among the candidates in $C \setminus \widetilde{C}$.

Note that a candidate $c \in C \setminus (\widetilde{C} \cup \{c^*, p\})$ has no effect on the answer to the problem instance. Indeed, if $\text{score}(p, V^\Gamma) \geq \text{score}(c^*, V^\Gamma)$, then the score of c is not relevant, and conversely, if $\text{score}(p, V^\Gamma) < \text{score}(c^*, V^\Gamma)$ then p loses anyway. Therefore, we can disregard each candidate in $C \setminus \widetilde{C}$ except for c^* and p.

The kernelization algorithm constructs an equivalent instance K as follows. In K, neither the budget, nor the preferred candidate will be changed. However, we will change the value of k to be $\beta + 1$, so the kernel instance K will contain a $(\beta + 1)$-approval election[1]. We define the set V_K of votes and the set C_K of candidates in K as follows.

First, the algorithm "truncates" each vote v, by deleting all its positions (together with the candidates in these positions) except for the 2β positions between $k - \beta + 1$ and $k + \beta$. Then again, we shall make use of dummy candidates (see the proof of Theorem 3); we will ensure $\text{score}(d, V^\Gamma) \leq 1$ for each such dummy d. Swapping a dummy with any other candidate will have cost 1 in K. Now, for each obtained truncated vote, the algorithm inserts a dummy candidate in the first position, so that the obtained votes have length $2\beta + 1$. In this step, the algorithm also determines the set $\widetilde{C}$ and the candidate c^*. This can be done in linear time. We denote the votes[2] obtained in this step by V_r. We do not change the costs of swapping candidates of $\widetilde{C} \cup \{c^*, p\}$ in some vote $v \in V_r$.

Next, to ensure that K is equivalent to the original instance, the algorithm constructs a set V_d of votes such that $\text{score}(c, V_r \cup V_d) = \text{score}(c, V)$ holds for each candidate c in $\widetilde{C} \cup \{p, c^*\}$. This can be done by constructing $\text{score}(c, V) - \text{score}(c, V_r)$ newly added votes where c is on the first position, and all the next 2β positions are taken by dummies. This way we ensure $\text{score}(c, V_d) = \text{score}(c, V_d^\Gamma)$ for any set Γ of swaps with total cost at most β.

If D is the set of dummy candidates created so far, then let $C_K = \widetilde{C} \cup \{p, c^*\} \cup D$. To finish the construction of the votes, it suffices to add for each vote $v \in V_r \cup V_d$ the candidates not yet contained in v, by appending them at the end (starting from the $(2\beta + 1)$-th position) in an arbitrary order. The obtained votes will be the votes V_K of the kernel.

The presented construction needs polynomial time. Using the above mentioned arguments, it is straightforward to verify that the constructed kernel instance is indeed equivalent to the original one. Thus, it remains to bound the size of K.

Clearly, $|\widetilde{C} \cup \{p, c^*\}| \leq 2|V|\beta + 2$. The number of dummies introduced in the first phase is exactly $|V_r| = |V|$. As the score of any candidate in V is at most $|V|$, the number of votes created in the second phase is at most $(2|V|\beta + 2)|V|$, which implies that the number of dummies created in this phase is at most $(2|V|\beta + 2)|V| \cdot 2\beta$. This shows $|C_K| \leq |V| + (2|V|\beta + 2)(2|V|\beta + 1) = O(|V|^2\beta^2)$, and also $|V_K| \leq (2|V|\beta + 3)|V| = O(|V|^2\beta)$. $\square$

Applying similar ideas, a kernel with $(\beta + k)|V|$ candidates is easy to obtain, which might be favorable to the above result in cases where k is small.

---

[1] We use $\beta + 1$ instead of β to avoid complications with the case $\beta = 0$.

[2] In fact, these vectors are not real votes yet in the sense that they do not contain each candidate.

5   Conclusion

We have taken the first step towards parameterized and multivariate investigations of SWAP BRIBERY under certain voting systems. We obtained W[1]-hardness for k-approval if the parameter is the budget β, while SWAP BRIBERY could be shown to be in FPT for a very large class of voting systems if the parameter is the number of candidates. This revaluates previous NP-hardness results: SWAP BRIBERY could be computed efficiently if the number of candidates is small, which is a common setting, e.g. in presidential elections.

However, we have shown this via an integer linear program formulation, using a result by Lenstra, which does not provide running times that are suitable in practice. Here, it would be interesting to give combinatorial algorithms that compute an optimal swap bribery.

As Elkind et al. [9] pointed out, it would be nice to characterize further natural cases of SWAP BRIBERY that are polynomial-time solvable. We provided one such example with Theorem 1 for k-approval in the case where costs are equal to 1. By contrast, as soon as we have two different costs, the problem becomes NP-complete for k-approval ($2 \leq k \leq m - 2$) and W[1]-hard if the parameter is the budget β.

There are plenty of possibilities to carry on our initiations. First, there are more parameterizations to be looked at, and in particular the study of combined parameters in the spirit of Niedermeier [20], see e.g. [1], is an interesting approach.

Also, we have focused our attention to k-approval, but the same questions could be studied for other voting systems, or for the special case of SHIFT BRIBERY which was shown to be NP-complete for several voting systems [9], or other variants of the bribery problem as mentioned in the introduction. For instance, we have only looked at *constructive* swap bribery, but the case of *destructive* swap bribery (when our aim is to achieve that a disliked candidate does *not* win) is worth further investigation as well.

Acknowledgments. We thank Rolf Niedermeier for an inspiring initial discussion.

References

1. Betzler, N.: On problem kernels for possible winner determination under the k-approval protocol. In: Hliněný, P., Kučera, A. (eds.) MFCS 2010. LNCS, vol. 6281, pp. 114–125. Springer, Heidelberg (2010)
2. Betzler, N., Dorn, B.: Towards a dichotomy for the Possible Winner problem in elections based on scoring rules. J. Comput. Syst. Sci. 76, 812–836 (2010)
3. Betzler, N., Hemmann, S., Niedermeier, R.: A multivariate complexity analysis of determining possible winners given incomplete votes. In: Proc. of IJCAI 2009, pp. 53–58 (2009)
4. Betzler, N., Uhlmann, J.: Parameterized complexity of candidate control in elections and related digraph problems. Theor. Comput. Sci. 410(52), 5425–5442 (2009)

5. Brams, S.J., Fishburn, P.C.: Voting procedures. In: Handbook of Social Choice and Welfare, vol. 1, pp. 173–236. Elsevier, Amsterdam (2002)
6. Christian, R., Fellows, M., Rosamond, F., Slinko, A.: On complexity of lobbying in multiple referenda. Review of Economic Design 11(3), 217–224 (2007)
7. Conitzer, V., Sandholm, T., Lang, J.: When are elections with few candidates hard to manipulate? J. ACM 54(3), 1–33 (2007)
8. Downey, R.G., Fellows, M.R.: Parameterized Complexity. Springer, Heidelberg (1999)
9. Elkind, E., Faliszewski, P., Slinko, A.: Swap bribery. In: Mavronicolas, M., Papadopoulou, V.G. (eds.) Algorithmic Game Theory. LNCS, vol. 5814, pp. 299–310. Springer, Heidelberg (2009)
10. Faliszewski, P.: Nonuniform bribery. In: Proc. of AAMAS 2008, pp. 1569–1572 (2008)
11. Faliszewski, P., Hemaspaandra, E., Hemaspaandra, L.A.: The complexity of bribery in elections. In: Proc. of AAAI 2006, pp. 641–646 (2006)
12. Faliszewski, P., Hemaspaandra, E., Hemaspaandra, L.A., Rothe, J.: Llull and Copeland voting computationally resist bribery and constructive control. J. Artif. Intell. Res. (JAIR) 35, 275–341 (2009)
13. Fellows, M.R., Hermelin, D., Rosamond, F.A., Vialette, S.: On the parameterized complexity of multiple-interval graph problems. Theor. Comput. Sci. 410(1), 53–61 (2009)
14. Flum, J., Grohe, M.: Parameterized Complexity Theory. Springer, Heidelberg (2006)
15. Hemaspaandra, E., Hemaspaandra, L.A.: Dichotomy for voting systems. J. Comput. Syst. Sci. 73(1), 73–83 (2007)
16. Konczak, K., Lang, J.: Voting procedures with incomplete preferences. In: Proc. of IJCAI 2005 Multidisciplinary Workshop on Advances in Preference Handling (2005)
17. Lenstra, H.: Integer programming with a fixed number of variables. Math. of OR 8, 538–548 (1983)
18. Liu, H., Feng, H., Zhu, D., Luan, J.: Parameterized computational complexity of control problems in voting systems. Theor. Comput. Sci. 410(27-29), 2746–2753 (2009)
19. Niedermeier, R.: Invitation to Fixed-Parameter Algorithms. Oxford University Press, Oxford (2006)
20. Niedermeier, R.: Reflections on multivariate algorithmics and problem parameterization. In: Proc. of STACS 2010, pp. 17–32 (2010)

Parameterizing by the Number of Numbers

Michael R. Fellows[1], Serge Gaspers[2], and Frances A. Rosamond[1]

[1] School of Engineering and IT, Charles Darwin University, NT 0909, Australia
{michael.fellows,frances.rosamond}@cdu.edu.au
[2] Institute of Information Systems, Vienna University of Technology, Vienna, Austria
gaspers@kr.tuwien.ac.at

Abstract. The usefulness of parameterized algorithmics has often depended on what Niedermeier has called "the art of problem parameterization". In this paper we introduce and explore a novel but general form of parameterization: *the number of numbers*. Several classic numerical problems, such as SUBSET SUM, PARTITION, 3-PARTITION, NUMERICAL 3-DIMENSIONAL MATCHING, and NUMERICAL MATCHING WITH TARGET SUMS, have multisets of integers as input. We initiate the study of parameterizing these problems by the number of distinct integers in the input. We rely on an FPT result for INTEGER LINEAR PROGRAMMING FEASIBILITY to show that all the above-mentioned problems are fixed-parameter tractable when parameterized in this way. In various applied settings, problem inputs often consist in part of multisets of integers or multisets of weighted objects (such as edges in a graph, or jobs to be scheduled). Such number-of-numbers parameterized problems often reduce to subproblems about transition systems of various kinds, parameterized by the size of the system description. We consider several core problems of this kind relevant to number-of-numbers parameterization. Our main hardness result considers the problem: given a non-deterministic Mealy machine M (a finite state automaton outputting a letter on each transition), an input word x, and a census requirement c for the output word specifying how many times each letter of the output alphabet should be written, decide whether there exists a computation of M reading x that outputs a word y that meets the requirement c. We show that this problem is hard for $W[1]$. If the question is whether there exists an input word x such that a computation of M on x outputs a word that meets c, the problem becomes fixed-parameter tractable.

1 Introduction

Parameterized complexity and algorithmics has been developing for more than twenty years. Some important progress of the field has depended on what Niedermeier has called "the art of problem parameterization" (see Chapter 5 of

---

[1] M.R.F. and F.A.R. acknowledge support from the Australian Research Council.
[2] S.G. acknowledges partial support from the European Research Council, grant reference 239962, from Conicyt Chile via project Basal-CMM, and from the Australian Research Council.

V. Raman and S. Saurabh (Eds.): IPEC 2010, LNCS 6478, pp. 123–134, 2010.

his monograph [14]). For example, it was Valerie King in 1994 who first suggested that the parameter might be $k = 1/\epsilon$ in the study of the complexity of approximation, leading eventually to the study of EPTASs.

Here we explore, for the first time (to our knowledge), a parameterization that seems widely relevant: *the number of numbers*. Many problems take as input information that consists (in part) of multisets of integers or multisets of weighted objects, such as weighted edges in a weighted graph, the time-requirements of jobs to be scheduled, or the sequence of molecular weights of a spectrographic dataset.

As an initial foray, we first show that a number of classic NP-hard problems about multisets of integers, when parameterized in this way, become fixed-parameter tractable. The proofs are easy, and the knowledgeable reader might anticipate them almost as exercises today — they use the relatively deep result that INTEGER LINEAR PROGRAMMING, parameterized by the number of variables, is FPT. Until recently, as noted in the 2006 monograph by Niedermeier [15], there were not so many interesting applications of this fundamental result (see [1,6,7,9] for some exceptions).

At a deeper level of engagement with this parameterization, we describe some examples of how number-of-numbers parameterized problems reduce to numerical problems about Mealy machines, parameterized by the size of the description of the machine. We show that one basic problem about Mealy machines, parameterized in this way, is FPT, and that another is $W[1]$-hard.

2   Preliminaries

In the INTEGER LINEAR PROGRAMMING FEASIBILITY problem (ILPF), the input is an $m \times n$ matrix $\mathbf{A}$ of integers and an m-vector $\mathbf{b}$ of integers, the parameter is n, and the question is whether there exists an n-vector $\mathbf{x}$ of integers satisfying the m inequalities $\mathbf{Ax} \leq \mathbf{b}$. ILPF was shown to be fixed-parameter tractable by Lenstra [12] and the running time has been improved by Kannan [10].

Let A be a multiset. The *cardinality* of A, denoted $|A|$, is the total number of elements in A, including repeated memberships. The *variety* of A, denoted $||A||$, is the number of distinct elements in A. Element a has *multiplicity* m in A if it occurs m times in A. We denote the set of integers from 1 to n by $[n]$.

Let $G = (V, E)$ be a graph, $v \in V$ be a vertex of G, and $S \subseteq V$ be a subset of vertices of G. The subgraph of G induced on S is the graph $G[S] = (S, E \cap \{uv : u, v \in S\})$. The set S is a *clique* of G if $G[S]$ is *complete*, i.e. there is an edge between every two distinct vertices of $G[S]$. The set S is an *independent set* of G if $G[S]$ is *empty*, i.e. $G[S]$ has no edge. The *neighborhood* of v is the set of vertices incident to v and denoted $N(v)$. Its *degree* is $d(v) = |N(v)|$. We also define $N_S(v) = N(v) \cap S$ and $d_S(v) = |N_S(v)|$.

Let Σ be an *alphabet*. The elements of Σ are called *letters*, and a *word* x of length $n = |x|$ is a sequence of n letters. The symbol λ denotes the empty letter. We denote the concatenation of two words $x_1, x_2 \in \Sigma^*$ by $x_1 x_2$. The ith *power* of a word x is denoted x^i or $(x)^i$ and represents the word $\underbrace{xx \dots x}_{i \text{ times}}$.

3   Subset Sum and Partition

We start with two classic problems on multisets and show that they are fixed-parameter tractable, parameterized by the number of numbers.

---

variety-SUBSET SUM (*var*-SUBSUM)

Input: a multiset A of integers and an integer s
Parameter: $k = ||A||$, the number of distinct integers in A
Question: Is there a multiset $X \subseteq A$ such that $\sum_{a \in X} a = s$?

---

variety-PARTITION (*var*-PART)

Input: a multiset A of integers
Parameter: $k = ||A||$
Question: Is there a multiset $X \subseteq A$ such that $\sum_{a \in X} a = \sum_{b \in A \setminus X} b$?

Theorem 1. *var*-SUBSUM *is fixed-parameter tractable.*

Proof. Given an instance (A, s) for *var*-SUBSUM, with $||A|| = k$, we create an equivalent instance of ILPF whose number of variables is upper bounded by a function of k. Let $a_1, \ldots, a_k$ denote the distinct elements of A and let $m_1, \ldots, m_k$ denote their respective multiplicities in A. The ILPF instance has the integer variables $x_1, \ldots, x_k$ and the following inequalities and equalities.

$$x_i \leq m_i \qquad\qquad \forall i \in [k]$$
$$x_i \geq 0 \qquad\qquad \forall i \in [k]$$
$$\sum_{i=1}^{k} x_i \cdot a_i = s.$$

For each $i \in [k]$, the variable x_i represents the number of times a_i occurs in X, the set summing to s in a valid solution. Using standard techniques in mathematical programming, these constraints can be transformed into the form $\mathbf{Ax} \leq \mathbf{b}$.   □

A very similar proof shows that *var*-PART is fixed-parameter tractable.

Theorem 2. *var*-PART *is fixed-parameter tractable.*

4   Numerical 3-Dimensional Matching

Using the ILPF machinery, we show in this section that several other problems, which are often used in NP-hardness proofs, become fixed-parameter tractable when parameterized by the number of numbers.

---

variety-NUMERICAL 3-DIMENSIONAL MATCHING (*var*-NUM3-DM)

Input: three multisets A, B, C of n integers each and an integer s
Parameter: $k = ||A \cup B \cup C||$
Question: Are there n triples $S_1, \ldots, S_n$, each containing one element from
            each of A, B, and C such that for every $i \in [n]$, $\sum_{a \in S_i} a = s$?

Theorem 3. *var*-NUM3-DM *is fixed-parameter tractable.*

Proof. Let (A, B, C, s) be an instance for *var*-NUM3-DM, with $k_1 = ||A||$, $k_2 = ||B||$, $k_3 = ||C||$, and $k = ||A \cup B \cup C||$. Let $a_1, \ldots, a_{k_1}$ denote the distinct elements of A, $b_1, \ldots, b_{k_2}$ denote the distinct elements of B, and $c_1, \ldots, c_{k_3}$ denote the distinct elements of C. Also, let $m_{1,a}, \ldots, m_{k_1,a}, m_{1,b}, \ldots, m_{k_2,b}, m_{1,c}, \ldots, m_{k_3,c}$ denote their respective multiplicities in A, B, and C. We create an instance of ILPF with at most k^3 integer variables $x_{i,j,\ell}$, $i \in [k_1], j \in [k_2], \ell \in [k_3]$:

$$x_{i,j,\ell} = 0 \qquad \text{for each } (i, j, \ell) \in [k_1] \times [k_2] \times [k_3]$$
$$\text{such that } a_i + b_j + c_\ell \neq s$$

$$\sum_{(j,\ell) \in ([k_2],[k_3])} x_{i,j,\ell} = m_{i,a} \qquad \forall i \in [k_1]$$

$$\sum_{(i,\ell) \in ([k_1],[k_3])} x_{i,j,\ell} = m_{j,b} \qquad \forall j \in [k_2]$$

$$\sum_{(i,j) \in ([k_1],[k_2])} x_{i,j,\ell} = m_{\ell,c} \qquad \forall \ell \in [k_3]$$

A variable $x_{i,j,\ell}$ represents the number of times the elements $a_i \in A$, $b_j \in B$ and $c_\ell \in C$ are used together to form a triple summing to s. The first constraint makes sure that such a triple is formed only if it sums to s. The remaining equalities make sure that each element of $A \cup B \cup C$ appears in a triple. Thus n such triples are formed, all summing to s if the integer program is feasible.   □

Note that the problem is also fixed-parameter tractable if parameterized by $||A \cup B||$ only: we face a NO-instance if $||C|| > ||\{a + b : a \in A, b \in B\}||$. A closely related, well known numerical problem, is the following.

---

variety-NUMERICAL MATCHING WITH TARGET SUMS (*var*-NMTS)
Input: three multisets A, B, S of n integers each
Parameter: $k = ||A \cup B \cup S||$
Question: Are there n triples $C_1, \ldots, C_n \in A \times B \times S$, such that the A-element and the B-element from each C_i sum to its S-element?

Corollary 1. *var*-NMTS *is fixed-parameter tractable.*

By the previous discussion, the natural parameterization by $||A \cup B||$ is also fixed-parameter tractable. A straightforward adaptation of the proof of Theorem 3 shows that *variety*-3-PARTITION is fixed-parameter tractable.

---

variety-3-PARTITION (*var*-3-PART)
Input: a multiset A of $3n$ integers
Parameter: $k = ||A||$
Question: Are there n triples $S_1, \ldots, S_n \subseteq A$, all summing to the same number?

Theorem 4. *var*-3-PART *is fixed-parameter tractable.*

5   Mealy Machines

Mealy machines [13] are finite-state transducers, generating an output based on their current state and input. A *deterministic Mealy machine* is a dual-alphabet state transition system given by a 5-tuple $M = (S, s_0, \Gamma, \Sigma, T)$:

- a finite set of states S,
- a start state $s_0 \in S$,
- a finite set Γ, called the input alphabet,
- a finite set Σ, called the output alphabet, and
- a transition function $T : S \times \Gamma \to S \times \Sigma$ mapping pairs of a state and an input letter to the corresponding next state and output letter.

In a *non-deterministic Mealy machine*, the only difference is that the transition function is defined $T : S \times \Gamma \to \mathcal{P}(S \times \Sigma)$ as for a given state and input letter, there may be more than one possibility for the next state and output letter. (Here $\mathcal{P}(X)$ denotes the powerset of a set X.)

A *census requirement* $c : \Sigma \to \mathbb{N}$ is a function assigning a non-negative integer to each letter of the output alphabet. It is used to constrain how many times each letter should appear in the output of a machine. A word $y \in \Sigma^*$ *meets* the census requirement if every letter $b \in \Sigma$ appears exactly $c(b)$ times in y.

Our first problem about Mealy machines asks whether there exists an input word and a computation of the Mealy machine such that the output word meets the census requirement.

---

variety-EXISTS WORD MEALY MACHINE (*var*-EWMM)
Input: a non-deterministic Mealy machine $M = (S, s_0, \Gamma, \Sigma, T)$, and a census requirement $c : \Sigma \to \mathbb{N}$
Parameter: $|S| + |\Gamma| + |\Sigma|$
Question: Does there exist a word $x \in \Gamma^*$ for which a computation of M on input x generates an output y that meets c?

Our proof that *var*-EWMM is fixed-parameter tractable is inspired by the proof from [5] showing that BANDWIDTH is fixed-parameter tractable when parameterized by the maximum number of leaves in a spanning tree of the input graph. We need the following definition and lemma from [5].

In a digraph D, two directed paths Δ and Δ' from a vertex s to a vertex t are *arc-equivalent*, if for every arc a of D, Δ and Δ' pass through a the same number of times.

Lemma 1 ([5]). *Any directed path Δ through a finite digraph D on n vertices from a vertex s to a vertex t of D is arc-equivalent to a directed path Δ' from s to t, where Δ' has the form:*

(1) Δ' consists of an underlying directed path ρ from s to t of length at most n^2,
(2) together with some number of short loops, where each such short loop l begins and ends at a vertex of ρ, and has length at most n.

Theorem 5. *var*-EWMM *is fixed-parameter tractable.*

Proof. Let $(M' = (S', s_0', \Gamma', \Sigma', T'), c)$ be an instance for *var*-EWMM with $k = |S'| + |\Gamma'| + |\Sigma'|$. As M' might have multiple transitions from one state to another, we first subdivide each transition in order to obtain a simple digraph underlying the Mealy machine (so we can use Lemma 1): create a new non-deterministic Mealy machine $M = (S, s_0, \Gamma, \Sigma, T)$ such that, initially, $S = S'$, $s_0 = s_0'$, $\Gamma = \Gamma' \cup \{\lambda\}$, and $\Sigma = \Sigma' \cup \{\lambda\}$; for each transition t of T' from a couple $(s_i, \langle i \rangle)$ to a couple $(s_o, \langle o \rangle)$, add a new state s_t to S and add the transition from $(s_i, \langle i \rangle)$ to $(s_t, \langle o \rangle)$ and the transition from (s_t, λ) to (s_o, λ) to T. Clearly, there is at most one transition between every two states in M.

Our algorithm goes over all transition paths in M of length at most $|S|^2$ that start from s_0. There are at most $|S|^{(|S|^2)}$ such transition paths and each such transition path has at most $|S|^{|S|}$ short loops, as they have length at most $|S|$ by Lemma 1. Let $P = (s_0, s_1, \ldots, s_{|P|})$ be such a transition path and $L = (\ell_0, \ell_1, \ldots, \ell_{|L|})$ be its short loops. It remains to check whether there exists a set of integers $X = \{x_1, x_2, \ldots, x_{|L|}\}$ such that a word output by a computation of M moving from s_0 to $s_{|P|}$ along the path P, and executing x_i times each short loop ℓ_i, $0 \le i \le |L|$, meets the census requirement. Note that if one such word meets the census requirement, then all such words meet the census requirement, as it does not matter in which order the short loops are executed. We verify whether such a set X exists by ILPF.

Let $\Sigma = \{\langle \ell, 1 \rangle, \langle \ell, 2 \rangle, \ldots, \langle \ell, |\Sigma| \rangle\}$. Define $m(i, j)$, $0 \le i \le |L|$, $1 \le j \le |\Sigma|$, to denote the number of times that M writes the letter $\langle \ell, j \rangle$ when it executes the loop ℓ_i once. Define $m(j)$, $1 \le j \le |\Sigma|$, to be the number of times that M writes the letter $\langle \ell, j \rangle$ when it transitions from s_0 to $s_{|P|}$ along the path P. Then, we only need to verify that there exist integers $x_1, x_2, \ldots, x_{|L|}$ such that

$$m(j) + \sum_{i=0}^{|L|} x_i \cdot m(i, j) = c(\langle \ell, j \rangle), \qquad \forall j \in [|\Sigma|].$$

By construction, $|S| \le |S'| + |T'| \le |S'| + |S'|^2 \cdot |\Gamma'| \cdot |\Sigma'| \le k + k^4$. As the number of integer variables of this program is at most $|L| \le |S|^{|S|} \le (k + k^4)^{k+k^4}$, and the number of transition paths that the algorithm considers is at most $|S|^{(|S|^2)} \le (k + k^4)^{k^2 + 2k^5 + k^8}$, *var*-EWMM is fixed-parameter tractable. $\square$

We note that the proof in [5] concerned a special case of a deterministic Mealy machine where the input and output alphabet are the same, and all transitions that read a letter $\langle \ell \rangle$ also write $\langle \ell \rangle$.

In our second Mealy machine problem, the question is whether, for a given input word, there is a computation of the Mealy machine which outputs a word that meets the census requirement.

---

variety-GIVEN WORD MEALY MACHINE (*var*-GWMM)

Input: a non-deterministic Mealy machine $M = (S, s_0, \Gamma, \Sigma, T)$, a word $x \in \Gamma^*$, and a census requirement $c : \Sigma \to \mathbb{N}$

Parameter: $|S| + |\Gamma| + |\Sigma|$

Question: Is there a computation of M on input x generating an output y that meets c?

Membership in XP is easily shown by dynamic programming.

Theorem 6. *var*-GWMM *is in* XP.

To show $W[1]$-hardness, we reduce from MCC, which is $W[1]$-hard [4].

MULTICOLORED CLIQUE (MCC)

Input: an integer k and a connected undirected graph $G = (V(1) \cup V(2) \ldots \cup V(k), E)$ such that for every $i \in [k]$, the vertices of $V(i)$ induce an independent set in G

Parameter: k

Question: Is there a clique of size k in G?

Clearly, a solution to this problem has one vertex from each color.

Theorem 7. *var*-GWMM *is* $W[1]$-*hard.*

Proof. Let $(k, G = (V(1) \cup V(2) \ldots \cup V(k), E))$ be an instance of MCC. Suppose $V(i) = \{v_{i,1}, v_{i,2}, \ldots, v_{i,|V(i)|}\}$ is the vertex set of color i, for each color class $i \in [k]$, $E = \{e_1, e_2, \ldots, e_{|E|}\}$, and $E(i,j) = \{e(i,j,1), e(i,j,2), \ldots, e(i,j,|E(i,j)|)\}$ is the subset of edges with one vertex in color class i, and the other in color class j, $i, j \in [k]$. Moreover, suppose $E(i,j)$ follows the same order as E, that is if $e_p = e(i,j,p')$, $e_q = e(i,j,q')$, and $p \le q$, then $p' \le q'$. For a vertex $v_{i,p}$ and two integers $j \in [k] \setminus \{i\}$ and $q \in [d_{V(j)}(v_{i,p}) + 1]$, we define $\mathsf{gap}(v_{i,p}, j, q) = t - s$, where $e(i,j,t)$ is the qth edge in $E(i,j)$ incident to $v_{i,p}$ (respectively, $t = |E(i,j)|$ if $q = d_{V(j)}(v_{i,p}) + 1$) and $e(i,j,s)$ is the $(q-1)^{\text{th}}$ edge in $E(i,j)$ incident to $v_{i,p}$ (respectively, $s = 0$ if $q = 1$).

We construct an instance $(M = (S, s_0, \Gamma, \Sigma, T), x, c)$ for *var*-GWMM. M's input alphabet, Γ, is $\{\langle i \rangle, \langle i, j \rangle, \langle \bar{e}, i, j \rangle, \langle e, i, j \rangle : i, j \in [k], i \ne j\}$. M's output alphabet, Σ, is $\{\lambda\} \cup \{\langle \ell, i, j \rangle, \langle \ell, \bar{e}, i, j \rangle : i, j \in [k], i \ne j\}$. The word x is defined

$$x := x_1 x_2 \ldots x_k$$

$$x_i := x_{i,0} x_{i,1} \ldots x_{i,i-1} x_{i,i+1} x_{i,i+2} \ldots x_{i,k} \langle i \rangle \qquad \forall i \in [k]$$

$$x_{i,0} := (\langle i, 1 \rangle \langle i, 2 \rangle \ldots \langle i, i-1 \rangle \langle i, i+1 \rangle \langle i, i+2 \rangle \ldots \langle i, k \rangle)^{|V(i)|} \qquad \forall i \in [k]$$

$$x_{i,j} := \langle i, j \rangle x_{i,j,1} \langle i, j \rangle x_{i,j,2} \ldots \langle i, j \rangle x_{i,j,|V(i)|} \langle i, j \rangle \qquad \forall i, j \in [k], i \ne j$$

$$x_{i,j,p} := \langle \bar{e}, i, j \rangle^{\mathsf{gap}(v_{i,p}, j, 1)} \langle e, i, j \rangle \langle \bar{e}, i, j \rangle^{\mathsf{gap}(v_{i,p}, j, 2)} \langle e, i, j \rangle$$

$$\ldots \langle \bar{e}, i, j \rangle^{\mathsf{gap}(v_{i,p}, j, d_{V(j)}(v_{i,p}))} \langle e, i, j \rangle \langle \bar{e}, i, j \rangle^{\mathsf{gap}(v_{i,p}, j, d_{V(j)}(v_{i,p})+1)}.$$

The census requirement c is, for every $i, j \in [k], i \ne j$,

$$c(\langle \ell, i, j \rangle) := |V(i)| + 1$$

$$c(\langle \ell, i, j \rangle) := |V(i)|$$

$$c(\langle \ell, \bar{e}, i, j \rangle) := |E(i,j)|.$$

The Mealy machine M consists of k parts. The ith part of M is depicted in Fig. 1. Its initial state is $s_{v,1}$. There is a transition from the last state of each part, $s^{(4)}_{e,i,k}$, to the first state of the following part, $s_{v,i+1}$ (from the kth part, there is

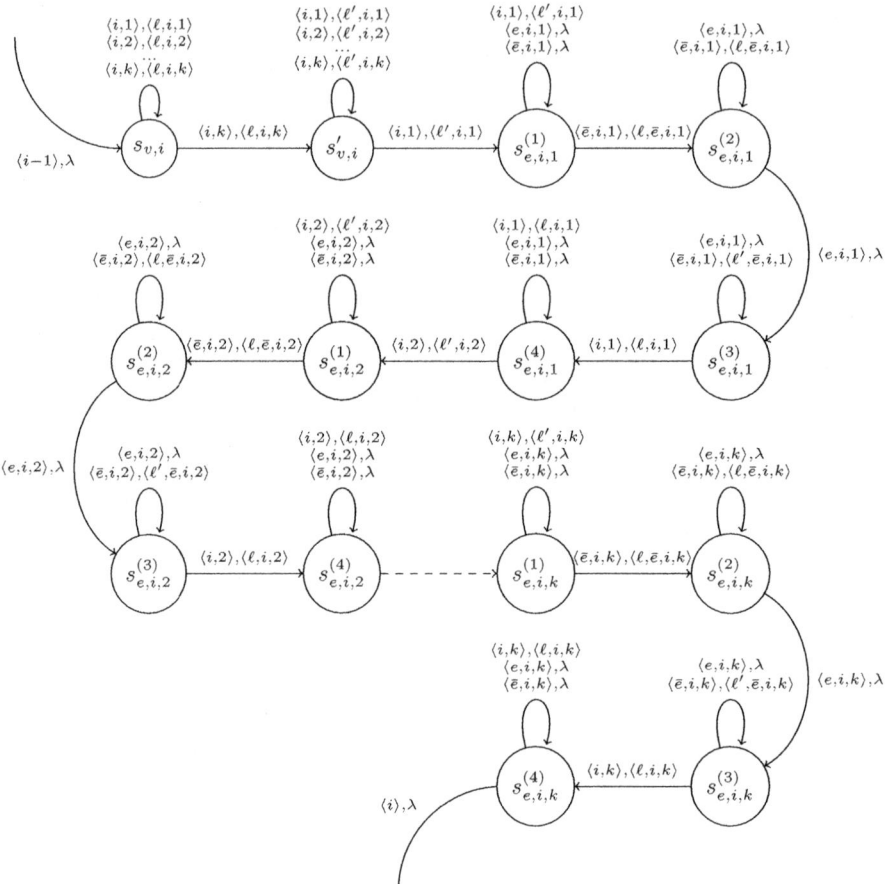

Fig. 1. The ith part of the Mealy machine M. It does not have the states $s_{e,i,i}^{(1)}$, $s_{e,i,i}^{(2)}$, $s_{e,i,i}^{(3)}$, and $s_{e,i,i}^{(4)}$; there is instead a transition from $s_{e,i,i-1}^{(4)}$ to $s_{e,i,i+1}^{(1)}$ reading $\langle i-1 \rangle$ and writing λ, and there is a transition from $s_{e,k,k-1}^{(4)}$ to a final state reading $\langle k \rangle$ and writing λ (drawing all this would have cluttered the figure too much).

a transition to a final state): it reads the letter $\langle i \rangle$ and writes the letter λ. We set $\langle \ell', \bar{e}, i, j \rangle = \langle \ell, \bar{e}, j, i \rangle$ for all $i \neq j \in [k]$.

First, suppose $(M = (S, s_0, \Gamma, \Sigma, T), x, c)$ is a YES-instance for *var*-GWMM.

We say that M *selects* a vertex $v_{i,p}$ if it makes a transition from state $s_{v,i}$ to state $s'_{v,i}$ reading $\langle i, k \rangle$ (respectively $\langle i, k-1 \rangle$ if $i = k$) for the pth time. In other words, in the ith part of M, it reads $p \cdot (k-1) - 1$ letters of $x_{i,0}$, staying in state $s_{v,i}$ and outputs the letter $\langle \ell, i, r \rangle$ for each letter $\langle i, r \rangle$ it reads; then it transitions to state $s'_{v,i}$ on reading $\langle i, k \rangle$ (respectively $\langle i, k-1 \rangle$) and outputs $\langle \ell, i, k \rangle$ (respectively $\langle \ell, i, k-1 \rangle$); in the state $s'_{v,i}$ it outputs the letter $\langle \ell', i, r \rangle$ for each letter $\langle i, r \rangle$ it reads.

We say that M *selects* an edge $e(i, j, q)$ if it makes a transition from state $s^{(2)}_{e,i,j}$ to state $s^{(3)}_{e,i,j}$ after having read the letter $\langle \bar{e}, i, j \rangle$ of $x_{i,j,p}$ exactly q times, where $v_{i,p}$ is the vertex of color i that $e(i, j, q)$ is incident on. In other words, in the ith part of M, it transitions from the state $s^{(1)}_{e,i,j}$ to the state $s^{(2)}_{e,i,j}$ on reading the first letter of $x_{i,j,p}$ (if it did this transition any later, the census requirement of $\langle \ell, \bar{e}, i, j \rangle$ could not be met); then it stays in the state $s^{(2)}_{e,i,j}$ until it has read q times the letter $\langle \bar{e}, i, j \rangle$ of $x_{i,j,p}$; then it transitions to the state $s^{(3)}_{e,i,j}$ on reading $\langle e, i, j \rangle$; it stays in this state and outputs $\langle \ell', \bar{e}, i, j \rangle$ for each letter $\langle \bar{e}, i, j \rangle$ it reads until transitioning to the state $s^{(4)}_{e,i,j}$ on reading the letter following $x_{i,j,p}$.

The following claims ensure that the edge-selection and the vertex-selection are compatible, i.e., that exactly one edge is selected from color i to color j, and that this edge is incident on the selected vertex of color i.

Claim 1. *Let i be a color and let $v_{i,p}$ be the vertex selected in the ith part of M. In its ith part, M selects one edge incident to $v_{i,p}$ and to a vertex of color j, for each $j \in [k] \setminus \{i\}$.*

Proof. After M has selected $v_{i,p}$, it has output p times each of the letters $\langle \ell, i, 1 \rangle$, $\langle \ell, i, 2 \rangle, \ldots, \langle \ell, i, i-1 \rangle, \langle \ell, i, i+1 \rangle, \langle \ell, i, i+2 \rangle, \ldots, \langle \ell, i, k \rangle$. For each $j \in [k] \setminus \{i\}$, the only other transitions that output $\langle \ell, i, j \rangle$ are the transition from $s^{(3)}_{e,i,j}$ to $s^{(4)}_{e,i,j}$ and a transition that loops on $s^{(4)}_{e,i,j}$. To meet the census requirement of $|V(i)|+1$ for $\langle \ell, i, j \rangle$, M selects an edge while reading $x_{i,j,p}$. This edge is incident on $v_{i,p}$ by construction. □

The following claim makes sure that the edge selected from color i to color j is the same as the edge selected from color j to color i.

Claim 2. *Suppose M selects the edge $e(i, j, q)$ in its ith part. Then, M selects the edge $e(j, i, q)$ in its jth part.*

By Claims 1 and 2, the k vertices that are selected by M form a multicolored clique. Thus, $(k, G = (V(1) \cup V(2) \ldots \cup V(k), E))$ is a YES-instance for MCC.

Now, suppose that $(M = (S, s_0, \Gamma, \Sigma, T), x, c)$ is a NO-instance for *var*-GWMM. For the sake of contradiction, suppose that $(k, G = (V(1) \cup V(2) \ldots \cup V(k), E))$ is a YES-instance for MCC. Let $\{v_{1,p_1}, v_{2,p_2}, \ldots, v_{k,p_k}\}$ be a multicolored clique in G. We will construct a word y meeting c such that a computation of M on input x generates y. For two adjacent vertices v_{i,p_i} and v_{j,p_j}, define $\mathsf{edge}(v_{i,p_i}, v_{j,p_j}) = t$ such that $e(i, j, t) = v_{i,p_i} v_{j,p_j}$. The word y is $y_1 y_2 \ldots v_k$, where $y_i, i \in [k]$ is

$$(\langle \ell, i, 1 \rangle \langle \ell, i, 2 \rangle \ldots \langle \ell, i, i-1 \rangle \langle \ell, i, i+1 \rangle \langle \ell, i, i+2 \rangle \ldots \langle \ell, i, k \rangle)^{p_i}$$

$$(\langle \ell', i, 1 \rangle \langle \ell', i, 2 \rangle \ldots \langle \ell', i, i-1 \rangle \langle \ell', i, i+1 \rangle \langle \ell', i, i+2 \rangle \ldots \langle \ell', i, k \rangle)^{|V(i)|-p_i}$$

$$y_{i,1} y_{i,2} \cdots y_{i,i-1} y_{i,i+1} y_{i,i+2} \cdots y_{i,k} \langle i \rangle$$

and $y_{i,j}, i \neq j \in [k]$ is

$$\langle \ell', i, j \rangle^{p_i} \langle \ell, \bar{e}, i, j \rangle^{\mathsf{edge}(v_{i,p_i}, v_{j,p_j})}$$

$$\langle \ell', \bar{e}, i, j \rangle^{|E(i,j)| - \mathsf{edge}(v_{i,p_i}, v_{j,p_j})} \langle \ell, i, j \rangle^{|V(i)| - p_i + 1}.$$

We note that there is a computation of M on input x that generates y and that y meets the census requirement c. This contradicts $(M = (S, s_0, \Gamma, \Sigma, T), x, c)$ being a No-instance.   □

6   Applications

In this section we sketch two examples that illustrate how number-of-numbers parameterized problems may reduce to census problems about Mealy machines, parameterized by the size of the machine. For another application, see [5].

Example 1: Heat-Sensitive Scheduling. In a recent paper Chrobak et al. [3] introduced a model for the issue of temperature-aware task scheduling for microprocessor systems. The motivation is that different jobs with the same time requirements may generate different heat loads, and it may be important to schedule the jobs so that some temperature threshold is not breached.

In the model, the input consists of a set of jobs that are all assumed to be of unit length, with each job assigned a numerical heat level. If at time t the processor temperature is T_t, and if the next job that is scheduled has heat level H, then the processor temperature at time $t + 1$ is

$$T_{t+1} = (T_t + H)/2$$

It is also allowed that perhaps no job is scheduled for time t (that is, *idle time* is scheduled), in which case $H = 0$ in the above calculation of the updated temperature.

The relevant decision problem is whether all of the jobs can be scheduled, meeting a specified deadline, in such a way that a given temperature threshold is never exceeded. This problem has been shown to be NP-hard [3] by a reduction from 3-Dimensional Matching. An image instance of the reduction, however, involves arbitrarily many distinct heat levels asymptotically close to $H = 2$, for a temperature threshold of 1.

In the spirit of the "deconstruction of hardness proofs" advocated by Komusiewicz et al. [11] (see also [2]), one might regard this problem as ripe for parameterization by the number of numbers, for example (scaling appropriately), a model based on $2k$ equally-spaced heat levels and a temperature threshold of k. Furthermore, if the heat levels of the jobs are only roughly classified in this way, it also makes sense to treat the temperature transition model similarly, as:

$$T_{t+1} = \lceil (T_t + H)/2 \rceil$$

The input to the problem can now be viewed equivalently as a census of how many jobs there are for each of the $2k+1$ heat levels, with the available potential units of idle time allowed to meet the deadline treated as "jobs" for which $H = 0$. Because of the ceiling function modeling the temperature transition, the problem now immediately reduces to *var*-EWMM, for a machine on $k + 1$ states (that represent the temperature of the processor) and an alphabet of size at most $2k + 1$. By Theorem 5, the problem is fixed-parameter tractable.

Example 2: A Problem in Computational Chemistry. The parameterized problem of WEIGHTED SPLITS RECONSTRUCTION FOR PATHS that arises in computational chemistry [8] reduces to a special case of *var*-GWMM. The input to the problem is obtained from time-series spectrographic data concerning molecular weights. The problem as defined in [8] is equivalent to the following two-processor scheduling problem. The input consists of

- a sequence x of positive integer *time gaps* taken from a set of positive integers Γ, and
- a census requirement c on a set of positive integers Σ of *job lengths*.

The question is whether there is a "winning play" for the following one-person two-processor scheduling game. At each step, first, *Nature* plays the next positive integer "gap" of the sequence of time gaps x — this establishes the next *immediate deadline*. Second, the *Player* responds by scheduling on one of the two processors, a job that begins at the last stop-time on that processor, and ends at the immediate deadline. The *Player* wins if there is a sequence of plays (against x) that meets the census requirement c on job lengths. Fig. 2 illustrates such a game.

Processor 1	4		3			3	
Processor 2		5		3	1		5
$x =$	4	1	2	1	1	1	4

Fig. 2. A winning game for the census: 1 (1), 3 (3), 4 (1), 5 (2)

This problem easily reduces to a special case of *var*-GWMM. Whether this special case is also $W[1]$-hard remains open.

7   Concluding Remarks

The practical world of computing is full of computational problems where inputs are "weighted" in a realistic model — weighted graphs provide a simple example relevant to many applications. Here we have begun to explore parameterizing on the *numbers of numbers* as a way of mitigating computational complexity for problems that are numerically structured. One might view some of the impulse here as *moving approximation issues into the modeling*, as illustrated by Example 1 in Section 6. We believe this line of attack may be widely applicable.

Finally, we remark that to date, there has been little attention to parameterized complexity in the context of cryptography. Number of numbers parameterization may provide some inroads into this underdeveloped area.

Acknowledgment. We thank Iyad Kanj for stimulating conversations about this work.

References

1. Alon, N., Azar, Y., Woeginger, G.J., Yadid, T.: Approximation schemes for scheduling on parallel machines. J. of Scheduling 1, 55–66 (1998)
2. Betzler, N., Fellows, M.R., Guo, J., Niedermeier, R., Rosamond, F.A.: Fixed-parameter algorithms for kemeny rankings. Theoret. Comput. Sci. 410(45), 4554–4570 (2009)
3. Chrobak, M., Dürr, C., Hurand, M., Robert, J.: Algorithms for temperature-aware task scheduling in microprocessor systems. In: Fleischer, R., Xu, J. (eds.) AAIM 2008. LNCS, vol. 5034, pp. 120–130. Springer, Heidelberg (2008)
4. Fellows, M.R., Hermelin, D., Rosamond, F.A., Vialette, S.: On the parameterized complexity of multiple-interval graph problems. Theoret. Comput. Sci. 410(1), 53–61 (2009)
5. Fellows, M.R., Lokshtanov, D., Misra, N., Mnich, M., Rosamond, F.A., Saurabh, S.: The complexity ecology of parameters: An illustration using bounded max leaf number. Theory Comput. Syst. 45(4), 822–848 (2009)
6. Fellows, M.R., Lokshtanov, D., Misra, N., Rosamond, F.A., Saurabh, S.: Graph layout problems parameterized by vertex cover. In: Hong, S.-H., Nagamochi, H., Fukunaga, T. (eds.) ISAAC 2008. LNCS, vol. 5369, pp. 294–305. Springer, Heidelberg (2008)
7. Fiala, J., Golovach, P.A., Kratochvíl, J.: Parameterized complexity of coloring problems: Treewidth versus vertex cover. In: TAMC. LNCS, vol. 5532, pp. 221–230. Springer, Heidelberg (2009)
8. Gaspers, S., Liedloff, M., Stein, M.J., Suchan, K.: Complexity of splits reconstruction for low-degree trees. CoRR, abs/1007.1733 (2010), Available on arXiv.org
9. Gramm, J., Niedermeier, R., Rossmanith, P.: Fixed-parameter algorithms for closest string and related problems. Algorithmica 37(1), 25–42 (2003)
10. Kannan, R.: Minkowski's convex body theorem and integer programming. Math. Oper. Res. 12(3), 415–440 (1987)
11. Komusiewicz, C., Niedermeier, R., Uhlmann, J.: Deconstructing intractability: A case study for interval constrained coloring. In: Kucherov, G., Ukkonen, E. (eds.) CPM 2009. LNCS, vol. 5577, pp. 207–220. Springer, Heidelberg (2009)
12. Lenstra, H.W.: Integer programming with a fixed number of variables. Math. Oper. Res. 8(4), 538–548 (1983)
13. Mealy, G.H.: A method for synthesizing sequential circuits. Bell System Technical Journal 34(5), 1045–1079 (1955)
14. Niedermeier, R.: The Art of Problem Parameterization. In: Invitation to Fixed-Parameter Algorithms. Oxford Lecture Series in Mathematics and Its Applications, ch. 5, pp. 41–49. Oxford University Press, Oxford (2006)
15. Niedermeier, R.: Invitation to Fixed-Parameter Algorithms. Oxford Lecture Series in Mathematics and Its Applications. Oxford University Press, Oxford (2006)

Are There Any Good Digraph Width Measures?[*]

Robert Ganian[1], Petr Hliněný[1], Joachim Kneis[2], Daniel Meister[2],
Jan Obdržálek[1], Peter Rossmanith[2], and Somnath Sikdar[2]

[1] Faculty of Informatics, Masaryk University, Brno, Czech Republic
{xganian1,hlineny,obdrzalek}@fi.muni.cz
[2] Theoretical Computer Science, RWTH Aachen University, Germany
{kneis,meister,rossmani,sikdar}@cs.rwth-aachen.de

Abstract. Several width measures for digraphs have been proposed in
the last few years. However, none of them possess all the "nice" properties
of treewidth, namely, (1) being *algorithmically useful*, that is, admitting
polynomial-time algorithms for a large class of problems on digraphs of
bounded width; and (2) having nice *structural properties* such as being
monotone under taking subdigraphs and some form of arc contractions.
As for (1), MSO_1 is the least common denominator of all reasonably expressive logical languages that can speak about the edge/arc relation on
the vertex set, and so it is quite natural to demand efficient solvability
of all MSO_1-definable problems in this context. (2) is a necessary condition for a width measure to be characterizable by some version of the
cops-and-robber game characterizing treewidth. More specifically, we introduce a notion of a *directed topological minor* and argue that it is the
weakest useful notion of minors for digraphs in this context. Our main
result states that any *reasonable* digraph measure that is algorithmically
useful and structurally nice cannot be substantially different from the
treewidth of the underlying undirected graph.

1 Introduction

An intensely investigated field in algorithmic graph theory is the design of
graph *width parameters* that satisfy two seemingly contradictory requirements:
(1) graphs of bounded width should have a reasonably rich structure; and, (2) a
large class of problems must be efficiently solvable on graphs of bounded width.
For undirected graphs, research into width parameters has been extremely successful with a number of algorithmically useful measures being proposed over the
years, chief among them being treewidth [17], clique-width [6], branchwidth [18]
and related measures (see also [3,10]). Many problems that are hard on general
graphs turned out to be tractable on graphs of bounded treewidth. These results
were combined and generalized by Courcelle's celebrated theorem which states
that a very large class of problems (MSO_2) is tractable on graphs of bounded
treewidth [4].

[*] This work is supported by the Deutsche Forschungsgemeinschaft (project RO 927/9)
and the Czech Science Foundation (project 201/09/J021).

However, there still do not exist *directed graph* width measures that are as successful as treewidth. This is because, despite many achievements and interesting results, most known digraph width measures do not allow for efficient algorithms for many problems. During the last decade, many digraph width measures were introduced, the prominent ones being directed treewidth [13], DAG-width [2,16], and Kelly-width [12]. These width measures proved useful for some problems. For instance, one can obtain polynomial-time (XP to be more precise) algorithms for HAMILTONIAN PATH on digraphs of bounded directed treewidth [13] and for PARITY GAMES on digraphs of bounded DAG-width [2] and Kelly-width [12]. But there is the negative side, too. HAMILTONIAN PATH, for instance, probably cannot be solved [15] on digraphs of directed treewidth, DAG-width, or Kelly-width at most k in time $O(f(k) \cdot n^c)$, where c is a constant independent of k. Note that HAMILTONIAN PATH *can* be solved in such a running time for undirected graphs of treewidth at most k [4].

Moreover, for the measures DAG-depth and Kenny-width[1] which are much more restrictive than DAG-width, problems such as DIRECTED DOMINATING SET, DIRECTED CUT, ORIENTED CHROMATIC NUMBER 4, MAX / MIN LEAF OUTBRANCHING, and k-PATH remain NP-complete on digraphs of constant width [8]. In contrast, clique-width and another recent digraph measure bi-rank-width [14] look more promising. A Courcelle-like [5] MSO$_1$ theorem exists for digraphs of bounded directed clique-width and bi-rank-width, and many other interesting problems can be solved in polynomial (XP) time on these [9,14]. For a recent exhaustive survey on complexity results for DAG-width, Kelly-width, bi-rank-width, and other digraph measures, see [8].

In this paper, we show that any *reasonable* digraph width measure that is *algorithmically useful* and is closed under a notion of *directed topological minors* upper-bounds the treewidth of the underlying undirected graph. In what follows, we formalize this statement. We start with the notion of algorithmic usefulness and note what is it that makes treewidth such a successful measure. Courcelle's theorem [4] states that all MSO$_2$-expressible problems are linear-time decidable on graphs of bounded treewidth. To us it seems that an algorithmically useful width measure must admit algorithms with running time $O(n^{f(k)})$, at least, for all MSO$_1$-expressible problems on n-vertex digraphs of width at most k, where f is some computable function (that is, XP running time). Algorithmically useful digraph width measures do indeed exist. Candidates include the number of vertices in the input graph and the treewidth of the underlying undirected graph. In the latter case we can apply the rich theory of (undirected) graphs of bounded treewidth, but we would not get anything substantially new for digraphs. As such, we are interested in digraph width measures that are *incomparable* to undirected treewidth.

To motivate our discussion of directed topological minors, we note that treewidth has an alternative cops-and-robber game characterization. In fact, several digraph width measures such as DAG-width [2,16], Kelly-width [12], and DAG-depth [8] admit some variants of this game-theoretic characterization.

---

[1] Kenny-width [8] is a different measure than Kelly-width [12].

While there is no formal definition of a cops-and-robber game-based width measure, all versions of the cops-and-robber game that have been considered share a basic property that shrinking induced paths does not help the robber. What we actually show is that a directed width measure that is "cops-and-robber game-based" must be closed under directed topological minors. On the other hand, we note that there exist algorithmically useful measures more general than undirected treewidth – digraph clique-width [6] and bi-rank-width [14] – which are not monotone even under taking subdigraphs.

Finally, the notion of a reasonable directed width measure is explained in Section 5 (see Definition 5.1). At this point, it suffices to say that this is simply a technicality that we make use of in the proof of our main theorem (Theorem 5.6). This theorem then states that a digraph width measure that admits XP-time algorithms for all MSO_1-problems wrt the width as parameter and is closed under directed topological minors must necessarily upper bound the treewidth of the underlying undirected graph. This implies that an algorithmically useful digraph width measure that is not treewidth-bounding cannot be characterized by a (version of) cops-and-robber game. We also show with examples that the prerequisities of our theorem cannot be weakened.

The paper is organized in four parts, starting with some core definitions in Section 2. Then in Section 3, we formally establish and discuss the (above outlined) properties an algorithmically useful digraph width measure should have. In Section 4, we introduce the notion of a directed topological minor, and discuss its properties and consider complexity issues. In particular, we show that it is hard to decide for a fixed (small) digraph whether it is a directed topological minor of a given digraph. In the last section, Section 5, we prove our main results which have already been outlined above. Due to lack of space, the proofs of results marked with a star ($\star$) are omitted.

2 Definitions and Notation

The graphs (both undirected and directed) that we consider in this paper are *simple*, i.e. they do not contain loops and parallel edges. Given a graph G, we let $V(G)$ denote its vertex set and $E(G)$ denote its edge set, if G is undirected. If G is directed, we let $A(G)$ denote its arc set. Given a directed graph D, the *underlying undirected graph* $U(D)$ of D is an undirected graph on the vertex set $V(D)$; and $\{u, v\}$ is an edge of $U(D)$ if and only if $(u, v) \in A(D)$ or $(v, u) \in A(D)$. A digraph D is an *orientation* of an undirected graph G if $U(D) = G$.

For a vertex pair u, v of a digraph D, a sequence $P = (u = x_0, \ldots, x_r = v)$ is called *directed (u, v)-path* of length $r > 0$ in D if the vertices $x_0, \ldots, x_r$ are pairwise distinct and $(x_i, x_{i+1}) \in A(G)$ for every $0 \leq i < r$. We also write $u \to_D^+ v$ if there exists a directed (u, v)-path in D, and $u \to_D^* v$ if either $u \to_D^+ v$ or $u = v$. A *directed cycle* is defined analogously with the modification that $x_0 = x_r$. A digraph D is *acyclic* (a DAG) if D contains no directed cycle.

A parameterized problem Q is a subset of $\Sigma \times \mathbb{N}_0$, where Σ is a finite alphabet. A parameterized problem Q is said to be *fixed-parameter tractable* if there is an algorithm that given $(x, k) \in \Sigma \times \mathbb{N}_0$ decides whether (x, k) is a yes-instance of Q in time $f(k) \cdot p(|x|)$ where f is some computable function of k alone, p is a polynomial and $|x|$ is the size measure of the input. The class of such problems is denoted by FPT. The class XP is the class of parameterized problems that admit algorithms with a run-time of $O(|x|^{f(k)})$ for some computable f, i.e. polynomial-time for every fixed value of k.

Monadic second-order (MSO in short) logic is a language particularly suited for description of problems on "tree-like structured" graphs. For instance, the celebrated result of Courcelle [4], and of Arnborg, Lagergren and Seese [1], states that all MSO_2 definable graph problems have linear-time FPT algorithms when parameterized by the undirected treewidth. The expressive power of MSO_2 is very strong, as it includes many natural graph problems.

Note 2.1. Check this newer description. In this paper we are, however, interested primarily in another logical dialect commonly abbreviated as MSO_1, whose expressive power is noticeably weaker than that of MSO_2. The weaker expressive power is not a handicap but an advantage for our paper since we are going to use it to prove negative results. Similarly to the previous, MSO_1 definable graph problems have FPT algorithms when parameterized by clique-width [5] and, consequently, by rank-width.

Definition 2.2. The language of MSO_1 contains the logical expressions that are built from the following elements:

- variables for elements (vertices) and their sets, and the predicate $x \in X$,
- the predicate $adj(u, v)$ with u and v vertex variables,
- equality for variables, the connectives $\land, \lor, \neg, \rightarrow$ and the quantifiers $\forall, \exists$.

Example 2.3. For an undirected graph to have the 3-colorability property is an MSO_1-expression:

$$\exists V_1, V_2, V_3 \left[\forall v (v \in V_1 \lor v \in V_2 \lor v \in V_3) \land \bigwedge_{i=1,2,3} \forall v, w (v \notin V_i \lor w \notin V_i \lor \neg adj(v, w)) \right]$$

A decision graph property $\mathcal{P}$ is MSO_1 *definable* if there exists an MSO_1 formula ϕ such that $\mathcal{P}$ holds for any graph G if, and only if, $G \models \phi$, i.e., ϕ is true on the model G. MSO_1 is analogously used for digraphs and their properties, where the predicate $arc(u, v)$ is used instead of $adj(u, v)$.

3   Desirable Digraph Width Measures

A *digraph width measure* is a function δ that assigns each digraph a non-negative integer. To stay reasonable, we expect that infinitely many non-isomorphic digraphs are of bounded width. We consider what properties a width measure is expected to have. Importantly, one must be able to solve a rich class of problems on digraphs of bounded width. But what does "rich" mean?

On one hand, looking at existing algorithmic results in the undirected case, it appears that a *good balance* between the richness of the class of problems we capture and the possibility of positive general algorithmic results is achieved by the class of MSO$_1$ expressible problems (Definition 2.2). On the other hand, if we consider any logical language $\mathcal{L}$ over digraphs that is powerful enough to deal with sets of singletons (i.e. of monadic second order) and that can identify the adjacent pairs of vertices of the digraph, then we see $\mathcal{L}$ can naturally interpret also the MSO$_1$ logic of the underlying graph. Hence the following specification appears to be the most natural common denominator in our context:

Definition 3.1. A digraph width measure δ is *powerful* if, for every MSO$_1$ definable undirected property $\mathcal{P}$, there is an XP algorithm deciding $\mathcal{P}$ on all digraphs D with respect to the parameter $\delta(D)$.

The traditional measures treewidth, branchwidth, clique-width, and more recent rank-width, are all powerful [4,5] for undirected graphs. For directed graphs, unfortunately, exactly the opposite holds. The width measures suggested in recent years as possible extensions of treewidth – including directed treewidth [13], D-width [20], DAG-width [16,2], and Kelly-width [12] – all are not powerful.

   Another concern is about "non-similarity" of our directed measure δ to the traditional treewidth of the underlying undirected graph; we actually want to obtain and study new measures that significantly differ from treewidth, in the negative sense of the following Definition 3.2. This makes sense because any measure δ which bounds the treewidth of the underlying graph would automatically be powerful but would not help to solve any more problem instances than we already can with traditional undirected measures.

Definition 3.2. A digraph width measure δ is called *treewidth-bounding* if there exists a computable function b such that, for every digraph D, $\delta(D) \leq k$ implies that the treewidth of $U(D)$ is at most $b(k)$.

To briefly outline the current state, we focus in the rest of this section on two of the treewidth-like directed measures which seem to attract most attention nowadays – DAG-width [16,2] and Kelly-width [12]; and on another two significantly more successful (in the algorithmic sense) measures – directed clique-width [6] and not-so-much-known bi-rank-width [14]. None of these measures are treewidth-bounding.

   Since the definitions of DAG-width and Kelly-width are not short, we skip them here and refer to [16,2,12] instead. Both DAG- and Kelly-width share some common properties important for us:

 – Acyclic digraphs (DAGs) have width 0 and 1, respectively.
 – If we replace each edge of a graph of treewidth k by a pair of opposite arcs, then the resulting digraph has DAG-width k and Kelly-width $k + 1$.
 – Both of the measures are characterized by certain cops-and-robber games.

Proposition 3.3 ($\star$). *If* P $\neq$ NP, *DAG-width and Kelly-width are not powerful.*

On the other hand, there are measures such as clique-width [6] which was originally defined for undirected graphs, but readily extends to digraphs. Another noticeable directed measure is *bi-rank-width* [14], which is related to clique-width in the sense that one is bounded on a digraph class iff the other one is. Due to restricted space we only refer to [14] or [9] for its definition and properties.

Theorem 3.4 (Courcelle, Makowsky, and Rotics [5]). *Directed clique-width, and consequently bi-rank-width, are powerful measures.*

For a better understanding of the situation, we note one important but elusive fact: Bounding the *undirected* clique-width or rank-width of the underlying undirected graph does not generally help solve directed graph problems.

Proposition 3.5 (⋆). *Undirected clique-width or rank-width are **not** powerful digraph measures unless $P = NP$.*

This is in a sharp contrast to the situation with treewidth where bounding the treewidth of the underlying undirected graph allows all the algorithmic machinery to work also on digraphs. As of now, there is no known non-trivial relationship between undirected measures and their directed generalizations.

After all, comparing Propositions 3.3 and 3.4, we clearly see the advantages of directed clique-width. There is, however, also the other side. Clique-width and bi-rank-width do not possess the nice structural properties common to the various treewidth-like measures, such as being subgraph- or contraction-monotone. This is due to symmetric orientations of complete graphs all having clique-width 2 while their subdigraphs include all digraphs, even those with arbitrarily high clique-width. This seems to be a drawback and a possible reason why clique-width- and rank-width-like measures are, unfortunately, not so widely accepted.

The natural question now is; can we take *the better of each of the two worlds*? In our search for the answer, we will not study specific digraph width measures but focus on general properties of possible width measures. The main result of this paper, Theorem 5.6, then answers this question negatively: One *cannot* have a "nice" digraph width measure which is powerful, not treewidth-bounding and, at the same time, monotone under taking subgraphs and directed topological minors (see in Section 4). This strong and conceptually new result holds modulo technical assumptions which prevent our digraph width measures from "cheating", such as in Theorems 5.7 and 5.8.

4   Directed Topological Minors

As for the second requirement we impose on a "good" digraph width measure – to possess some nice structural properties similar to those we often see in undirected graph measures – we argue as follows in this section.

Many width measures (e.g. treewidth) for undirected graphs are *monotone* under taking minors. In other words, the measure of a minor is not larger than the measure of the graph itself. Graph H is a *minor* of a graph G if it can be obtained by a sequence of applications of three operations: vertex deletion, edge deletion and edge contraction. (See e.g. [7].) It is therefore only natural to expect that a "good" digraph measure should also be closed under some notion of a directed minor. However, there is currently no widely agreed definition of a directed minor in general. One published, but perhaps too restrictive on subdivisions (as we will see later), notion is the *butterfly minor* [13].

To deal with directed minors, we first need a formal notion of an *arc contraction* for digraphs:

Definition 4.1. Let D be a digraph and $a = (x, y) \in A(D)$ be an arc. The digraph obtained by *contracting arc* a, denoted by D/a, is the digraph on the vertex $(V(D) \setminus \{x, y\}) \cup \{v_a\}$ where v_a is a new vertex, and the arc set A' such that $(u, v) \in A'$ iff one of the following holds

$$(u, v) \in A(D \setminus \{x, y\}), \text{ or}$$
$$v = v_a, \text{ and } (u, x) \in A \text{ or } (u, y) \in A, \text{ or}$$
$$u = v_a, \text{ and } (x, v) \in A \text{ or } (y, v) \in A.$$

See Fig. 1 for an example of a contraction. Note that contraction always produces simple digraphs (that is, no arcs of the form (x, x)). The result of a contraction does not depend on the orientation of the contracted arc, and we treat contraction of a pair of opposite arcs (x, y) and (y, x) as a contraction of a single bidirectional arc.

Fig. 1. Arc contraction: digraphs D (left) and D/a

An important decision point when defining a minor is; which arcs do we *allow to contract*? In the case of undirected graph minors, any edge can be contracted. However, the situation is not so obvious in the case of digraphs. Look again at Figure 1. If we contract the arc a, we actually introduce a new directed path $u \to^+ v$, whereas in undirected graphs no new (undirected) path is ever created by the edge contraction. On the other hand, simply never introducing a new directed path (e.g. the *butterfly minor* of [13]) is not a good strategy either – since one can easily construct, see Figure 2, digraphs in which no arc can be contracted without introducing a new directed path. Yet such digraphs can be "very simple" with respect to usual cops-and-robber games, and arc contractions do not help the robber in the depicted situation.

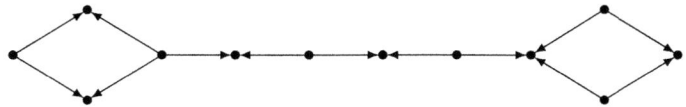

Fig. 2. Any arc contraction in this digraph introduces a new directed path

In order to deal with the mentioned issue of contractibility of arcs, and to remain as general as possible at the same time, we consider *directed topological minors* where we allow only those arc contractions that do not introduce any new directed path between vertices of degree at least three (cf. Figure 2 again). (Note that our definition is different from the one given in Hunter's thesis [11], where an arc is contractible iff at least one endvertex has both out- and in-degree one.) Subsequent claims then justify our choice. For reference we denote by $V_3(D) \subseteq V(D)$ the subset of vertices having at least three neighbors in D.

Definition 4.2. An arc $a = (u, v) \in A(D)$ is *2-contractible* in a digraph D if

- u or v has exactly two neighbors, and
- $(v, u) \in A(D)$ or there is no pair of vertices $x, y \in V_3(D)$ such that $x \rightarrow^*_{(D/a)} v$ and $u \rightarrow^*_{(D/a)} y$.

A digraph H is a *directed topological minor* of D if there exists a sequence of digraphs $D_0, \ldots, D_r \cong H$ such that D_0 is a subgraph of D, and for all $0 \le i \le r - 1$, D_{i+1} is obtained from D_i by contracting a 2-contractible arc.

Proposition 4.3 ($\star$). *Let D be a digraph and D' be a digraph obtained from D by a sequence of vertex deletions, arc deletions and contractions of 2-contractible arcs. Then D' is a directed topological minor of D.*

A useful notion in reasoning about directed topological minors is that of a 2-path. Let D be a digraph and $P = (x_0, \ldots, x_k)$ a sequence of vertices of D. Then P is a *2-path* (of length k) in D if P is a path in the underlying graph $U(D)$ and all internal vertices x_i for $0 < i < k$ have exactly two neighbors in D. Obviously not every 2-path is a directed path. The following lemma explains a key property of our directed topological minors – that they behave analogously to ordinary topological minors, being able to contract any long 2-paths.

Lemma 4.4 ($\star$). *Let $S = (x_0, \ldots, x_k)$ a 2-path of length $k > 2$ in a digraph D. Then there exists a sequence of 2-contractions of arcs of S in D turning S into*
  a) a 2-path of length two, or
  b) a single arc if S was a directed path in D.

The obvious question is which of the traditional digraph measures are closed under taking directed topological minors. Here we give the answer:

Proposition 4.5 ($\star$). *DAG-width and Kelly-width are monotone under taking directed topological minors unless the width is 0 or 1, respectively. Directed clique-width and bi-rank-width are not.*

We note that the problem of deciding whether a given digraph is a directed topological minor of another digraph is NP-complete.

Theorem 4.6 ($\star$). *There exists a digraph H such that the H-DIRECTED TOPO-LOGICAL MINOR problem (given a digraph D, decide whether H is directed topological minor of D) is NP-complete.*

The last result can be generalized to:

Theorem 4.7. *The following decision problem is NP-complete: given acyclic digraphs D and H, decide whether H is directed topological minor of D.*

5   An (Almost) Optimal Closure Property Result for Digraph Width Measures

In this section we finally prove some "almost optimal" negative answers to the questions raised in the Introduction and at the end of Section 3. To recapitulate, we have asked whether it is possible to define a digraph width measure that is closed under some reasonable notion of a directed minor (e.g., Definition 4.2) and that is still powerful (Definition 3.1) analogously to ordinary treewidth. We also recall the property of being treewidth-bounding (which we want to avoid) from Definition 3.2.

Besides the aforementioned several properties we suggest one more technical property that a desired good directed width measure should possess, in order to avoid "cheating" such as in the example of Theorem 5.8. Informally, we do not want to allow the measure to keep "computationally excessive" information in the orientation of edges:

Definition 5.1. A digraph width measure δ is *efficiently orientable* if there exist a computable function h, and a polynomial-time computable function $r : \mathscr{G} \to \mathscr{D}$ (from the class of all graphs to that of digraphs), such that for every undirected graph $G \in \mathscr{G}$, it is $U(r(G)) = G$ and

$$\delta(r(G)) \leq h(\min\{\delta(D) : D \text{ a digraph s.t. } U(D) = G\}) .$$

Proposition 5.2 ($\star$). *DAG-width, Kelly-width, digraph clique-width, and bi-rank-width are all efficiently orientable.*

Our main proof also relies on some deep ingredients from Graph Minors:

Theorem 5.3 ([19]). *If H is a planar undirected graph then there exists a number n_H such that for every G of treewidth at least n_H, H is a minor of G.*

Proposition 5.4 (folklore). *If H is a minor of G and the maximum degree of H is three, then H is a topological minor of G.*

Finally we need the following result whose proof we omit. For a set S of natural numbers, an *S-regular graph* is a graph having every vertex degree in S.

Theorem 5.5 ($\star$). *For any simple undirected graph H and every MSO_1 formula φ, there exist a $\{1,3\}$-regular planar graph G and an MSO_1 formula ψ, such that*

a) *$H \models \varphi \iff G \models \psi$, and*
b) *for every subdivision G_1 of G, we have $G_1 \models \psi \iff G \models \psi$.*
c) *Moreover, ψ depends only on φ, $|\psi| = \mathcal{O}(|\varphi|)$, and both G and ψ are computable in polynomial time from H and φ, respectively.*

With all the ingredients at hand, we state and prove our main result:

Theorem 5.6. *Let δ be a digraph width measure with the following properties*

a) *δ is not treewidth-bounding;*

b) δ is monotone under taking directed topological minors;
c) δ is efficiently orientable.

Then P = NP, or δ is not powerful.

Proof. We assume that δ is powerful and show that for every MSO_1 definable property φ of undirected graphs there exists a polynomial-time algorithm that decides, given as input an undirected graph G, whether $G \models \varphi$. Since, by Example 2.3, there are MSO_1 properties φ such that deciding whether $G \models \varphi$ is NP-hard, this would imply that P = NP.

Given an MSO_1-formula φ and an undirected graph H, we construct a $\{1,3\}$-regular planar graph G and an MSO_1-formula ψ as in Theorem 5.5. Let G_1 be the 1-subdivision of G (i.e. replacing every edge of G with a path of length two). We claim that, under assumptions (a) and (b), there exists an orientation D of G_1 such that $\delta(D) \leq k$, for some constant k dependent on δ.

We postpone the proof of this claim, and show its implications first. Since δ is efficiently orientable, by Definition 5.1, we can efficiently construct an orientation D_1 of G_1 such that $\delta(D_1) \leq h(k)$, for some computable function h. Note that since k is a constant, the width of D_1 is at most a constant. Let ψ_1 be the (directed) MSO_1 formula obtained from ψ by replacing $adj(u,v)$ with $(arc(u,v) \vee arc(v,u))$. Then, by Theorem 5.5, $H \models \varphi$ iff $D_1 \models \psi_1$, and hence we have a polynomial reduction of the problems $H \models \varphi$ onto $D_1 \models \psi_1$. Since δ is assumed to be powerful, the latter problem can be solved by an XP algorithm wrt the constant parameter $h(k)$, that is, in polynomial time.

We now return to our claim. Since δ is not treewidth-bounding, there is $k \geq 0$ such that the class $\mathcal{U} = \{U(D) \colon \delta(D) \leq k\}$ has unbounded treewidth. By Theorem 5.3, there exists D_0 such that $\delta(D_0) \leq k$ and $U(D_0)$ contains a G_1-minor. Since the maximum degree of G_1 is three, by Proposition 5.4, G_1 is a topological minor of $U(D_0)$ and hence some subdivision G_2 of G_1 is a subgraph of $U(D_0)$. Therefore there exists D_2, a subdigraph of D_0, with $U(D_2) = G_2$. Finally, by Lemma 4.4 one can contract 2-paths in D_2, if necessary, to obtain a digraph D_3 with $U(D_3) = G_1$. Clearly D_3 is a directed topological minor of D_0 and since δ is closed under taking directed topological minors, we have $\delta(D_3) \leq k$. This completes the proof of our claim and the theorem.   □

Due to Theorem 5.6, a powerful digraph width measure essentially "cannot be stronger" than ordinary undirected treewidth, unless P = NP. Our result requires two assumptions about the width measure δ in consideration: δ should be closed under taking directed topological minors, and it should be efficiently orientable. An interesting question is whether these conditions are necessary, or, put differently, whether the result of Theorem 5.6 can be strengthened by weakening these two essential assumptions.

We address this question in the remainder of this section – we show that Theorem 5.6 is almost strongest possible in the following Theorems 5.7 and 5.8. Specifically, we show that if one relaxes either of these two conditions b), c), then one can construct directed measures which definitely do not "look nice". In the first round, we relax the condition b) just to subdigraphs, and get:

Theorem 5.7 ($\star$). *There exists a **powerful** digraph width measure δ s.t.*

a) δ is not treewidth-bounding;
b) δ is monotone under taking subdigraphs;
c) δ is efficiently orientable.

Moreover, the same remains true if we replace b) with b') δ is monotone under taking butterfly minors.

The way the measure of Theorem 5.7 attains its "power" is by subdividing every edge with a tower-exponential number of new vertices. This is certainly not a nice behavior of a desired measure, and hence such a measure δ *should be dismissed.*

In the second round, we take a closer look at the condition that δ is efficiently orientable. It is not unreasonable to assume a digraph width measure to be efficiently orientable since most known digraph measure are, e.g. Proposition 5.2. Furthermore, efficient orientability prevents digraph measures from "keeping excessive information" in the orientation of arcs, such as (Theorem 5.8) the information about 3-colorability of the underlying graph.

Theorem 5.8 ($\star$). *There exists a digraph width measure δ such that*

a) δ is not treewidth-bounding;
b) δ is monotone under taking directed topological minors;
c) for every $k \geq 1$, on any digraph D with $\delta(D) \leq k$, one can decide in time $\mathcal{O}(3^k \cdot n^2)$ whether $U(D)$ is 3-colorable, and find a 3-coloring if it exists;
d) and for every 3-colorable graph G there exists an orientation D, $U(D) = G$ such that $\delta(D) = 1$.

6   Conclusions

The main result of this paper shows that an algorithmically useful digraph width measure that is substantially different from treewidth cannot be closed under taking directed topological minors. Since "standard" cops-and-robber games remain invariant on directed topological minors, we can conclude that a digraph width measure that allows efficient decisions of MSO_1-definable digraph properties on classes of bounded width cannot be defined using such games. This gives more weight to the argument [8] that bi-rank-width [14] is the best (though not optimal) currently known candidate for a *good* digraph width measure.

Our main result also leaves room for other ways of overcoming the problems with the currently existing digraph width measures. We have asked for width measures that are powerful, i.e., all MSO_1-definable digraph properties are decidable in polynomial time on digraphs of bounded width. What happens if we relax this requirement? We can ask for more time, like subexponential running time, or we can ask for restricted classes of MSO_1-definable digraph properties. Currently, we are not aware of any noticeable progress in this direction. Another interesting direction for future research is a closer study of efficient orientability and directed topological minors.

Finally, we believe that the results and suggestions contained in our paper will lead to new ideas and research directions in the area of digraph width measures – an area that seems to be stuck at this moment.

References

1. Arnborg, S., Lagergren, J., Seese, D.: Easy problems for tree-decomposable graphs. J. Algorithms 12(2), 308–340 (1991)
2. Berwanger, D., Dawar, A., Hunter, P., Kreutzer, S.: DAG-width and parity games. In: Durand, B., Thomas, W. (eds.) STACS 2006. LNCS, vol. 3884, pp. 524–536. Springer, Heidelberg (2006)
3. Bodlaender, H.: Treewidth: Characterizations, Applications, and Computations. In: Fomin, F.V. (ed.) WG 2006. LNCS, vol. 4271, pp. 1–14. Springer, Heidelberg (2006)
4. Courcelle, B.: The monadic second order logic of graphs I: Recognizable sets of finite graphs. Inform. and Comput. 85, 12–75 (1990)
5. Courcelle, B., Makowsky, J.A., Rotics, U.: Linear time solvable optimization problems on graphs of bounded clique-width. Theory Comput. Syst. 33(2), 125–150 (2000)
6. Courcelle, B., Olariu, S.: Upper bounds to the clique width of graphs. Discrete Appl. Math. 101(1-3), 77–114 (2000)
7. Diestel, R.: Graph Theory. Graduate texts in mathematics, vol. 173. Springer, New York (2005)
8. Ganian, R., Hliněný, P., Kneis, J., Langer, A., Obdržálek, J., Rossmanith, P.: On digraph width measures in parametrized algorithmics. In: IWPEC 2009. LNCS, vol. 5917, pp. 161–172. Springer, Heidelberg (2009)
9. Ganian, R., Hliněný, P., Obdržálek, J.: Unified approach to polynomial algorithms on graphs of bounded (bi-)rank-width (2009) (submitted)
10. Hliněný, P., Oum, S., Seese, D., Gottlob, G.: Width parameters beyond tree-width and their applications. The Computer Journal 51(3), 326–362 (2008)
11. Hunter, P.: Complexity and Infinite Games on Finite Graphs. PhD thesis, University of Cambridge (2007)
12. Hunter, P., Kreutzer, S.: Digraph measures: Kelly decompositions, games, and orderings. Theor. Comput. Sci. 399(3), 206–219 (2008)
13. Johnson, T., Robertson, N., Seymour, P.D., Thomas, R.: Directed tree-width. J. Combin. Theory Ser. B 82(1), 138–154 (2001)
14. Kanté, M.: The rank-width of directed graphs. arXiv:0709.1433v3 (March 2008)
15. Lampis, M., Kaouri, G., Mitsou, V.: On the algorithmic effectiveness of digraph decompositions and complexity measures. In: Hong, S.-H., Nagamochi, H., Fukunaga, T. (eds.) ISAAC 2008. LNCS, vol. 5369, pp. 220–231. Springer, Heidelberg (2008)
16. Obdržálek, J.: DAG-width – connectivity measure for directed graphs. In: SODA 2006, pp. 814–821. ACM-SIAM (2006)
17. Robertson, N., Seymour, P.D.: Graph minors. II. Algorithmic aspects of tree-width. J. Algorithms 7(3), 309–322 (1986)
18. Robertson, N., Seymour, P.D.: Graph minors. X. Obstructions to tree-decomposition. J. Combin. Theory Ser. B 52(2), 153–190 (1991)
19. Robertson, N., Seymour, P.D., Thomas, R.: Quickly excluding a planar graph. J. Combin. Theory Ser. B 62, 323–348 (1994)
20. Safari, M.: D-width: A more natural measure for directed tree-width. In: Jedrzejowicz, J., Szepietowski, A. (eds.) MFCS 2005. LNCS, vol. 3618, pp. 745–756. Springer, Heidelberg (2005)

On the (Non-)existence of Polynomial Kernels for P_l-free Edge Modification Problems[⋆]

Sylvain Guillemot[1], Christophe Paul[2], and Anthony Perez[2]

[1] Lehrstuhl für Bioinformatik, Friedrich-Schiller Universität Jena
[2] Université Montpellier II - CNRS, LIRMM

Abstract. Given a graph $G = (V, E)$ and an integer k, an edge modification problem for a graph property Π consists in deciding whether there exists a set of edges F of size at most k such that the graph $H = (V, E \,\triangle\, F)$ satisfies the property Π. In the Π *edge-completion problem*, the set F of edges is constrained to be disjoint from E; in the Π *edge-deletion problem*, F is a subset of E; no constraint is imposed on F in the Π *edge-editing problem*. A number of optimization problems can be expressed in terms of graph modification problems which have been extensively studied in the context of parameterized complexity. When parameterized by the size k of the edge set F, it has been proved that if Π is an hereditary property characterized by a finite set of forbidden induced subgraphs, then the three Π edge-modification problems are FPT [4]. It was then natural to ask [4] whether these problems also admit a polynomial size kernel. Using recent lower bound techniques, Kratsch and Wahlström answered this question negatively [15]. However, the problem remains open on many natural graph classes characterized by forbidden induced subgraphs. Kratsch and Wahlström asked whether the result holds when the forbidden subgraphs are paths and pointed out that the problem is already open in the case of P_4-free graphs (i.e. cographs). This paper provides positive and negative results in that line of research. We prove that parameterized cograph edge modification problems have cubic vertex kernels whereas polynomial kernels are unlikely to exist for P_l-free and C_l-free edge deletion problems for large enough l.

1 Introduction

An edge modification problem aims at changing the edge set of an input graph $G = (V, E)$ in order to get a certain property Π satisfied (see [16] for a recent study). Edge modification problems cover a broad range of graph optimization problems among which completion problems (*e.g.* MINIMUM FILL-IN, *a.k.a* CHORDAL GRAPH COMPLETION [19,21]), editing problems (*e.g.* CLUSTER EDITING [20]) and edge deletion problems (*e.g.* MAXIMUM PLANAR SUBGRAPH [10]). In a completion problem, the set F of modified edges is constrained to be disjoint from E; in an edge deletion problem, F has to be a subset of E; and in an editing problem, no restriction applies to F. These problems are fundamental in

---

[⋆] Research supported by the AGAPE project (ANR-09-BLAN-0159).

V. Raman and S. Saurabh (Eds.): IPEC 2010, LNCS 6478, pp. 147–157, 2010.

graph theory and play an important role in computational complexity theory (indeed they represent a large number of the earliest NP-Complete problems [10]). Edge modification problems are also relevant in the context of applications as graphs are often used to model data sets which may contain errors. Adding or deleting an edge thereby corresponds to fixing some false negatives or false positives (see *e.g.* [20] in the context of CLUSTER EDITING). Different variants of edge modification problems have been studied in the literature such as graph sandwich problems [11]. Most of the edge modification problems turns out to be NP-Complete [16] and approximation algorithms exist for some known graph properties (see *e.g.* [14,22]). But for those who want to compute an exact solution, fixed parameter algorithms [5,8,17] are a good alternative to cope with such hard problems. In the last decades, edge modification problems have been extensively studied in the context of fixed parameterized complexity (see [4,7,13]).

A parameterized problem Q is *fixed parameter tractable* (FPT for short) with respect to parameter k whenever it can be solved in time $f(k) \cdot n^{O(1)}$, where $f(k)$ is an arbitrary computable function [5,17]. In the context of edge modification problems, the size k of the set F of modified edges is a natural parameterization. The generic question is thereby whether a given edge modification problem is FPT for this parameterization. More formally:

PARAMETERIZED Π EDGE–MODIFICATION PROBLEM
Input: An undirected graph $G = (V, E)$.
Parameter: An integer $k \geqslant 0$.
Question: Is there a subset $F \subseteq V \times V$ with $|F| \leqslant k$ such that the graph $H = (V, E \triangle F)$ satisfies Π?

A classical result of parameterized complexity states that a parameterized problem Q is FPT if and only if it admits a *kernelization*. A *kernelization* of a parameterized problem Q is a polynomial time algorithm $\mathcal{K}$ that given an instance (x, k) computes an equivalent instance $\mathcal{K}(x, k) = (x', k')$ such that the sizes of x' and k' are bounded by a computable function $h()$ depending only on the parameter k. The reduced instance (x', k') is called a *kernel* and we say that Q admits a *polynomial kernel* if the function $h()$ is a polynomial. The equivalence between the existence of an FPT algorithm and the existence of a kernelization only yields kernels of (at least) exponential size. Determining whether an FPT problem has a polynomial (or even linear) size kernel is thus an important challenge. Indeed, the existence of such polynomial time reduction algorithm (or pre-processing algorithm or *reduction rules*) really speed-up the resolution of the problem, especially if it is interleaved with other techniques [18]. However, recent results proved that not every fixed parameter tractable problem admits a polynomial kernel [1].

Cai [4] proved that if Π is an hereditary graph property characterized by a finite set of forbidden subgraphs, then the PARAMETERIZED Π MODIFICATION problems (edge-completion, edge-deletion and edge-editing) are FPT. It was then natural to ask [4] whether these Π edge-modification problems also admit a polynomial size kernel. Using recent lower bound techniques, Kratsch and Wahlström answered negatively this question [15]. However, the problem remains open on many natural graph classes characterized by forbidden induced subgraphs. Kratsch and Wahlström asked whether the result holds when the forbidden subgraphs are paths or cycles and pointed out that the problem is already open in the case of P_4-free graphs (i.e. cographs). In this paper, we prove that PARAMETERIZED COGRAPH EDGE MODIFICATION problems have cubic vertex kernels whereas polynomial kernels are unlikely to exist for P_l-FREE and C_l-FREE EDGE DELETION problems for large enough l. The NP-Completeness of the cograph edge-deletion and edge-completion problems have been proved in [6].

Outline of the paper. We first establish structural properties of optimal edge-modification sets with respect to modules of the input graph (Section 2). These properties allow us to design general reduction rules (Section 3.1). We then establish cubic kernels using an extra sunflower rule (Section 3.2 and 3.3). Finally, we show it is unlikely that the C_l-FREE and the P_l-FREE EDGE-DELETION problems have polynomial kernels (Section 4).

2   Preliminaries

We only consider finite undirected graphs without loops nor multiple edges. Given a graph $G = (V, E)$, we denote by xy the edge of E between the vertices x and y of V. We set $n = |V|$ and $m = |E|$ (subscripts will be used to avoid possible confusion). The neighbourhood of a vertex x is denoted by $N(x)$. If S is a subset of vertices, then $G[S]$ is the subgraph induced by S (i.e. any edge $xy \in E$ between vertices $x, y \in S$ belongs to $E_{G[S]}$). Given a set of pairs of vertices F and a subset $S \subseteq V$, $F[S]$ denotes the pairs of F with both vertices in S. Given two sets S and S', we denote by $S \triangle S'$ their symmetric difference.

2.1   Fixed Parameter Complexity and Kernelization

We let Σ denote a finite alphabet and $\mathbb{N}$ the set of natural numbers. A *(classical) problem* Q is a subset of Σ^*, and a string $x \in \Sigma^*$ is an *input* of Q. A *parameterized problem* Q over Σ is a subset of $\Sigma^* \times \mathbb{N}$. The second component of an input (x, k) of a parameterized problem is called the *parameter*. Given a parameterized problem Q, one can derive its unparameterized (or classical) version $\tilde{Q}$ by $\tilde{Q} = \{x \# 1^k : (x, k) \in Q\}$, where $\#$ is a symbol that does not belong to Σ.

A parameterized problem Q is *fixed parameter tractable* (FPT for short) if there is an algorithm which given an instance $(x, k) \in \Sigma^* \times \mathbb{N}$ decides whether $(x, k) \in Q$ in time $f(k) \cdot n^{O(1)}$ where $f(k)$ is an arbitrary computable function (see [5,8,17]). A *kernelization* of a parameterized problem Q is a polynomial time algorithm $\mathcal{K} : \Sigma^* \times \mathbb{N} \to \Sigma^* \times \mathbb{N}$ which given an instance $(x, k) \in \Sigma^* \times \mathbb{N}$ outputs an instance $(x', k') \in \Sigma^* \times \mathbb{N}$ such that

1. $(x, k) \in Q \Leftrightarrow (x', k') \in Q$ and
2. $|x'|, k' \leqslant h(k)$ for some computable function $h : \mathbb{N} \to \mathbb{N}$.

The reduced instance (x', k') is called a *kernel* and we say that Q admits a *polynomial kernel* if the function $h()$ is a polynomial.

It is well known that a parameterized problem Q is FPT if and only if it has a kernelization [17]. But this equivalence only yields (at least) exponential size kernels. Recent results proved that it is unlikely that every fixed parameter tractable problem admits a polynomial kernel [1]. These results rely on the notion of *(or-)composition algorithms* for parameterized problems, which together with a polynomial kernel would imply a collapse on the polynomial hierarchy [1].

An *or-composition algorithm* for a parameterized problem Q is an algorithm that receives as input a sequence of instances $(x_1, k) \ldots (x_t, k)$ with $(x_i, k) \in \Sigma^* \times \mathbb{N}$ for $1 \leqslant i \leqslant t$, runs in time polynomial in $\sum_{i=1}^{t} |x_i| + k$ and outputs an instance (y, k') of Q such that:

1. $(y, k') \in Q \Leftrightarrow (x_i, k) \in Q$ for some $1 \leqslant i \leqslant t$ and
2. k' is polynomial in k.

A parameterized problem admitting an *or-composition algorithm* is said to be *or-compositional*.

Theorem 1. *[1,9] Let Q be an or-compositional parameterized problem whose unparameterized version $\tilde{Q}$ is NP-complete. The problem Q does not admit a polynomial kernel unless $NP \subseteq coNP/Poly$.*

Let P and Q be parameterized problems. A *polynomial time and parameter transformation* from P to Q is a polynomial time computable function $\mathcal{T} : \Sigma^* \times \mathbb{N} \to \Sigma^* \times \mathbb{N}$ which given an instance $(x, k) \in \Sigma^* \times \mathbb{N}$ outputs an instance $(x', k') \in \Sigma^* \times \mathbb{N}$ such that:

1. $(x, k) \in P \Leftrightarrow (x', k') \in Q$ and
2. $k' \leqslant p(k)$ for some polynomial p.

Theorem 2. *[2] Let P and Q be parameterized problems and let $\tilde{P}$ and $\tilde{Q}$ be their unparameterized versions. Suppose that $\tilde{P}$ is NP-complete and $\tilde{Q}$ belongs to NP. If there is a polynomial time and parameter transformation from P to Q and if Q admits a polynomial kernel, then P also admits a polynomial kernel.*

2.2   Modular Decomposition and Cographs

A *module* in a graph $G = (V, E)$ is a set of vertices $M \subseteq V$ such that for any $x \notin M$ either $M \subseteq N(x)$ or $M \cap N(x) = \emptyset$. Clearly if $M = V$ or $|M| = 1$, then M is a *trivial* module. A graph without any non-trivial module is called *prime*. For two disjoint modules M and M', either all the vertices of M are adjacent to all the vertices of M' or none of the vertices of M is adjacent to any vertex of M'. A partition $\mathcal{P} = \{M_1, \ldots, M_k\}$ of the vertex set $V(G)$ whose parts are modules is a *modular partition*. A *quotient graph* $G_{/\mathcal{P}}$ is associated with any modular partition $\mathcal{P}$: its vertices are the parts of $\mathcal{P}$ and there is an edge between M_i and M_j iff M_i and M_j are adjacent in G.

A module M is *strong* if for any module M' distinct from M, either $M \cap M' = \emptyset$ or $M \subset M'$ or $M' \subset M$. It is clear from definition that the family of strong modules arranges in an inclusion tree, called the *modular decomposition tree* and denoted $MD(G)$. Each node N of $MD(G)$ is associated with a quotient graph G_N whose vertices correspond to the children $N_1, \ldots N_k$ of N. We say that a node N of $MD(G)$ is *parallel* if G_N has no edge, *series* if G_N is complete, and *prime* otherwise. For a survey on modular decomposition theory, refer to [12].

Definition 1. *Let $G_1 = (V_1, E_1)$ and $G_2 = (V_2, E_2)$ be two vertex disjoint graphs. The* series composition *of G_1 and G_2 is the graph $G_1 \otimes G_2 = (V_1 \cup V_2, E_1 \cup E_2 \cup V_1 \times V_2)$. The* parallel composition *of G_1 and G_2 is the graph $G_1 \oplus G_2 = (V_1 \cup V_2, E_1 \cup E_2)$.*

Parallel and series nodes in the modular decomposition tree respectively correspond to a parallel and series composition of their children.

Cographs are commonly known as P_4-free graphs (a P_4 is an induced path on four vertices). An equivalent definition states that a graph is a cograph if it can be constructed from single vertex graphs by a sequence of parallel and series composition [3]. In particular, this means that the modular decomposition tree of a *cograph* does not contain any prime node. It follows that cographs are also known as the totally decomposable graphs for the modular decomposition.

3   Polynomial Kernels for Cograph Edge-Modification Problems

3.1   Modular Decomposition Based Reduction Rules

Since cographs correspond to P_4-free graphs, cograph edge-modification problems consist in adding or deleting at most k edges to the input graph in order to make it P_4-free. The use of the modular decomposition tree in our algorithms follows from the following observation: given a module M of a graph $G = (V, E)$ and four vertices $\{a, b, c, d\}$ inducing a P_4 of G, then $|M \cap \{a, b, c, d\}| \leqslant 1$ or $\{a, b, c, d\} \subseteq M$. This means that given a modular partition $\mathcal{P}$ of a graph G, any induced P_4 of G is either contained in a part of $\mathcal{P}$ or intersects the parts of $\mathcal{P}$ in at most one vertex. This observation allows us to show that a cograph edge-modification problem can be solved independently on modules of the partition $\mathcal{P}$ and on the quotient graph $G_{/\mathcal{P}}$, as stated by the following results:

Observation 1 [1]. *Let M be a non-trivial module of a graph $G = (V, E)$. Let F_M be an optimal edge-deletion (resp. edge-completion, edge-edition) set of $G[M]$ and let F_{opt} be an optimal edge-deletion (resp. edge-completion, edge-edition) set of G. Then*

$$F = (F_{opt} \setminus F_{opt}[M]) \cup F_M$$

is an optimal edge-deletion (resp. edge-completion, edge-edition) set of G.

---

[1] Due to space limits, most proofs are deferred to a full version of this paper.

Lemma 2. *Let $G = (V, E)$ be an arbitrary graph. There exists an optimal edge-deletion (resp. edge-completion, edge-edition) set F such that every module M of G is module of the cograph $H = (V, E \triangle F)$.*

We now present three reduction rules which apply to the three cograph edge-modification problems we consider. The second reduction rule is not required to obtain a polynomial kernel for each of these problems. However, it will ease the analysis of the structure of a reduced graph.

Rule 1. *Remove the connected components of G which are cographs.*

Rule 2. *If $C = G_1 \otimes G_2$ is a connected component of G, then replace C by $G_1 \oplus G_2$.*

Rule 3. *If M is a non-trivial module of G which is strictly contained in a connected component and is not an independent set of size at most $k+1$, then return the graph $G' \oplus G[M]$ where G' is obtained from G by deleting M and adding an independent set of size $\min\{|M|, k+1\}$ having the same neighbourhood than M.*

Observe that if $G[M]$ is a cograph, adding a disjoint copy to the graph is useless since it will then be removed by Rule 1. The soundness of Rule 3 follows from Lemma 2: there always exists an optimal solution that updates all or none of the edges between any two disjoint modules. Thereby if a module M has size greater than $k + 1$, none of the edges (or non-edges) xy with $x \in M$, $y \notin M$ can be edited in any solution. Observe also that these three reduction rules preserve the parameter. However, Rule 3 increases the number of vertices of the instance. Nevertheless, we will be able to bound the number of vertices of a reduced instance.

The analysis of the size of the kernel relies on the following structural property of the modular decomposition tree of an instance reduced under Rule 1, Rule 2 and Rule 3.

Observation 3. *Let G be a graph reduced under Rule 1, Rule 2 and Rule 3. If C is a non-prime connected component of G, then the modules of C are independent sets of size at most $k + 1$.*

Finally, computing a reduced graph requires polynomial time:

Lemma 4. *The reduction rules 1, 2 and 3 are safe and can be carried out in polynomial time. Moreover, given a graph $G = (V, E)$, computing a graph reduced under Rules 1 to 3 requires polynomial time.*

3.2   Cograph Edge-Deletion (and Edge-Completion)

In addition to the previous reduction rules, we need the classical *sunflower* rule to obtain a polynomial kernel for the parameterized cograph edge-deletion problem.

Rule 4. *If e is an edge of G that belongs to a set $\mathcal{P}$ of at least $k + 1$ P_4's such that e is the only common edge of any two distinct P_4's of $\mathcal{P}$, then remove e and decrease k by one.*

Observation 5. *The reduction rule 4 is safe and can be carried out in polynomial time.*

To analyse the size of a reduced graph $G = (V, E)$, we study the structure of the cograph $H = (V, E \bigtriangleup F)$ resulting from the removal of an optimal (of size at most k) edge-deletion set F. The modular decomposition tree (or cotree) is the appropriate tool for this analysis.

Theorem 3. *The parameterized cograph edge-deletion (and edge completion) problems have a cubic vertex kernel.*

Proof. Let $G = (V, E)$ be a graph reduced under Rule 1, Rule 2, Rule 3 and Rule 4 that can be turned into a cograph by deleting at most k edges. Let F be an optimal edge-deletion set and denote by $H = (V, E \bigtriangleup F)$ the cograph resulting from the deletion of F and by T its cotree. We will count the number of leaves of T (or equivalently of vertices of G and H).

Observe that since a set of k edges covers at most $2k$ vertices, T contains at most $2k$ affected leaves (i.e. leaves corresponding to a vertex incident to a removed edge). We say that an internal node of the cotree T is *affected* if it is the least common ancestor of two affected leaves. Notice that there are at most $2k$ affected nodes.

We first argue that the root of T is a parallel node and is affected. Assume that the root of T is a series node: since no edges are added to G, this would imply that G is not reduced under Rule 2, a contradiction. Moreover, since G is reduced under Rule 1, none of its connected components is a cograph. It follows that every connected component of G contains a vertex incident to a removed edge, and thus that every subtree attached to the root contains an affected leaf as a descendant. Hence the root of T is an affected node.

Claim 1. *Let p be an affected leaf or an affected node different from the root, and q be the least affected ancestor of p. The path between p and q has length at most $2k + 3$.*

Since there are at most $2k$ affected nodes and $2k$ affected leaves, T contains at most $(4k - 1)(2k + 3) + 2k$ internal nodes. As G is reduced, Observation 3 implies that each of these $O(k^2)$ nodes is attached to a set of at most $k + 1$ leaves or a parallel node with $k + 1$ children. It follows that T contains at most $2k + (k + 1)[(4k - 1)(2k + 3) + 2k] \leqslant 8k^3 + 20k^2 + 11k$ leaves, which correspond to the number of vertices of G.

We now conclude with the time complexity needed to compute the kernel. Since the application of Rule 4 decreases the value of the parameter (which is not changed by the other rules), Rule 4 is applied at most $k \leqslant n^2$ times. It then follows from Lemma 4 that a reduced instance can be computed in polynomial time.   □

3.3 Cograph Edge-Editing

The lines of the proof for the cubic kernel of the edge-editing problem are essentially the same as for the edge-deletion problem. But since edges can be added and deleted, the reduction Rule 4 has to be refined in order to avoid that a single edge addition breaks an arbitrary large set of P_4's.

Rule 5. *If $\{x, y\}$ is a pair of vertices of G that belongs to a set S of $t \geqslant k + 1$ quadruples $P_i = \{x, y, a_i, b_i\}$ such that for $1 \leqslant i \leqslant t$, every P_i induces a P_4 and for any $1 \leqslant i < j \leqslant t$, $P_i \cap P_j = \{x, y\}$, then change E into $E \bigtriangleup \{xy\}$ and decrease k by one.*

As for reduction Rule 4, it is clear that reduction Rule 5 is safe and can be applied in polynomial time. The kernelization algorithm of cograph edge-editing consists of an exhaustive application of Rules 1, 2, 3 and 5.

Theorem 4. *The parameterized cograph edge-editing problem has a cubic vertex kernel.*

For the deletion (resp. editing) problem there exists a graph reduced under Rule 1, Rule 2, Rule 3 and Rule 4 (resp. Rule 5) that achieves the cubic bound.

4 Kernel Lower Bounds for P_l-free Graph Edge-Deletion Problems

In [15], Kratsch and Wahlström show that the NOT-1-IN-3-SAT problem has no polynomial kernelization under a complexity-theoretic assumption ($NP \nsubseteq coNP/poly$). We observe that their argument still applies to a graph restriction of NOT-1-IN-3-SAT where the constraints arise from the triangles of an input graph.

4.1 A Graphic Version of the NOT-1-IN-3-SAT Problem

For a graph $G = (V, E)$, an edge-bicoloring is a function $B : E \rightarrow \{0, 1\}$. A *partial edge-bicoloring* of G is an edge-bicoloring of a subset of edges of E. An edge colored 1 (resp. 0) is called a 1-edge (resp. 0-edge). We say that the edge-bicoloring B' *extends* a partial edge-bicoloring B if for every $e \in E$ colored by B, then $B(e) = B'(e)$. The weight of an edge-bicoloring is the number $\omega(B)$ of 1-edges. We consider the following problem:

TRIPARTITE NOT-1-IN-3-EDGE-TRIANGLE
Input: An undirected graph $G = (V, E)$ s.t. $V = V_1 \cup V_2 \cup V_3$ where V_i is an independent set, $i = 1, 2, 3$, and a partial edge-bicoloring $B : E \rightarrow \{0, 1\}$.
Parameter: An integer $k \in \mathbb{N}$.
Question: Can we extend B to a valid edge-bicoloring B' of weight at most k? An edge-bicoloring is *valid* if every triangle of G contains either zero, two or three 1-edges.

Lemma 6. *The* TRIPARTITE-NOT-1-IN-3-EDGE-TRIANGLE *problem does not admit a polynomial kernel unless* $NP \subseteq coNP/poly$.

The proof of this result relies on a polynomial time and parameter transformation from the more general NOT-1-IN-3-EDGE-TRIANGLE problem (where no constraint is imposed to G). This problem can be shown to be both NP-complete and or-compositional, and Theorem 2 then allows us to conclude.

4.2   Negative Results for Γ-free Edge Deletion Problems

In this section, we show that unless $NP \subseteq coNP/poly$, the C_l-FREE EDGE-DELETION and the P_l-FREE EDGE-DELETION problems have no polynomial kernel for large enough $l \in \mathbb{N}$. To that aim, we provide polynomial time and parameter transformations from TRIPARTITE-NOT-1-IN-3-EDGE-TRIANGLE to the ANNOTATED C_l-FREE EDGE-DELETION problem and to the ANNOTATED P_l-FREE EDGE-DELETION problem. For a graph Γ, the ANNOTATED Γ-FREE EDGE-DELETION problem is defined as follows:

ANNOTATED Γ-FREE EDGE-DELETION
Input: An undirected graph $G = (V, E)$ and a subset S of vertices.
Parameter: An integer $k \in \mathbb{N}$.
Question: Is there a subset $F \subseteq E \cap (S \times S)$ such that $H = (V, E \setminus F)$ is Γ-free?

Observe that the ANNOTATED Γ-FREE EDGE-DELETION problem reduces to the (unannotated) Γ-FREE EDGE-DELETION problem whenever Γ is closed under true (resp. false) twin addition: it suffices to add for every vertex $v \in V \setminus S$ a set of $k+1$ true (resp. false) twin vertices, a true (resp. false) twin vertex of v being a vertex u adjacent (resp. non-adjacent) to v and having the same neighbourhood than v. Clearly this transformation also preserves the parameter.

Theorem 5. *The C_l-FREE EDGE-DELETION problem has no polynomial kernel for any $l \geqslant 12$, unless $NP \subseteq coNP/poly$.*

Proof (sketch). We use the fact that we can restrict the TRIPARTITE-NOT-1-IN-3-EDGE-TRIANGLE problem to instances (G, B, k) not containing any 0-edge (*i.e.* $B(e) = 1$ whenever it is defined). The same argument was used in [15] for the NOT-1-IN-3-SAT problem. We now describe a polynomial time and parameter transformation from this restriction of TRIPARTITE-NOT-1-IN-3-EDGE-TRIANGLE (i.e. without 0-edges) to ANNOTATED C_l-FREE EDGE-DELETION. The statement then follows from Theorem 2 and the above discussion.

Let (G, B, k) be an instance of the TRIPARTITE-NOT-1-IN-3-EDGE-TRIANGLE problem. The construction of the instance (H, S, k') of ANNOTATED C_l-FREE EDGE-DELETION works as follows. First the sets V_1, V_2 and V_3 are turned into cliques and the 1-edges of G are removed. In addition to V, the graph H contains a set U of new vertices. For each pair $t = (e, v)$ with $e = uw$ an edge of G and v a vertex of G, such that $\{u, v, w\}$ induces a triangle in G, we create a path P_t of length $l - 1$ between u and w in H (the internal vertices of P_t are added to U). Notice that each triangle of G generates three such paths in H. It remains to add some *safety* edges incident to the vertices of U. Every two vertices x and y of U that do not belong to the same path are made adjacent. In every path P_t, we select an internal vertex c_t, called its *centre*, at distance $(l - 1)/2$ from u. Every centre vertex c_t is made adjacent to $V \setminus \{u, v, w\}$. We denote by $H = (V_H, E_H)$ the resulting graph. To complete the description of (H, S, k') we set $S = V$ and the parameter $k' = k - k_1$ where k_1 is the number of 1-edges of (G, B, k). The correctness of the transformation can be shown using the following result:

Claim 2. *A subset of vertices $C \in V_H$ induces a cycle of length l iff G contains a triangle uvw, with $e = uw$ a 1-edge and uv, vw uncolored edges, such that $C = P_t \cup \{v\}$ with $t = (e, v)$.* □

To adapt this result to the P_l-FREE EDGE-DELETION problem we need to modify the transformation: instead of creating one path of length $l - 1$ from u to w for every triangle $\{u, v, w\}$ of G, the gadget now consists in two paths of length $(l - 1)/3$ (starting from u) and $2(l - 1)/3$ (starting from w). Paths are still made adjacent to each other, and the centre vertex is now located on the second path, at distance $(l - 1)/3$ from w. Similar arguments then apply to show the soundness of the construction.

Theorem 6. *The P_l-FREE EDGE-DELETION problem has no polynomial kernel for any $l \geqslant 13$, unless $NP \subseteq coNP/poly$.*

References

1. Bodlaender, H., Downey, R., Fellows, M., Hermelin, D.: On problems without polynomial kernels. In: Aceto, L., Damgård, I., Goldberg, L.A., Halldórsson, M.M., Ingólfsdóttir, A., Walukiewicz, I. (eds.) ICALP 2008, Part I. LNCS, vol. 5125, pp. 563–574. Springer, Heidelberg (2008)
2. Bodlaender, H.L., Thomassé, S., Yeo, A.: Kernel bounds for disjoint cycles and disjoint paths. In: Fiat, A., Sanders, P. (eds.) ESA 2009. LNCS, vol. 5757, pp. 635–646. Springer, Heidelberg (2009)
3. A. Brandstädt, V. B. Le, and J. P. Spinrad. Graph Classes: A Survey. SIAM Monographs on Discrete Mathematics and Applications (1999)
4. Cai, L.: Fixed-parameter tractability of graph modification problems for hereditary properties. Information Processing Letters 58(4), 171–176 (1996)
5. Downey, R.G., Fellows, M.R.: Parameterized complexity. Springer, Heidelberg (1999)
6. El-Mallah, E.S., Colbourn, C.: The complexity of some edge deletion problems. IEEE Transactions on Circuits and Systems 35(3), 354–362 (1988)
7. Fellows, M.R., Langston, M., Rosamond, F., Shaw, P.: Efficient parameterized preprocessing for cluster editing. In: Csuhaj-Varjú, E., Ésik, Z. (eds.) FCT 2007. LNCS, vol. 4639, pp. 312–321. Springer, Heidelberg (2007)
8. Flum, J., Grohe, M.: Parameterized complexity theorey. Texts in Theoretical Computer Science. Springer, Heidelberg (2006)
9. Fortnow, L., Santhanam, R.: Infeasibility of instance compression and succinct PCPs for NP. In: STOC, pp. 133–142 (2008)
10. Garey, M., Johnson, S.: Computers and intractability: a guide to the theory of NP-completeness. Freeman, New York (1978)
11. Golumbic, M., Kaplan, H., Shamir, R.: Graph sandwich problems. Journal of Algorithms 19, 449–473 (1995)
12. Habib, M., Paul, C.: A survey on algorithmic aspects of modular decomposition. Computer Science Review 4(1), 41–59 (2010)
13. Heggernes, P., Paul, C., Telle, J.A., Villanger, Y.: Interval completion with few edges. In: STOC, pp. 374–381 (2007)
14. Kenyon-Mathieu, C., Schudy, W.: How to rank with few errors. In: Annual ACM Symposium on Theory of Computing (STOC), pp. 95–103 (2007)

15. Kratsch, S., Wahlström, M.: Two edge modification problems without polynomial kernels. In: IWPEC 2009. LNCS, vol. 5917, pp. 264–275. Springer, Heidelberg (2009)
16. Natanzon, A., Shamir, R., Sharan, R.: Complexity classification of some edge modification problems. Discrete Applied Mathematics 113(1), 109–128 (2001)
17. Niedermeier, R.: Invitation to fixed parameter algorithms. Oxford Lectures Series in Mathematics and its Applications, vol. 31. Oxford University Press, Oxford (2006)
18. Niedermeier, R., Rossmanith, P.: A general method to speed up fixed-parameter-tractable algorithms. Information Processing Letters 73(3-4), 125–129 (2000)
19. Rose, D.J.: A graph-theoretic study of the numerical solution of sparse positive systems of linear equations. In: Graph Theory and Computing, pp. 183–217 (1972)
20. Shamir, R., Sharan, R., Tsur, D.: Cluster graph modification problems. Discrete Applied Mathematics 144(1-2), 173–182 (2004)
21. Tarjan, R., Yannakakis, M.: Simple linear-time algorithms to test chordality of graphs, test acyclicity of hypergraphs, and selectively reduce acyclic hypergraphs. SIAM Journal of Computing 13, 566–579 (1984)
22. van Zuylen, A., Williamson, D.P.: Deterministic algorithms for rank aggragation and other ranking and clustering problems. In: Kaklamanis, C., Skutella, M. (eds.) WAOA 2007. LNCS, vol. 4927, pp. 260–273. Springer, Heidelberg (2008)

Parameterized Complexity Results for General Factors in Bipartite Graphs with an Application to Constraint Programming

Gregory Gutin[1], Eun Jung Kim[1], Arezou Soleimanfallah[1],
Stefan Szeider[2], and Anders Yeo[1]

[1] Department of Computer Science, Royal Holloway, University of London, Egham, Surrey
TW20 0EX, UK
{gutin,eunjung,arezou,anders}@cs.rhul.ac.uk
[2] Institute of Information Systems, Vienna University of Technology, A-1040 Vienna, Austria
stefan@szeider.net

Abstract. The NP-hard general factor problem asks, given a graph and for each vertex a list of integers, whether the graph has a spanning subgraph where each vertex has a degree that belongs to its assigned list. The problem remains NP-hard even if the given graph is bipartite with partition $U \uplus V$, and each vertex in U is assigned the list $\{1\}$; this subproblem appears in the context of constraint programming as the consistency problem for the extended global cardinality constraint. We show that this subproblem is fixed-parameter tractable when parameterized by the size of the second partite set V. More generally, we show that the general factor problem for bipartite graphs, parameterized by $|V|$, is fixed-parameter tractable as long as all vertices in U are assigned lists of length 1, but becomes W[1]-hard if vertices in U are assigned lists of length at most 2. We establish fixed-parameter tractability by reducing the problem instance to a bounded number of acyclic instances, each of which can be solved in polynomial time by dynamic programming.

1 Introduction

To find in a given graph a spanning subgraph (or *factor*) that satisfies certain degree constraints is a fundamental task in combinatorics that entails several classical polynomial-time solvable problems such as PERFECT MATCHING (the factor is 1-regular), r-FACTOR (the factor is r-regular), and (a,b)-FACTOR (the degree of each vertex v in the factor lies in a given interval (a_v, b_v)). Lovász [7,8] introduced the following NP-hard problem which generalizes all mentioned factor problems:

GENERAL FACTOR
Instance: A graph $G = (V, E)$ and a mapping K that assigns to each vertex $v \in V$ a set $K(v) \subseteq \{0, \ldots, d(v)\}$ of integers.
Question: Is there a subset $F \subseteq E$ such that for each vertex $v \in V$ the number of edges in F incident with v is an element of $K(v)$?

The problem remains NP-hard even for bipartite graphs $G = (U \uplus V, E)$ where $K(u) = \{1\}$ for all $u \in U$ and $K(v) = \{0, 3\}$ for all $v \in V$. Cornuéjols [4] obtained a

V. Raman and S. Saurabh (Eds.): IPEC 2010, LNCS 6478, pp. 158–169, 2010.

dichotomy result that classifies the complexity of all GENERAL FACTOR problems that are formed by restricting the sets $K(v)$ to a fixed class $\mathcal{C}$ of sets of integers. For each class $\mathcal{C}$ the corresponding problem is either polynomial or NP-complete.

In this paper we study the parameterized complexity of GENERAL FACTOR for bipartite graphs $G = (U \uplus V, E)$ parameterized by the size of V. Our main results can be summarized as follows.

> The problem GENERAL FACTOR for bipartite graphs $G = (U \uplus V, E)$, parameterized by the size of V, is
> (1) fixed-parameter tractable if $|K(u)| \leq 1$ for all $u \in U$;
> (2) W[1]-hard if $|K(u)| \leq 2$ for all $u \in U$.

We establish result (1) by a novel combination of concepts from polynomial-time algorithmics (alternating cycles) with concepts from fixed-parameter algorithmics (data reduction and annotation).

Next we briefly discuss an application of our fixed-parameter tractability result. Constraint Programming (CP) is a general-purpose framework for combinatorial problems that can be solved by assigning values to variables such that certain restrictions on the combination of values are satisfied; the restrictions are formulated by a combination of so-called *global constraints*. For example the global constraint ALLDIFFERENT enforces that certain variables must all be assigned to mutually different values. The Catalog of Global Constraints [1] lists hundreds of global constraints that are used to model various real-world problems. For several global constraints the consistency problem (i.e., deciding whether there exists an allowed value assignment) is NP-complete [3]. It is an interesting line of research to study such global constraints under the framework of parameterized complexity. We think that global constraints are an excellent platform for deploying efficient fixed-parameter algorithms for real-world applications.

An important global constraint is the *extended global cardinality constraint* (or *EGC constraint*, for short) [14]. Let X be a finite set of variables, each variable $x \in X$ given with a finite set $D(x)$ of possible values. An EGC constraint over X is specified by a mapping that assigns to each value $d \in D := \bigcup_{x \in X} D(x)$ a set $K(d)$ of non-negative integers. The constraint is consistent if one can assign each variable $x \in X$ a value $\alpha(x) \in D(x)$ such that $|\alpha^{-1}(d)| \in K(d)$ holds for all values $d \in D$. The consistency problem for EGC constraints can clearly be expressed as an instance (G, K') of GENERAL FACTOR where G, the *value graph* of the constraint [14], is the bipartite graph $(X \uplus D, \{ xd : d \in D(x) \})$ and K' is the degree list assignment defined by $K'(x) = \{1\}$ for all $x \in X$ and $K'(d) = K(d)$ for all $d \in D$. Hence our result (1) renders the consistency problem for EGC constraints fixed-parameter tractable when parameterized by the number $|D|$ of values.

1.1 Related Work

The parameterized complexity of EGC constraints was first studied by Samer and Szeider [12] using the treewidth of the value graph as the parameter. For value graphs of bounded degree it is easy to see that the consistency problem is fixed-parameter tractable for this parameter, as one can express the restrictions imposed by the sets $K(v)$ in monadic second-order logic, and use Courcelle's Theorem. However, for graphs of unbounded degree the problem is W[1]-hard. That instances of unbounded degree but bounded treewidth are solvable in non-uniform polynomial time (i.e., the consistency problem is in XP) can be shown by means of an extension of Courcelle's Theorem [13]. A further parameterization of GENERAL FACTOR was considered by Mathieson and Szeider [9], taking as the parameter the number of edges that need to be deleted to obtain the general factor. The problem is W[1]-hard in general but fixed parameter tractable for graphs of bounded degree.

The parameterized complexity of other global constraints were recently studied by Bessiere et al. [2], considering, among others, the global constraints NVALUE, DIS-JOINT, and ROOTS.

In the context of parameterized complexity it is interesting to mention the results of van Hoeve et al. [15] who compare various algorithms for the SEQUENCE constraint (a global constraint that is important for various scheduling problems). Although the consistency problem is polynomial for this constraint, it turns out that a fixed-parameter algorithm outperforms the polynomial-time algorithm on several realistic instances.

1.2 Notation and Preliminaries

Unless otherwise stated, all graphs considered are finite, simple, and undirected. We denote a graph G with vertex set V and edge set E by $G = (V, E)$ and write $V(G) = V$ and $E(G) = E$. We denote an edge between two vertices u and v by uv or equivalently vu. For a set F of edges and a vertex v we write $N_F(v) = \{ u : uv \in F \}$ and we write $d_F(v)$ for the number of edges in F that are incident with v. For a graph G we also write $N_G(v) = N_{E(G)}(v)$ and $d_G(v) = d_{E(G)}(v)$, and we omit the subscripts if the context allows.

A *degree list assignment* K is a mapping that assigns to each vertex $v \in V(G)$ a set $K(v) \subseteq \{0, \ldots, d_G(v)\}$. A set $F \subseteq E(G)$ is a *general K-factor* of G if $d_F(v) \in K(v)$ holds for each $v \in V(G)$. Sometimes it is convenient to identify a set $F \subseteq E(G)$ with the spanning subgraph $(V(G), F)$ of G.

An instance of a *parameterized problem* L is a pair (I, k) where I is the *main part* and k is the *parameter*; the latter is usually a non-negative integer. L is *fixed-parameter tractable* if there exist a computable function f and a constant c such that instances (I, k) can be solved in time $O(f(k)n^c)$ where n denotes the size of I. FPT is the class of all fixed-parameter tractable decision problems.

A *parameterized reduction* is a many-one reduction where the parameter for one problem maps into the parameter for the other. More specifically, problem L reduces to problem L' if there is a mapping R from instances of L to instances of L' such that (i) (I, k) is a YES-instance of L if and only if $(I', k') = R(I, k)$ is a YES-instance of L', (ii) $k' = g(k)$ for a computable function g, and (iii) R can be computed in time $O(f(k)n^c)$ where f is a computable function, c is a constant, and n is the size of I. The parameterized complexity classes $W[1] \subseteq W[2] \subseteq \cdots \subseteq XP$ are defined as the closure of certain parameterized problems under parameterized reductions. There is strong theoretical evidence that parameterized problems that are hard for classes $W[i]$ are not fixed-parameter tractable.

For more background on parameterized complexity we refer to other sources [5,6,10].

2   Fixed-Parameter Tractability

This section is devoted to the proof of our fixed-parameter tractability result. Let BI-PARTITE GENERAL FACTOR WITH SINGLETONS denote the problem GENERAL FACTOR restricted to instances (G, K) where $G = (U \uplus V, E)$ is bipartite and $|K(u)| \leq 1$ for all $u \in U$. We will show the following:

Theorem 1. BIPARTITE GENERAL FACTOR WITH SINGLETONS *parameterized by the size of V is fixed parameter tractable.*

Let (G, K) be an instance of BIPARTITE GENERAL FACTOR WITH SINGLETONS with $G = (U \uplus V, E)$ and $V = \{v_1, \ldots, v_k\}$. Clearly we may assume that $K(v) \notin \{\emptyset, \{0\}\}$ for all $v \in U \uplus V$: if $K(v) = \emptyset$ then G has no general K-factor, and if $K(v) = \{0\}$ then we can delete v from G. Thus, in particular for each $u \in U$ we have $K(u) \in \{\{1\}, \ldots, \{k\}\}$.

2.1   General Factors of Edge-Weighted Graphs

Key to our algorithm for BIPARTITE GENERAL FACTOR WITH SINGLETONS is the transformation to a more general "annotated" problem on edge-weighted graphs that allows a more succinct representation.

Let G be a graph. A (non-negative integral) *edge-weighting* ρ of G is a mapping that assigns to each edge $e \in E(G)$ a non-negative integer $\rho(e)$. We refer to a pair (G, ρ) as an *edge-weighted graph*. For a vertex v of G we define $d_\rho(v)$ as the sum of $\rho(e)$ over all edges incident with v (or 0 if v has no incident edges). $d_G(v)$ denotes as usual the number of edges incident with v, the degree of v. Let K be a degree list assignment of G. We define a general K-factor of an edge-weighted graph (G, ρ) by using $\rho(e)$ as the "capacity" of an edge e. More precisely, we say that an edge-weighting φ is a *general K-factor* of the edge-weighted graph (G, ρ) if (i) $\varphi(e) \leq \rho(e)$ holds for all edges e of G and (ii) $d_\varphi(v) \in K(v)$ for all $v \in V(G)$. Evidently this definition generalizes the above definition of general K-factors for unweighted graphs (by considering an unweighted graph as an edge-weighted graph where each edge has weight 1, and a set F of edges as an edge-weighting that assigns each edge in F the weight 1, and all other edges the weight 0). By GENERAL FACTOR FOR EDGE-WEIGHTED GRAPHS we refer to the obvious generalization of the decision problem GENERAL FACTOR to edge-weighted graphs.

In the following we will present several *reduction rules* that take as input an instance $I = (G, \rho, K)$ of GENERAL FACTOR FOR EDGE-WEIGHTED GRAPHS and produce as output an instance $I' = (G', \rho', K')$ of the same problem (or rejects I as a no-instance). We say that a reduction rule is *sound* if it always holds that either both I and I' are no-instances or both are yes-instances (or in case of rejection, I is indeed a no-instance). A reduction rule is *polynomial* if we can decide in polynomial time whether it applies to I and we can compute I' in polynomial time if the rule applies.

2.2   Contractions of Modules

Let (G, ρ, K) be an instance of GENERAL FACTOR FOR EDGE-WEIGHTED GRAPHS. For an integer $c \geq 1$ we call a subset $M \subseteq V(G)$ a *c-module* if

1. M is nonempty and independent;
2. $K(v) = \{c\}$ for all $v \in M$;
3. all vertices in M have exactly the same neighbors;
4. $\rho(e) = 1$ holds for all edges $e \in E(G)$ with one end in M.

Reduction Rule 1. *Let M be a c-module of (G, ρ, K). Obtain a new instance (G', ρ', K') by replacing M with a new vertex u_M that is adjacent with the same vertices as the vertices in M. Set $\rho'(e) = |M|$ for all edges e incident with u_M and $\rho'(e) = \rho(e)$ for all other edges. Set $K'(u_M) = \{c|M|\}$ and $K'(v) = K(v)$ for all other vertices.*

Lemma 1. *Reduction Rule 1 is sound and polynomial.*

Proof. Let φ be a general K-factor of (G, ρ). We define $\varphi'(u_M w) = \sum_{v \in M} \varphi(vw)$ for edges $u_M w$ that are incident with u_M and $\varphi'(e) = \varphi(e)$ for all other edges. Observe that $\varphi'(u_M w) \leq |M| = \rho'(u_M w)$ and $d_{\varphi'}(u_M) = \sum_{v \in M} d_\varphi(v) = c|M| \in K'(u_M)$, hence φ' is a general K'-factor of (G', ρ').

Conversely, let φ' be a general K'-factor of (G', ρ'). Let $M = \{u_1, \ldots, u_s\}$ and let $N = \{v_1, \ldots, v_t\}$ be the set of neighbors of u_M.

We define an edge-weighting φ of G. For $0 \leq i \leq t$ let $S_i = \sum_{i'=1}^{i} \varphi'(u_M v_{i'})$; thus $S_0 = 0$ and $S_t = cs$. For $1 \leq i \leq t$ and $1 \leq j \leq s$ we set

$$\varphi(v_i u_j) = \begin{cases} 1 & \text{if } j \equiv S_{i-1} + l \pmod{s} \text{ for some } 1 \leq l \leq \varphi'(u_M v_i); \\ 0 & \text{otherwise.} \end{cases}$$

For $e \in E(G) \cap E(G')$ we set $\varphi(e) = \varphi'(e)$. Since $\varphi'(u_M v_i) \leq s$ for $1 \leq i \leq t$ this definition is correct. To see that φ is a general K-factor of G we note that $d_\varphi(v_i) = d_{\varphi'}(v_i)$ for all $1 \leq i \leq t$, and $d_\varphi(u_j) = d_\varphi(u_M)/s = c \in K(u_j)$ for all $1 \leq j \leq s$.

As it is obvious that the rule is polynomial, the lemma follows. □

2.3   Acyclic General Factors

Let φ be a general K-factor of an edge-weighted graph (G, ρ). We say that an edge $e \in E(G)$ is *full* in φ if $\varphi(e) = \rho(e)$ and $\varphi(e) > 0$, an edge $e \in E(G)$ is *empty* in φ if $\varphi(e) = 0$.

The *skeleton* of φ is the spanning subgraph G_φ of G with $E(G_\varphi) = \{e \in E(G) : 0 < \varphi(e) < \rho(e)\}$; i.e., $E(G_\varphi)$ consists of all edges that are neither full nor empty. The *full skeleton* of φ is the spanning subgraph G_φ^+ of G with $E(G_\varphi^+) = \{e \in E(G) : 0 < \varphi(e) \leq \rho(e)\}$; i.e., $E(G_\varphi^+)$ consists of all edges that are not empty. We say that φ is *acyclic* if its skeleton G_φ contains no cycles (i.e., is a forest), φ is *fully acyclic* if its full skeleton G_φ^+ contains no cycles.

Lemma 2. *If a bipartite edge-weighted graph (G, ρ) has a general K-factor, then it also has an acyclic general K-factor.*

Proof. Let $G = (U \uplus V, E)$ where $U = \{u_1, \ldots, u_p\}$ and $V = \{v_1, \ldots, v_k\}$. For an edge-weighting φ of G and a pair v_i, u_j with $v_i u_j \notin E(G)$ we define $\varphi(v_i u_j) = 0$.

With each edge-weighting φ of G we associate a vector $A(\varphi)$ defined as follows:

$$A(\varphi) = (\varphi(v_1 u_1), \varphi(v_1 u_2), \ldots, \varphi(v_1 u_p),$$
$$\varphi(v_2 u_1), \varphi(v_2 u_2), \ldots, \varphi(v_2, u_p),$$
$$\ldots$$
$$\varphi(v_k u_1), \varphi(v_k u_2), \ldots, \varphi(v_k u_p)).$$

Let φ be a general K-factor of G such that $A(\varphi)$ is lexicographically maximal among all vectors of general K-factors of G. (A vector $(a_1, \ldots, a_n)$ is lexicographically larger than $(b_1, \ldots, b_n)$ if for some i, $a_i > b_i$ and $a_j = b_j$ for all $j < i$.) We are going to show that φ is acyclic.

Suppose to the contrary that the skeleton G_φ contains a cycle $C = v_{j_1} u_{i_1} \ldots v_{j_t} u_{i_t} v_{j_1}$. Without loss of generality, we may assume that $j_2 = \min\{j_1, \ldots, j_t\}$. Moreover we may assume that $i_1 < i_2$ since otherwise we can consider the reverse of C instead.

We define a new general K-factor φ' of (G, ρ) by setting $\varphi'(v_{j_l} u_{i_{l-1}}) = \varphi(v_{j_l} u_{i_{l-1}}) + 1$ and $\varphi'(v_{j_l} u_{i_l}) = \varphi(v_{j_l} u_{i_l}) - 1$ for $1 \leq l \leq t$ (computing indices modulo t), and $\varphi'(e) = \varphi(e)$ for all other edges. That φ' is indeed a general K-factor follows from the following observations:

(i) $d_{\varphi'}(w) \in K(w)$ holds for all $w \in U \uplus V$ since $d_{\varphi'}(w) = d_\varphi(w) \in K(w)$;

(ii) For each $e \in E(C)$ we have $\varphi'(e) \leq \varphi(e) + 1 \leq \rho(e)$ since $e \in E(C) \subseteq E(G_\varphi)$ and therefore e is not full in φ.

(iii) For each $e \in E(C)$ we have $0 \leq \varphi(e) - 1 \leq \varphi'(e)$ since $e \in E(C) \subseteq E(G_\varphi)$ and therefore e is not empty in φ.

Furthermore, we observe that $\varphi'(v_{j_2} u_{i_1}) > \varphi(v_{j_2} u_{i_1})$, but all entries in the vector $A(\varphi')$ before $\varphi(v_{j_2} u_{i_1})$ remain the same as in $A(\varphi)$. Hence $A(\varphi')$ is lexicographically larger than $A(\varphi)$, a contradiction to our assumption that $A(\varphi)$ is lexicographically maximal. This proves that G_φ is indeed acyclic. $\qquad\square$

Let $I = (G, \rho, K)$ be an instance of GENERAL FACTOR FOR EDGE-WEIGHTED GRAPHS and $X \subseteq E(G)$. For a vertex v of G let $d_X(v)$ be the sum of $\rho(e)$ over all edges $e \in X$ that are incident with v. Let $I - X$ denote the instance (G^X, ρ^X, K^X) obtained from I by deleting X, decreasing the capacities ρ of all edges not in X by one, and updating the degree list assignment assuming the edges in X are full. More precisely, we set $G^X = G - X$, $\rho^X(e) = \max(\rho(e) - 1, 0)$ for all $e \in E(G^X)$, and for all $v \in V(G^X)$ we set

$$K^X(v) = \{c - d_X(v) : c \in K(v), c - d_X(v) \geq 0\}.$$

If φ' is a general K^X-factor of (G^X, ρ^X) we denote by $\varphi' + X$ the edge-weighting of (G, ρ) defined by $(\varphi' + X)(e) = \varphi(e)$ for $e \in E(G^X)$ and $(\varphi' + X)(e) = \rho(e)$ for $e \in X$.

The following lemma is an easy consequence of this definition.

Lemma 3. *Let* $I = (G, \rho, K)$ *be an instance of* GENERAL FACTOR FOR EDGE-WEIGHTED GRAPHS, $X \subseteq E(G)$, *and* $I - X = (G^X, \rho^X, K^X)$.

1. *If φ is a general K-factor of (G, ρ) such that X is precisely the set of full edges of φ, then the restriction of φ to G^X is a general K^X-factor of (G^X, ρ^X).*
2. *Conversely, if φ' is a general K^X-factor of (G^X, ρ^X), then $\varphi' + X$ is a general K-factor of (G, ρ) where X is the set of full edges of $\varphi' + X$.*

Lemma 4. *Let (G, ρ) be a bipartite edge-weighted graph and K a degree list assignment such that for each edge uv of G we have $K(u) = \{\rho(uv)\}$ or $K(v) = \{\rho(uv)\}$. If (G, ρ) has a general K-factor, then it also has a fully acyclic general K-factor.*

Proof. Assume that (G, ρ) has a general K-factor. By Lemma 2, (G, ρ) has an acyclic general K-factor φ. Let X be the set of full edges of φ. Let $I-X = (G^X, \rho^X, K^X)$ and φ^X the restriction of φ to G^X. By Lemma 3, φ^X is a general K^X-factor of (G^X, ρ^X), and since φ is acyclic, φ^X is fully acyclic (observe that the full skeleton of φ^X equals the skeleton of φ, i.e., $G^+_{\varphi^X} = G_\varphi$). For each edge $uv \in X$ we have $K^X(u) = \{0\}$ or $K^X(v) = \{0\}$, hence at least one of the ends of any $uv \in X$ is of degree 1 in the full skeleton G^+_φ. Since G^+_φ can be obtained by adding the edges in X to the forest G_φ, it follows that also G^+_φ is a forest, i.e., φ is fully acyclic.   □

2.4   Eliminating Vertices of Low Degree

Reduction Rule 2. *Assume that G has a vertex v of degree 0. If $0 \notin K(v)$, then reject the instance; if $0 \in K(v)$ then delete v from G and let $G' = G - v$, $\rho' = \rho$, and K' the restrictions of K to $V(G')$.*

Reduction Rule 3. *Assume G has a vertex v of degree 1. Let u be the neighbor of v. We let $G' = G - v$, ρ' the restriction of ρ to $E(G')$, and we put*

$$K'(u) = \{ c_u - c_v : c_u \in K(u), \ c_v \in K(v), \ c_v \leq \min(\rho(uv), c_u) \}$$

and $K'(w) = K(w)$ for all $w \in V(G') \setminus \{u\}$.

The proof of the following lemma is obvious.

Lemma 5. *Reduction Rules 2 and 3 are sound and polynomial.*

Proposition 1. GENERAL FACTOR FOR EDGE-WEIGHTED GRAPHS *can be solved in polynomial time for edge-weighted forests.*

Proof. Let $I = (G, \rho, K)$ be an instance of GENERAL FACTOR FOR EDGE-WEIGHTED GRAPHS such that G is a forest. If $V(G) \neq \emptyset$, G has a vertex of degree ≤ 1, and hence Reduction Rule 2 or 3 applies, and we obtain in polynomial time an equivalent instance with one vertex less which is again a forest (or we reject the instance). By at most $|V(G)|$ applications of the rules we either reject the instance (I is a no-instance) or we eliminate all the vertices (I is a yes-instance). Thus the result follows by repeated application of Lemma 5.   □

2.5   The Algorithm

It remains to put together the above results to show that BIPARTITE GENERAL FACTOR WITH SINGLETONS parameterized by the size of V is fixed-parameter tractable.

Let (G, K) with $V(G) = U \uplus V$ be the given instance of the problem. As explained above we can consider (G, K) as an instance $I = (G, \rho, K)$ of GENERAL FACTOR FOR EDGE-WEIGHTED GRAPHS letting $\rho(e) = 1$ for all $e \in E(G)$. Let $|V| = k$.

1. We partition U into maximal sets $U_1, \ldots, U_p$ such that each U_i is a c-module for some $1 \le c \le k$.
2. We apply Reduction Rule 1 with respect to the modules $U_1, \ldots, U_p$ and obtain an instance $I^M = (G^M, \rho^M, K^M)$ of GENERAL FACTOR FOR EDGE-WEIGHTED GRAPHS where $G^M = (U^M \uplus V, E^M)$ is a bipartite graph with $p + k$ vertices.
3. We guess a set $X \subseteq E^M$ of edges and consider the instance $I^X = I^M - X = (G^X, \rho^X, K^X)$.
4. We guess a spanning forest T of G^X and consider the instance $I^T = (T, \rho^T, K^T)$ where ρ^T is the restriction of ρ^X to T and $K^T = K^X$. We check if (T, ρ^T) has a general K^T-factor using Proposition 1 (i.e., applying the Reduction Rules 2 and 3). If (T, ρ^T) has general K^T-factor, then we stop and output YES.
5. If none of the guesses for X and T produces the answer YES, we stop and output NO.

In the statement of the following theorem we use the O^*-notation which suppresses polynomial factors [16].

Theorem 2. *Given a bipartite graph $G = (U \uplus V, E)$, $k = |V|$, and a degree list assignment K with $|K(u)| \le 1$ for all $u \in U$, we can decide whether G has a general K-factor in time $O^*(2^{k^2 2^k + k^2}(k + 1)^{k2^k + k})$.*

Thus, BIPARTITE GENERAL FACTOR WITH SINGLETONS parameterized by the size of V is fixed-parameter tractable.

Proof. We show that the above algorithm decides correctly and in the claimed time bound whether G has a general K-factor.

Due to Lemma 1, (G, ρ) has a general K-factor if and only if (G^M, ρ^M) has a general K^M-factor. By Lemma 2, (G^M, ρ^M) has a general K^M-factor if and only if (G^M, ρ^M) has an *acyclic* general K^M-factor. By Lemma 3, (G^M, ρ^M) has an *acyclic* general K^M-factor if and only if there is some X such that (G^X, ρ^X) has a fully acyclic general K^X-factor. The latter is exactly the case if some spanning forest T of G^X has a general K^T-factor. The correctness of the algorithm thus follows from Proposition 1, it remains to bound its running time.

We may assume that U contains no isolated vertices as such vertices can be ignored without changing the problem. Each U_i is defined by some degree constraint $c \in \{0, \ldots, k\}$ and a nonempty subset of V, hence $p \le k2^k - 1$. Since G^M has at most $k^2 2^k$ edges, there are at most $2^{k^2 2^k}$ possible choices for X. In a spanning forest T, each vertex has at most one parent. Each vertex in U has $k + 1$ alternatives for its parent (including not having one), and each vertex in V has at most $k2^k$

alternatives for its parent (including not having one). Thus, each G^X has at most $O((k+1)^{k2^k}(k2^k)^k)$ possible spanning forests T. It follows that we apply Proposition 1 at most $O(2^{k^2 2^k}(k+1)^{k2^k}(k2^k)^k) = O(2^{k^2 2^k + k^2}(k+1)^{k2^k + k})$ times. $\square$

Corollary 1. *Given a bipartite graph* $G = (U \uplus V, E)$, $k = |V|$, *and a degree list assignment* K *with* $K(u) = \{1\}$ *for all* $u \in U$, *we can decide in time* $O^*(2^{k^2}(k+1)^{k2^k + k})$ *whether* G *has a general* K-*factor.*

Proof. We use a simplified version of the above algorithm. In view of Lemma 4 we do not need to guess a set X of full edges in order to be able to restrict our scope to fully acyclic general K-factors. Thus we may skip step 3 of the algorithm and save a factor of $2^{k^2 2^k}$ in the running time. $\square$

3   W[1]-Hardness

This section is devoted to establishing the W[1]-hardness result. Let BIPARTITE GENERAL FACTOR WITH PAIRS denote the problem GENERAL FACTOR restricted to instances (G, K) where $G = (U \uplus V, E)$ is bipartite and $K(u) \in \{\{1\}\} \cup \{\{0, r\} : 1 \leq r \leq |V|\}$ holds for all $u \in U$. We will show the following:

Theorem 3. BIPARTITE GENERAL FACTOR WITH PAIRS *parameterized by the size of* V *is* W[1]-*hard.*

We give a parameterized reduction from PARTITIONED CLIQUE, which asks whether a given k-partite graph $G = (V_1 \uplus \ldots \uplus V_k, E)$, where $|V_i| = n$ for all $1 \leq i \leq k$, has a k-clique (a complete subgraph on k vertices). This problem is known to be W[1]-complete [11] for parameter k.

For this reduction we need to ensure that exactly one vertex v_i is selected from each partite set V_i, $1 \leq i \leq k$, and that v_i and v_j are adjacent for all $1 \leq i < j \leq k$. For the first part we shall use the following gadget construction.

Given a set A of non-negative integers with $M = \max(A)$ and a number $r > 0$, we construct a bipartite graph $G_{A,r} = (U' \uplus V', E')$ and a degree list assignment K as follows. We let $U' = U \cup W$ where $U = \{u_1, \ldots, u_M\}$, $W = \{w_1, \ldots, w_M\}$, and $V' = \{x, y\} \cup Z$ where $Z = \{z_1, \ldots, z_r\}$. We specify the edge set E' in terms of neighborhoods: $N(x) = U$, $N(y) = U'$, and $N(z) = W$ for all $z \in Z$. We call the vertices of Z the *outputs* of the gadget. We define the mapping K as follows: $K(u) = \{1\}$ for each $u \in U$, $K(w) = \{0, r+1\}$ for each $w \in W$, $K(x) = A$, $K(y) = \{M\}$. We do not apply any degree restrictions to the outputs $z \in Z$ and therefore we put $K(z) = \{0, \ldots, M\} = \{0, \ldots, d(z)\}$. We call the graph $G_{A,r}$ together with the degree list assignment K a *selection gadget*.

Lemma 6. *If a set* F *of edges forms a general* K-*factor of a selection gadget* $G_{A,r}$ *then all outputs are incident to the same number* α *of edges in* F, *and* $\alpha \in A$. *Conversely, for each* $\alpha \in A$ *there exists a general* K-*factor* F *of* $G_{A,r}$ *such that each output is incident with exactly* α *edges of* F.

Proof. Suppose that F is a general K-factor of $G_{A,r}$. Let $d_F(x) = \alpha \in A$. Since $N(x) = U$ and $K(u) = \{1\}$ for all $u \in U$, it follows that exactly α vertices from U are adjacent with x via edges of F. Hence the other $M - \alpha$ vertices of U must be adjacent with y via edges of F. Since $K(y) = \{M\}$, y has to be adjacent with α vertices in W via edges in F. Since $K(w) = \{0, r+1\}$ for every $w \in W$, these α vertices in W must also be incident to r vertices in Z via edges of F, i.e., to all vertices in Z. It follows that $d_F(z) = \alpha$ for all $z \in Z$.

Conversely, let $\alpha \in A$. We define a general factor F. We choose $W_\alpha \subseteq W$ and $U_\alpha \subseteq U$ with $|U_\alpha| = |W_\alpha| = \alpha$. We put $F = \{ wz : w \in W_\alpha, z \in Z \cup \{y\} \} \cup \{ uy : u \in U \setminus U_\alpha \} \cup \{ ux : u \in U_\alpha \}$. It is easy to verify that F is indeed a general K-factor and $d_F(z) = \alpha$ for each $z \in Z$. □

Let A be a set of non-negative integers, $N = \max(A) + 1$, $A' = \{ N\alpha : \alpha \in A \}$ and $r, r' \geq 0$ two numbers. We take two vertex disjoint selection gadgets $G_{A,r+1}$ and $G_{A',r'+1}$ and identify one output z of the first with one output z' of the second gadget. Let us call this identified vertex q. We define $K(q) = \{ a + Na : a \in A \}$. We call this new gadget a *double selection gadget* $G_{A,r,r'}$. We consider the outputs of $G_{A,r+1}$ and $G_{A',r'+1}$ except z and z' as the outputs of $G_{A,r,r'}$. We call the r outputs that originate from $G_{A,r+1}$ the *lower outputs*, and the r' outputs that originate from $G_{A',r'+1}$ the *upper outputs* of $G_{A,r,r'}$. If $U \uplus V$ denotes the vertex set of $G_{A,r,r'}$, then $|V| = r + r' + 5$.

Lemma 7. *If a set F of edges is a general K-factor of a double selection gadget then all lower outputs are incident to the same number α of edges in F, all upper outputs are incident to the same number β of edges in F, and we have $\alpha \in A$ and $\beta = \alpha N$. Conversely, for each $\alpha \in A$ there is a general K-factor F such that all lower outputs are incident to α edges in F, and all upper outputs are incident to αN edges in F.*

Proof. Let $G_{A,r,r'}$ be a double selection gadget constructed from two selection gadgets $G_{A,r+1}$ and $G_{A',r'+1}$, and let F be a general factor of $G_{A,r,r'}$. Let $c = |N_F(q) \cap V(G_{A,r+1})|$ and $c' = |N_F(q) \cap V(G_{A',r'+1})|$. Clearly $c + c' = d_F(q) \in K(q)$. By Lemma 6 we have $c \in A$, $c' \in A'$, $d_F(z) = c$ for all $z \in Z$ and $d_F(z') = c'$ for all $z' \in Z'$, thus the first part of the lemma is shown. The second part follows easily by using the second part of Lemma 6 twice. □

Next we describe the parameterized reduction from PARTITIONED CLIQUE to BIPARTITE GENERAL FACTOR WITH PAIRS that uses the double selection gadgets. Let $G = (V_1 \uplus \ldots \uplus V_k, E)$ be an instance of PARTITIONED CLIQUE, and assume $n = |V_i|$ for $1 \leq i \leq k$. We write $V_i = \{v_1^i, \ldots, v_n^i\}$. For every $1 \leq i \leq k$, we take a copy H_i of the double selection gadget $G_{A,r,r'}$ where $A = \{1, \ldots, n\}$, $r = i - 1$ and $r' = k - i$. For each pair $1 \leq i < j \leq k$ we identify an upper output of H_i and a lower output of H_j. We denote the identified vertex as $h_{i,j}$. We can choose the identified pairs in such a way that finally each output is identified with exactly one other output. Let $H = (U_H \uplus V_H, E_H)$ be the bipartite graph constructed in this way. We define a degree list assignment K where each identified vertex $h_{i,j}$, $1 \leq i < j \leq k$, gets assigned the list $\{ N\alpha + \beta : v_\alpha^i v_\beta^j \in E(G), \alpha, \beta \in \{1, \ldots, n\} \}$, and all other vertices inherit the list assigned to them in the definition of a double selection gadget. Thus (H, K) is an instance of GENERAL FACTOR that satisfies the properties as stated in Theorem 3 (in fact, for all $u \in U_H$ we have $K(u) \in \{\{1\}\} \cup \{\{0, r\} : 2 \leq r \leq k + 1\}$). Furthermore, we have $|V_H| = \binom{k}{2} + 5k$ as V_H contains $\binom{k}{2}$ identified vertices and each H_i, $1 \leq i \leq k$, contributes 5 more vertices to V_H. Therefore the new parameter $k' = |V_H|$ of the BIPARTITE GENERAL FACTOR WITH PAIRS instance is indeed a function of the old parameter k of the PARTITIONED CLIQUE instance. Furthermore, it is easy to check that $|U_H| = k \cdot 2n(n + 2)$ and clearly (H, K) can be obtained from G in polynomial time. It remains to show that the reduction is correct:

Lemma 8. *H has a general K-factor if and only if G has a k-clique.*

Proof. Let F be a general K-factor of H. For $1 \leq i \leq k$ let $F_i = F \cap E(H_i)$ and observe that F_i is a general factor of H_i. Thus, by the first part of Lemma 7 there is some $a_i \in A = \{1, \ldots, n\}$ such that $d_{F_i}(z) = a_i$ for each lower output of H_i and $d_{F_i}(z') = Na_i$ for each upper output z' of H_i. Let $1 \leq i < j \leq k$ and consider the identified vertex $h_{i,j}$. We have $d_F(h_{i,j}) = Na_i + a_j$. Since $K(h_{i,j}) = \{N\alpha + \beta : v_\alpha^i v_\beta^j \in E(G)\}$, it follows that $v_{a_i}^i v_{a_j}^j \in E(G)$. Hence $C = \{v_{a_i}^i : 1 \leq i \leq k\}$ induces a clique in G.

Conversely, assume that $C \subseteq V(G)$ induces a k-clique in G. Since G is k-partite, C contains exactly one vertex $v_{x_i}^i$ from each set V_i, $1 \leq i \leq k$. By the second part of Lemma 7, each H_i, $1 \leq i \leq k$, has a general factor F_i such that $d_{F_i}(z) = x_i$ for each lower output z and $d_{F_i}(z') = Nx_i$ for each upper output z'. Let $F = \bigcup_{i=1}^{k} F_i$. Since for each pair $1 \leq i < j \leq k$ we have $v_{x_i} v_{x_j} \in E(G)$, it follows that $d_F(h_{i,j}) = x_j + Nx_i \in K(h_{i,j})$, hence F is indeed a general K-factor of H. □

With Lemma 8 we have shown that our reduction is correct, thus Theorem 3 is established.

4 Conclusion

We have studied the parameterized complexity of general factor problems for bipartite graphs $G = (U \uplus V, E)$ where the size of the sets $K(u)$ for $u \in U$ is bounded by a small constant and where $|V|$ is the parameter. There are various further variants of general factor problems whose parameterized complexities would be interesting to explore, for example, one could consider $|U|$ instead of $|V|$ as the parameter. A further possibility is to restrict $K(v)$ for all vertices v of one or both partite sets to a fixed class $\mathcal{C}$ of sets of integers, similar to Cornuejols' dichotomy result [4]. It would be interesting to reveal fixed-parameter tractable general factor problems that are W[1]-hard without the restriction of $K(v)$ to a fixed class $\mathcal{C}$ and NP-hard without the parameterization.

Acknowledgements. Research of Gutin, Kim and Yeo was supported in part by an EPSRC grant. Research of Gutin was supported in part by the IST Programme of the European Community, under the PASCAL 2 Network of Excellence. Research of Soleimanfallah was supported in part by the Overseas Research Students Award Scheme (ORSAS). Research of Szeider was supported by the European Research Council, grant reference 239962.

References

1. Beldiceanu, N., Carlsson, M., Rampon, J.-X.: Global constraint catalog. Technical Report T2005:08, SICS, SE-16 429 Kista, Sweden (August 2006), http://www.emn.fr/x-info/sdemasse/gccat/
2. Bessiere, C., Hebrard, E., Hnich, B., Kiziltan, Z., Quimper, C.-G., Walsh, T.: The parameterized complexity of global constraints. In: Proceedings of AAAI 2008, AAAI Conference on Artificial Intelligence, pp. 235–240. AAAI Press, Menlo Park (2008)
3. Bessiére, C., Hebrard, E., Hnich, B., Walsh, T.: The tractability of global constraints. In: Wallace, M. (ed.) CP 2004. LNCS, vol. 3258, pp. 716–720. Springer, Heidelberg (2004)
4. Cornuéjols, G.: General factors of graphs. J. Combin. Theory Ser. B 45(2), 185–198 (1988)
5. Downey, R.G., Fellows, M.R.: Parameterized Complexity. Monographs in Computer Science. Springer, New York (1999)
6. Flum, J., Grohe, M.: Parameterized Complexity Theory. Texts in Theoretical Computer Science. An EATCS Series, vol. XIV. Springer, Berlin (2006)
7. Lovász, L.: The factorization of graphs. In: Proceedings of Combinatorial Structures and their Applications 1969, pp. 243–246. Gordon and Breach, New York (1970)
8. Lovász, L.: The factorization of graphs. II. Acta Math. Acad. Sci. Hungar. 23, 223–246 (1972)
9. Mathieson, L., Szeider, S.: Editing graphs to satisfy degree constraints: A parameterized approach (May 2009) (submitted)
10. Niedermeier, R.: Invitation to Fixed-Parameter Algorithms. Oxford Lecture Series in Mathematics and its Applications. Oxford University Press, Oxford (2006)
11. Pietrzak, K.: On the parameterized complexity of the fixed alphabet shortest common supersequence and longest common subsequence problems. J. of Computer and System Sciences 67(4), 757–771 (2003)
12. Samer, M., Szeider, S.: Tractable cases of the extended global cardinality constraint. In: Proceedings of CATS 2008, Computing: The Australasian Theory Symposium. CRPIT, vol. 77, pp. 67–74. Australian Computer Society (2008); Full version to appear in Constraints: an International Journal
13. Szeider, S.: Monadic second order logic on graphs with local cardinality constraints. In: Ochmański, E., Tyszkiewicz, J. (eds.) MFCS 2008. LNCS, vol. 5162, pp. 601–612. Springer, Heidelberg (2008); Full version to appear in ACM Transactions on Computational Logic
14. van Hoeve, W.-J., Katriel, I.: Global constraints. In: Rossi, F., van Beek, P., Walsh, T. (eds.) Handbook of Constraint Programming, ch. 6. Elsevier, Amsterdam (2006)
15. van Hoeve, W.J., Pesant, G., Rousseau, L.-M., Sabharwal, A.: Revisiting the sequence constraint. In: Benhamou, F. (ed.) CP 2006. LNCS, vol. 4204, pp. 620–634. Springer, Heidelberg (2006)
16. Woeginger, G.J.: Exact algorithms for NP-hard problems: A survey. In: Jünger, M., Reinelt, G., Rinaldi, G. (eds.) Combinatorial Optimization - Eureka, You Shrink! LNCS, vol. 2570, pp. 185–208. Springer, Heidelberg (2003)

On the Grundy Number of a Graph

Frédéric Havet[*] and Leonardo Sampaio[**]

Projet Mascotte, I3S (CNRS, UNSA) and INRIA, 2004 route des lucioles, BP 93,
06902 Sophia-Antipolis Cedex, France
{fhavet,leonardo.sampaio_rocha}@inria.sophia.fr
http://www-sop.inria.fr/mascotte/

Abstract. The *Grundy number* of a graph G, denoted by $\Gamma(G)$, is the largest k such that G has a *greedy k-colouring*, that is a colouring with k colours obtained by applying the greedy algorithm according to some ordering of the vertices of G. Trivially $\Gamma(G) \leq \Delta(G) + 1$. In this paper, we show that deciding if $\Gamma(G) \leq \Delta(G)$ is NP-complete. We then show that deciding if $\Gamma(G) \geq |V(G)| - k$ is fixed parameter tractable with respect to the parameter k.

Keywords: Colouring, Online Colouring, Greedy Colouring, NP-completeness, Fixed Parameter Complexity.

1 Introduction

A *k-colouring* of a graph $G = (V, E)$ is a surjective mapping $c : V \to \{1, 2, \ldots, k\}$ such that $c(u) \neq c(v)$ for any edge $uv \in E$. A *k-colouring* may also be seen as a partition of the vertex set of G into k disjoint *stable sets* $S_i = \{v \mid c(v) = i\}$, $1 \leq i \leq k$. For convenience (and with a slight abuse of terminology), by colouring we mean either the mapping c or the partition $(S_1, \ldots, S_k)$. The elements of $\{1, \ldots, k\}$ are called *colours*. We may sometimes refer to S_i by its colour i. A graph is k-colourable if it admits a k-colouring. The *chromatic number* is $\chi(G) = min\{k \mid G$ is k-colourable$\}$.

The most basic and most widespread on-line algorithm producing colourings is the greedy algorithm or first-fit algorithm. A *greedy colouring* relative to a vertex ordering $\sigma = v_1 < v_2 < \ldots < v_n$ of $V(G)$ is obtained by colouring the vertices in the order $v_1, \ldots, v_n$, assigning to v_i the smallest positive integer not already used on its lowered-indexed neighbours. A greedy colouring has the following property:

For every $j < i$, every vertex v in S_i has a neighbour in S_j. $\hspace{1em}(P)$

Otherwise v would have been coloured by an integer not greater than j. Conversely, a colouring satisfying Property (P) is a greedy colouring relative to any vertex ordering in which the vertices of S_i precede those of S_j when $i < j$. The *Grundy number* $\Gamma(G)$ is the largest k such that G has a greedy k-colouring.

---

[*] Partially supported by ANR Blanc AGAPE.
[**] Partially supported by CAPES/Brazil and ANR Blanc AGAPE.

V. Raman and S. Saurabh (Eds.): IPEC 2010, LNCS 6478, pp. 170–179, 2010.

Hence the Grundy number and its ratio with the chromatic number measure how bad the greedy algorithm may perform on a graph.

Easily, $\chi(G) \leq \Gamma(G) \leq \Delta(G) + 1$.

Determining the chromatic number of a graph is well-known to be an NP-hard problem. Similarly, determining the Grundy number of a graph is NP-hard [14]. But deciding if the chromatic number of a graph is at most k is NP-complete [7] for $k \geq 3$, whereas Zaker [14] showed that for any fixed k, it is decidable in polynomial time if a given graph has Grundy number at most k. To show this, he proved that there is a finite number of graphs called k-atoms such that if $\Gamma(G) \geq k$ then G contains a k-atom as an induced subgraph.

Brooks [3] proved that, for any connected graph G, $\chi(G) \leq \Delta(G)$ unless G is a complete graph or an odd cycle. This implies that it can be checked in polynomial time if $\chi(G) \leq \Delta(G)$. Extensions of Brooks' Theorem have also been considered. A well-known conjecture of Borodin and Kostochka [2] states that every graph of maximal degree $\Delta \geq 9$ and chromatic number at least Δ has a Δ-clique. Reed [13] proved that this is true when Δ is sufficiently large, thus settling a conjecture of Beutelspacher and Herring [1]. Further information about this problem can be found in the monograph of Jensen and Toft [10, Problem 4.8]. Generalisation of this problem has also been studied by Farzad, Molloy and Reed [6] and Molloy and Reed [11]. In particular, it is proved [11] that determining whether a graph with large constant maximum degree Δ is $(\Delta-q)$-colourable can be done in linear time if $(q + 1)(q + 2) \leq \Delta$. This threshold is optimal by a result of Emden-Weinert, Hougardy and Kreuter [5], since they proved that for any two constants Δ and $q \leq \Delta - 3$ such that $(q+1)(q+2) > \Delta$, determining whether a graph of maximum degree Δ is $(\Delta - q)$-colourable is NP-complete.

A natural question is then to ask if an analog of Brooks' Theorem exists for the Grundy number. One may ask if it is decidable in polynomial time whether a graph G satisfies $\Gamma(G) = \Delta(G) + 1$. In Section 2, we answer this question in the negative by showing that this problem is NP-complete even when restricted to bipartite graphs. In particular, it implies that it is NP-hard to compute the Grundy number of a bipartite graph.

We then investigate some parameterised version of the Grundy number problem. For an introduction to parameterised algorithms and complexity, we refer the reader to [4] or [12]. Telle (See [4], Exercise 3.2.7) proved that the following problem is Fixed Parameter Tractable (FPT).

DUAL OF COLOURING
Instance: A graph G and an integer k.
Parameter: k.
Question: $\chi(G) \leq |V(G)| - k$?

In Section 3, we show that the following analog for greedy colouring is also FPT.

DUAL OF GREEDY COLOURING
Instance: A graph G and an integer k.
Parameter: k.
Question: $\Gamma(G) \geq |V(G)| - k$?

In fact, Telle showed DUAL OF COLOURING is FPT by showing that it has a quadratic kernel. Being FPT is equivalent to having a kernel but not necessarily polynomial. Then, a natural question is the following.

Problem 1. Does DUAL OF GREEDY COLOURING have a polynomial kernel?

2   NP-Hardness Results

Before proving some complexity results, we need a preliminary lemma.

Lemma 1. *Let G be a graph and x a vertex of G. If there is a greedy colouring c such that x is coloured p, then, for any $1 \leq i \leq p$, there is a greedy colouring such that x is coloured i.*

Proof. For $1 \leq i \leq p$, let S_i be the set of vertices coloured i by c. Then for any $1 \leq i \leq p$, $(S_{p-i+1}, \ldots, S_p)$ is a greedy i-colouring of $G[\bigcup_{j=p-i+1}^{p} S_j]$ in which x is coloured i. This partial greedy colouring of G may be extended into a greedy colouring of G in which x is coloured i.   □

We now show that no Brooks type theorem exists for the Grundy number.

Theorem 1. *It is NP-complete to decide if a bipartite graph G satisfies $\Gamma(G) = \Delta(G) + 1$.*

Proof. The problem is trivially in NP. To show that it is also NP-complete, we present a reduction from 3-edge-colourability of 3-regular graphs, which is known to be NP-complete [8].

Let G be a 3-regular graph with $t - 3$ vertices. Set $V(G) = \{v_4, v_5, \ldots, v_t\}$ and $E(G) = \{e_1, \ldots, e_p\}$. Let I be the vertex-edge incidence graph of G, that is the bipartite graph with vertex set $V(I) = V(G) \cup E(G)$ in which an edge of G is adjacent to its two endvertices. Also, let $M_{p,p}$ denote the graph obtained from the complete bipartite graph $K_{p,p}$ by removing a perfect matching. It can be easily checked that $\Gamma(M_{p,p}) = p$. We construct from I a new bipartite graph H as follows. For each vertex $e_i \in E(G)$, we add a copy $M_{3,3}(e_i)$ of $M_{3,3}$ and identify one of its vertices with e_i. We add a new vertex r adjacent to all the vertices of $V(G)$. We add copies $M_{1,1}^r$, $M_{2,2}^r$, $M_{3,3}^r$, $M_{t+1,t+1}^r$ of K_1, K_2, $M_{3,3}$, $M_{t+1,t+1}$ and we choose arbitrary vertices v_1, v_2, v_3, v_{t+1} respectively from each copy and add the edges $v_1 r$, $v_2 r$, $v_3 r$, $v_{t+1} r$. Finally, for every $4 \leq i \leq t$, we do the following: for every $4 \leq j \leq i - 1$, we add a copy $M_{j,j}^i$ of $M_{j,j}$, choose an arbitrary vertex v_j^i of it and add the edge $v_i v_j^i$. See Figure 1.

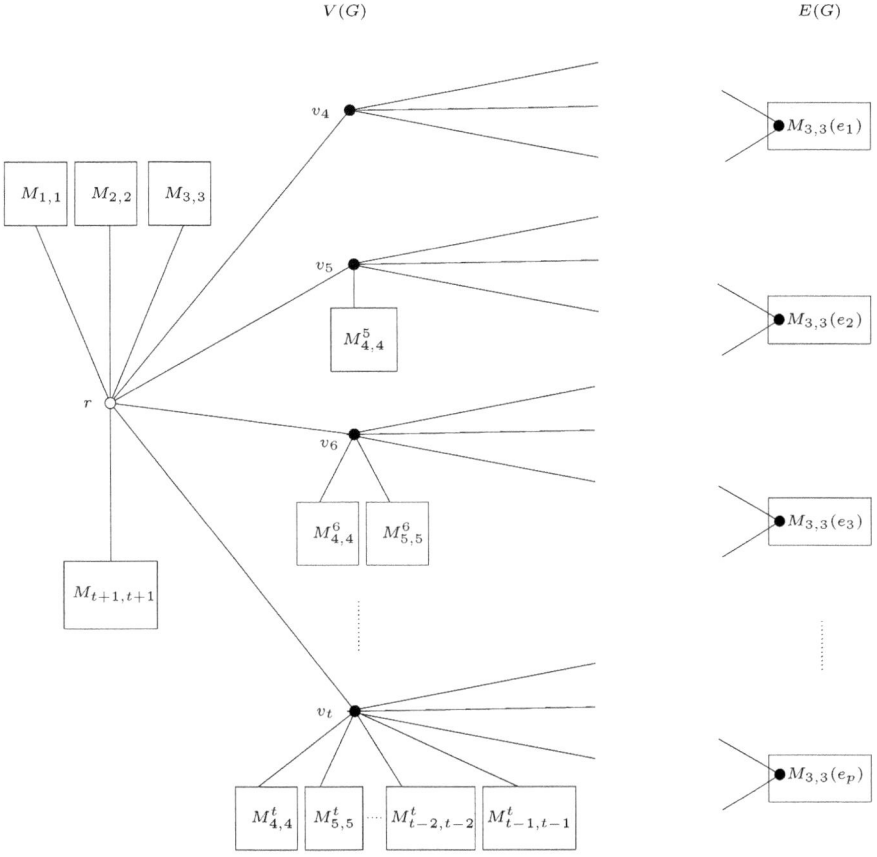

Fig. 1. The graph H of the reduction in Theorem 1

Observe that:

(i) $d_H(r) = t + 1$,

(ii) $d_H(v_i) = 1 + (i - 1) = i$, for $4 \leq i \leq t + 1$.

(iii) $d_H(e_j) = 4$, for $1 \leq j \leq p$, since e_i has two neighbours in I and two in $M_{3,3}(e_i)$.

(iv) $d_H(v_j^i) = j$, for $5 \leq i \leq t$ and $4 \leq j < i$, since a vertex in $M_{j,j}^i$ has degree $j - 1$ and v_j^i is adjacent to v_i.

(v) $\Delta(H) = t + 1$ and the only vertices with degree $t + 1$ are r and v_{t+1}.

Let us show that $\Gamma(H) = \Delta(H) + 1 = t + 2$ if and only if G is 3-edge-colourable.

Assume first G has a 3-edge-colouring ϕ. By Lemma 1, for every $1 \leq i \leq p$, there is a greedy colouring of the copy of $M_{3,3}(e_i)$ associated with e_i where e_i is coloured $\phi(e_i)$. Then in I every vertex in $V(G)$ has one neighbour of each colour in $\{1, 2, 3\}$. There is a greedy colouring of $M_{j,j}^i$, $4 \leq j < i \leq t$ so that v_j^i is coloured j.

Then we greedily extend the union of these colourings to v_i, $4 \leq i \leq t$, so that v_i is coloured i. We also greedily colour $M_{1,1}^r$, $M_{2,2}^r$, $M_{3,3}^r$, $M_{t+1,t+1}^r$ in such a way that r has one neighbour coloured i, $i \in \{1, 2, 3, t+1\}$. Finally, r has one neighbour of each colour j, $1 \leq j \leq t+1$. So we colour it with $t+2$.

Hence $\Gamma(H) \geq t+2$ and so $\Gamma(H) = t+2$ because $\Delta(H) = t+1$.

Let us now show that if $\Gamma(H) = t+2$ then G is 3-edge-colourable. Assume that c is a greedy $(t+2)$-colouring of H.

Claim 1. $\{c(r), c(v_{t+1})\} \subseteq \{t+1, t+2\}$

Proof. Let u be a vertex such that $c(u) = t+2$. Then, u must have one neighbour coloured with each of the other $t+1$ colours and then $d(u) \geq t+1$. Hence, by Observation (v), u is either r or v_{t+1}.

Case 1: $u = r$.
Then, $c(v_{t+1}) = t+1$, since the only neigbours of r with degree at least t are v_t and v_{t+1}, $d(v_t) = t$, and v_t is adjacent to r which is already coloured $t+2$.
Case 2: $u = v_{t+1}$. The only neighbour of v_{t+1} that could be coloured $t+1$ is r, since all its neighbours in $M_{t+1,t+1}^r$ have degree t and are adjacent to v_{t+1} which is coloured $t+2$. □

Claim 2. For $1 \leq i \leq t$, $c(v_i) = i$.

Proof. By Claim 1, $\{c(r), c(v_{t+1})\} \subseteq \{t+1, t+2\}$. Since $d_H(r) = t+1$, r has one neighbour coloured i, for each $1 \leq i \leq t$. A neighbour of r which is coloured t must have degree at least t. So, by Observation (ii), it must be v_t. And so on, by induction, we show that $c(v_i) = i$, for $1 \leq i \leq t$. □

We now prove that c induces a proper 3-edge-colouring of G.

Consider vertex v_i, $4 \leq i \leq t$. By Claim 2, it is coloured i, and by Observation(ii) it has degree equal to $i+1$. Since it is adjacent to r, which by Claim 1 has a colour greater than i, there qre only $i-1$ vertices remaining for the other $i-1$ colours. So, v_i has exactly one neighbour coloured j, for each $1 \leq j \leq i-1$. Hence the three edges incident to v_i in G, which are adjacent to v_i in I, have different colours. □

As a corollary to Theorem 1, it is NP-hard to compute the Grundy number of a bipartite graph.

Corollary 1. *Given a bipartite graph G and an integer k, it is NP-complete to decide if $\Gamma(G) \geq k$.*

Theorem 1 also implies the following.

Corollary 2. *Let $k \geq 0$ be a fixed integer. It is NP-complete to decide if a bipartite graph G satisfies $\Gamma(G) \geq \Delta(G) + 1 - k$.*

Proof. We present a reduction from the problem of deciding if $\Gamma(G) = \Delta(G)+1$ for a bipartite graph G, that we just proved to be NP-complete. Let G be a graph of maximum degree Δ and H the disjoint union of G and the star $S_{\Delta+k+1}$ on $\Delta+k+1$ vertices ($S_{\Delta+k+1}$ has $\Delta+k$ edges incident to a vertex). Then $\Delta(H) = \Delta+k$ and $\Gamma(G) = \Gamma(H)$ because $\Gamma(S_{\Delta+k+1}) = 2$. Hence $\Gamma(H) \geq \Delta(H)+1-k$ if and only if $\Gamma(G) = \Delta(G)+1$. □

3   Fixed Parameter Tractability

The aim of this section is to prove the following theorem which shows that DUAL OF GREEDY COLOURING is FPT.

Theorem 2. DUAL OF GREEDY COLOURING can be solved in time $O\left((2k)^{2k} \cdot |E| + 2^{2k} k^{3k+5/2}\right)$.

A *vertex cover* in a graph G is a set $C \subseteq V(G)$ such that for every $e \in E(G)$, at least one of the endvertices of e is in C. A vertex cover is said to be *minimal* if there is no vertex cover $C' \subset C$. The *complement* of a graph G is the graph $\overline{G}$ with the same vertex set and such that $uv \in E(\overline{G})$ if and only if $uv \notin E(G)$.

The proof of Theorem 2 may be outlined as follows. We first show that a graph $G = (V, E)$ has Grundy number at least $|V| - k$ if and only if its complement has a vertex cover with certain properties and in particular size at most $2k$. We then give an algorithm in $O\left(k^{2k} \cdot |E| + k^{3k+5/2}\right)$ that decides if a given minimal vertex cover of $\overline{G}$ is contained in a vertex cover having such properties.

There are at most 2^{2k} minimal vertex covers of size at most $2k$ and we can enumerate them in time $O\left(2^{2k} \cdot |V|\right)$ using a search tree (see for example Section 8.2 of [12]). Hence applying the above-mentioned algorithm for each minimal vertex cover yields an algorithm in time $O\left((2k)^{2k} \cdot |E| + 2^{2k} k^{3k+5/2}\right)$ for DUAL OF GREEDY COLOURING.

Lemma 2. Let $G = (V, E)$ be a graph and $k \geq 0$ an integer. Then, $\Gamma(G) \geq |V| - k$ if and only if there is a vertex cover C of $\overline{G}$ such that $G[C]$ admits a greedy colouring $(C_1, C_2, \ldots, C_{k'})$ with the following properties:

P1: $|C| - k \leq k' \leq k$;
P2: $|C_i| \geq 2$, for every $1 \leq i \leq k'$;
P3: For each $v \in V \backslash C$ and for every $1 \leq i \leq k'$, there is $u \in C_i$ such that $uv \in E$.

Proof. ($\Rightarrow$) Assume that $\Gamma(G) \geq |V| - k$ and consider a greedy $\Gamma(G)$-colouring c. Let C be the set of vertices that are in a colour class with more than one vertex. Then $V \backslash C$ is the set of vertices that are alone in their colour classes.

Let u and v be two vertices in $V \backslash C$. Without loss of generality we may assume that $c(u) > c(v)$. Then, as c is a greedy colouring, u has a neighbour coloured $c(v)$, which must be v. So $uv \in E$. Hence $V \backslash C$ is clique in G and so an independent set in $\overline{G}$. Consequently, C is a vertex cover in $\overline{G}$.

Let c' be the greedy colouring $(C_1, C_2, \ldots, C_{k'})$ of $G[C]$ induced by c. Then, $|C_i| \geq 2$, for every $1 \leq i \leq k'$, so $|C| \geq 2k'$. By definition of C, we have $\Gamma(G) = |V| - |C| + k' \leq |V| - k'$. Since $\Gamma(G) \geq |V| - k$, we obtain $k \geq k' \geq |C| - k$.

Finally, let $v \in V \backslash C$ and $1 \leq i \leq k'$. If the colour of the vertices of C_i in c is smaller than $c(v)$, then v is adjacent to at least one vertex of C_i because c is greedy. If not then every vertex of C_i is adjacent to v because it is the sole vertex coloured $c(v)$. In both cases, v is adjacent to at least one vertex in C_i, so c' also has Property P3.

($\Leftarrow$) Let C be a vertex cover of $\overline{G}$ such that there is a greedy colouring $c' = \left(C_1, C_2, \ldots, C_{k'}\right)$ of $G[C]$ having Properties P1, P2 and P3. One can extend c' to the entire graph G by assigning $|V| - |C|$ distinct colours to the vertices of $V \backslash C$. As a consequence of P3 and the fact that $V \backslash C$ is an independent set in $\overline{G}$ and therefore a clique in G, the obtained colouring is greedy. Because of P1, it uses $k' + |V| - |C| \geq (|C| - k) + |V| - |C| = |V| - k$ colours.   □

Let C be a vertex cover of $\overline{G}$. A greedy colouring $(C_1, C_2, \ldots, C_{k'})$ of $G[C]$ having the Properties P1, P2 and P3 of Lemma 2 is said to be *good*. C is *suitable* if $G[C]$ has a good greedy colouring. Observe that Property P1 implies that a suitable vertex cover has cardinality at most $2k$.

Proposition 1. *Let C be a suitable vertex cover of $\overline{G}$, $(C_1, C_2, \ldots, C_{k'})$ a good greedy colouring of $G[C]$ and $C_{min} \subseteq C$ a minimal vertex cover. Then for all $1 \leq i \leq k'$, $|C_i \setminus C_{min}| \leq 1$.*

Proof. Each colour class C_i, $1 \leq i \leq k'$, is an independent set of size at least 2 in G. So it is a clique of size at least 2 in $\overline{G}$. Since C_{min} is a vertex cover in $\overline{G}$, $|C_i \cap C_{min}| \geq |C_i| - 1$, so $|C_i \setminus C_{min}| \leq 1$.   □

Lemma 3. *Let k be an integer, $G = (V, E)$ a graph and C_{min} a minimal vertex cover of $\overline{G}$ of size at most $2k$. It can be determined in time $O\left(k^{2k} \cdot |E| + k^{3k+5/2}\right)$ if C_{min} is contained in a suitable vertex cover C.*

Proof. In order to determine if C_{min} is contained in a suitable vertex cover, we enumerate all possible proper colourings of $G[C_{min}]$ with k' colours, $|C_{min}| - k \leq k' \leq k$. For each of them, we then check in time $O(|E| + k^{k+5/2})$ if it can be extended into a good greedy colouring of a suitable vertex cover. There are at most $k^{|C_{min}|} \leq k^{2k}$ proper colourings of C_{min} with at most k colours and they can be enumerated in time $O(k^{2k})$. Hence the running time of our algorithm is $O\left(k^{2k} \cdot |E| + k^{3k+5/2}\right)$.

Let us now detail an algorithm that, given a proper colouring $c = (C_1, C_2, \ldots, C_{k'})$ of $G[C_{min}]$, decides if it can be extended into a good greedy colouring of a suitable vertex cover in time $O(|E| + k^{k+5/2})$. By Proposition 1, for such an extension at most one vertex of $V \setminus C_{min}$ is added in each colour class.

If c is a good colouring of C_{min} then we are done. So we may assume that it is not. We say that a colour class S_i is *defective* with respect to a colouring $s = (S_1, \ldots, S_l)$ of $S \subseteq V$ if at least one of the following holds:

(i): $|S_i| < 2$;
(ii): For some $j > i$, there is $v \in S_j$ with no neighbour in S_i;
(iii): There is $v \in V \setminus S$ such that v has no neighbours in S_i.

Let S_i be a defective colour class with respect to s. An *i-candidate* with respect to s is a vertex $v \in V \setminus \bigcup_{j=1}^{l} S_j$ such that $S_i \cup \{v\}$ is an independent set which is not defective with respect to the colouring $(S_1, \ldots, S_{i-1}, S_i \cup \{v\}, S_{i+1}, \ldots, S_l)$. We denote by $X_s(j)$ the set of j-candidates with respect to s and D_s the set of defective colour classes with respect to s. If $|X_s(j)| \geq k$, we say that j is a colour class of *type 1*. Otherwise, we say that it is of *type 2*. It is easy to see that the set of defective colour classes of c and their candidates can be computed in time $O(|E|)$.

Clearly, if c can be extended into a good colouring, it means that we can place candidates to some of its colour classes and obtain a colouring without defective colour classes. Because of Proposition 1, we are only allowed to place at most one vertex in each colour class. As we will show later, the only defective colour classes that may not receive a candidate in the extension of c to a good colouring are those of type 2.

Claim 3. *Let s be a k'-colouring of C_{min}, i one of its defective colour classes and v an i-candidate. Let s' be the extension of s where we place v in colour class i. Then, for every colour class $j \neq i$, $X_{s'}(j) = X_s(j)\backslash\{v\}$.*

Proof. First, assume that j is not defective in s. If it does become defective in s' it is due to condition (ii), since it cannot satisfy (i) or (iii) after the insertion of v in colour i. But then, since j does not satisfy (ii) in s, v is the only vertex in i that may have no neighbours coloured j, which implies that j satisfies condition (iii) in s, a contradiction.

Now assume that j is defective in s. We first prove that $X_s(j)\backslash\{v\}\subseteq X_{s'}(j)$. In this case, again we have that (i) and (iii) remain unchanged after the insertion of v, in the sense that if a vertex different from v is a candidate for j in s because of one of these conditions, the same will happen in s'. Regarding condition (iii), the only thing that may change in s' is that v is now one of the vertices with no neighbours in j. Since v is not in C_{min}, it is adjacent to every j-candidate. Then, every vertex distinct from v that was a candidate for j in s remains a candidate for j in s'.

The converse, that is that every vertex in $X_{s'}(j)$ is also in $X_s(j)\backslash\{v\}$, is trivial. □

In particular, Claim 3 shows that the insertion of a candidate in a defective colour classes does not create a new defective colour class.

Claim 4. *Let $s = (S_1, \ldots, S_{k'})$ be a k'-colouring of C_{min} and assume it can be extended into a good colouring $s' = (S'_1, \ldots, S'_{k'})$ of a suitable vertex cover. If j is a defective colour class in s and $|S'_j\backslash S_j| = 0$, then j is of type 2 in s. Moreover, j is defective only because of (iii).*

Proof. Let j be such that $S'_j = S_j$. If j satisfies (i) then there is only one vertex coloured j in s', and thus s' cannot be a good colouring, a contradiction. If j satisfies (ii) then there is a vertex v coloured $j' > j$ with no neighbours in S_j. Since no vertices were added to S_j in s', vertex v also has no neighbours coloured j in s', a contradiction. Hence j can be defective only because of condition (iii). But then, if $|X_s(j)| \geq k$, as we add at most one vertex to each colour class when extending s to s', at least one vertex in $X_s(j)$ is not in any colour class of s'. Hence, there is a vertex that is not coloured in s' and has no neighbours with colour j, which implies that j is a defective colour class in s', a contradiction. □

In order to determine if c can be properly extended, in a first step we consider all possible extensions of the type 2 colour classes. For such a colour class of type 2, we can choose to add to it either one of its candidates or none. By Claim 4 this later case is possible if the colour class satisfies only (iii). There are k_1 defective colour classes of type 2, where $k_1 \leq k$. Moreover each of these colour classes has k_2 candidates with $k_2 < k$, and so has $k_2 + 1$ possible ways of extension: adding one of the k_2 candidates or adding none. Hence, we can enumerate all the possible extensions of the type 2 colour classes in time $O(k_1^{k_2+1}) = O(k^k)$. In the second step, for each possible extension, we check if the type 1 colour classes could be extended in order to obtain a good greedy colouring.

Let c' be one possible extension of c as considered in the last paragraph. If a colour class S_i of type 2 has not been extended, it may still be defective. If it is defective because it satisfies (i) or (ii), then it will remain defective after the second step in which we add some candidate to colour classes of type 1. Hence, we can stop, it will never lead to a good colouring. If it is defective because it satisfies (iii) (and only (iii)) then all the vertices $v \in V \setminus \bigcup_{j=1}^{l} S_j$ such that v has no neighbours in S_i must be placed into some type 1 colour class in any extension to a good colouring. In particular, they need to be candidates of at least one colour class of type 1. We call such vertices v *necessary candidates*.

Let $D_1 \subseteq D_{c'}$ be the set of defective colour classes in c' that are of type 1 in c. Also let N be the set of necessary candidates in c' and C the vertex cover given by the vertices coloured in c'. Remember that a suitable vertex cover C must satisfy $|C| - k \leq k'$, and so if $|C| + |D_1| - k > k'$ there is no way of properly extending c', since by Claim 4 we need to place one candidate in each of the $|D_1|$ defective colour classes of type 1. The number of colour classes in c is at most k, and after a candidate is placed in a defective colour class, the colour class is no longer defective. So, since the type 1 colour classes have at least k candidates in c and because of Claim 3, there are at least $|D_{c'}|$ candidates for each of the $|D_1|$ defective colour classes of type 1 in c. As a consequence, there are enough candidates to place in each colour class in D_1. But we also have to ensure that every necessary candidate is placed in a defective colour class. This is equivalent to finding a matching in the bipartite graph H with vertex set $V(H) = D_1 \cup N$ and with edge set $E(H) = \{(i \in D_1, v \in N) \mid$ vertex v is an i-candidate$\}$ such that every vertex in N is saturated. This can be done in time $O\left((|V(H)| + |E(H)|)\sqrt{|V(H)|}\right) = O(k^{5/2})$ by the algorithm of Hopcroft and Karp [9].

If such a matching does not exist, then we cannot properly extend c' by adding candidates to the vertices in D_1, and so we may reject c'. If such a matching exists, since each type 1 colour class has more than $|D_1|$ candidates, we can greedily extend c' to a good colouring.

Hence one can check in time $O(|E| + k^{k+5/2})$ if a proper colouring of $G[C_{min}]$ can be extended into a good greedy colouring of G. □

We are now able to prove Theorem 2.

Proof (of Theorem 2).

Let G be an instance of the problem. To answer the question, we enumerate all minimal vertex covers of $\overline{G}$, and check, for each one, if it is contained in a suitable vertex cover. To enumerate all minimal vertex covers takes time $O(2^{2k})$. For each of these at most 2^{2k} minimal vertex covers, we check if it is contained in a suitable vertex cover. By Lemma 3, it can be done in $O\left(k^{2k} \cdot |E| + k^{3k+5/2}\right)$. Hence the total running time is $O\left((2k)^{2k} \cdot |E| + 2^{2k}k^{3k+5/2}\right)$.   □

References

1. Beutelspacher, A., Hering, P.-R.: Minimal graphs for which the chromatic number equals the maximal degree. Ars Combin. 18, 201–216 (1984)
2. Borodin, O.V., Kostochka, A.V.: On an upper bound of a graph's chromatic number, depending on the graph's degree and density. J. Combinatorial Theory Ser. B 23(2-3), 247–250 (1977)
3. Brooks, R.L.: On colouring the nodes of a network. Proc. Cambridge Phil. Soc. 37, 194–197 (1941)
4. Downey, R.G., Fellows, M.R.: Parameterized Complexity. Monographs in Computer Science, 1st edn. Springer, New York (1999)
5. Emden-Weinert, T., Hougardy, S., Kreuter, B.: Uniquely colourable graphs and the hardness of colouring graphs of large girth. Combin. Probab. Comput. 7(4), 375–386 (1998)
6. Farzad, B., Molloy, M., Reed, B.: $(\Delta - k)$-critical graphs. J. Combinatorial Theory Ser. B 93(2), 173–185 (2005)
7. Garey, M.R., Johnson, D.S., Stockmeyer, L.J.: Some simplified NP-complete graph problems. Theoret. Comput. Sci. 1, 237–267 (1976)
8. Holyer, I.: The NP-completeness of edge-coloring. SIAM J. Computing 2, 225–231 (1981)
9. Hopcroft, J.E., Karp, R.M.: An $O(n^{5/2})$ algorithm for maximum matchings in bipartite graphs. SIAM J. Comput. 2(4), 225–231 (1973)
10. Jensen, T.R., Toft, B.: Graph coloring problems. Wiley-Interscience Series in Discrete Mathematics and Optimization. John Wiley & Sons, Inc., New-York (1995)
11. Molloy, M., Reed, B.: Colouring graphs when the number of colours is nearly the maximum degree. In: Proceedings of the Thirty-Third Annual ACM Symposium on Theory of Computing, pp. 462–470 (electronic). ACM, New York (2001)
12. Niedermeier, R.: Invitation to Fixed-Parameter Algorithms. Oxford University Press, Oxford (2006)
13. Reed, B.: A strengthening of Brooks' theorem. J. Combinatorial Theory Ser. B 76(2), 136–149 (1999)
14. Zaker, M.: Results on the Grundy chromatic number of graphs. Discrete Math. 306(23), 3166–3173 (2006)

Exponential Time Complexity of
Weighted Counting of Independent Sets

Christian Hoffmann[*]

Donghua University Shanghai, China
`christian.hoffmann2010@googlemail.com`

Abstract. We consider weighted counting of independent sets using a
rational weight x: Given a graph with n vertices, count its independent
sets such that each set of size k contributes x^k. This is equivalent to
computation of the partition function of the lattice gas with hard-core
self-repulsion and hard-core pair interaction. We show the following conditional lower bounds: If counting the satisfying assignments of a 3-CNF
formula in n variables (#3SAT) needs time $2^{\Omega(n)}$ (i.e. there is a $c > 0$
such that no algorithm can solve #3SAT in time 2^{cn}), counting the independent sets of size $n/3$ of an n-vertex graph needs time $2^{\Omega(n)}$ and
weighted counting of independent sets needs time $2^{\Omega(n/\log^3 n)}$ for all rational weights $x \neq 0$.

We have two technical ingredients: The first is a reduction from 3SAT
to independent sets that preserves the number of solutions and increases
the instance size only by a constant factor. Second, we devise a combination of vertex cloning and path addition. This graph transformation
allows us to adapt a recent technique by Dell, Husfeldt, and Wahlén
which enables interpolation by a family of reductions, each of which increases the instance size only polylogarithmically.

1   Introduction

Finding independent sets with respect to certain restrictions is a fundamental
problem in theoretical computer science. Perhaps the most studied version is the
maximum independent set problem: Given a graph, find an independent set[1] of
maximum size. This problem is closely related to the clique and vertex cover
problems. The decision versions of these are among the 21 problems considered
by Karp in 1972 [18], and they are used as examples in virtually every exposition
of the theory of NP-completeness [10, Section 3.1], [21, Section 9.3], [4, Section
34.5]. Exact algorithms for the independent set problem have been studied since
the 70s of the last century [27, 17, 22] and there is still active research [9, 19, 3].

Besides finding a maximum independent set, algorithms that count the number of independent sets have also been developed [6]. If the counting process

---

[*] This work has been done while the author was a research assistant at Saarland
University, Germany.

[1] A subset A of the vertices of a graph is *independent* iff no two vertices in A are
joined by an edge of the graph.

V. Raman and S. Saurabh (Eds.): IPEC 2010, LNCS 6478, pp. 180–191, 2010.
© Springer-Verlag Berlin Heidelberg 2010

is done in a weighted manner (as in (1) below), we arrive at a problem from statistical physics: computation of the partition function of the lattice gas with hard-core self-repulsion and hard-core pair interaction [24]. In graph theoretic language, this is the following problem: Given a graph $G = (V, E)$ and a weight $x \in \mathbb{Q}$, compute

$$I(G; x) = \sum_{\substack{A \subseteq V \\ A \text{ independent set}}} x^{|A|}. \tag{1}$$

$I(G; x)$ is also known as the independent set polynomial of G [12, 11]. Luby and Vigoda mention that "equivalent models arise in the Operations Research community when considering properties of stochastic loss systems which model communication networks" [20]. Evaluation of $I(G; x)$ has received a considerable amount of attention, mainly concerning approximability if x belongs to a certain range depending on Δ, the maximum degree of G [20, 8, 29, 30].

In this paper, we give evidence that exact evaluation of $I(G; x)$ needs almost exponential time (Theorem 2). We do this by reductions from the following problem:

Name: #d-SAT
Input: Boolean formula φ in d-CNF with m clauses in n variables
Output: Number of satisfying assignments for φ

All lower bounds of this work are based on the following assumption, which is a counting version of the exponential time hypothesis (ETH) [15, 7]:

#ETH (Dell, Husfeldt, Wahlén 2010). *There is a constant $c > 0$ such that no deterministic algorithm can compute #3-SAT in time $\exp(c \cdot n)$.*

Our first result concerns the following problem:

Name: #$\frac{1}{3}$-IS
Input: Graph with n vertices
Output: Number of independent sets of size exactly $n/3$ in G

Theorem 1. *#$\frac{1}{3}$-IS requires time $\exp(\Omega(n))$ unless #ETH fails.*

Theorem 1 gives an important insight for the development of exact algorithms counting independent sets: Let us consider algorithms that count independent sets of a particular kind. (For example: algorithms that count the independent sets of maximum size. Another example: algorithms that simply count all independent sets). Using only slight modifications, many of the actual algorithms that have been suggested for these problems can be turned into algorithms that solve #$\frac{1}{3}$-IS. Theorem 1 tells us that there is some $c > 1$ such that every such algorithm has worst-case running time $\geq c^n$—unless #ETH fails. In other words: There is a universal c^n barrier for counting independent sets that can only be broken 1) if substantial progress on counting SAT is made or 2) by approaches that are custom-tailored to the actual version of the independent set problem such that they can not be used to solve #$\frac{1}{3}$-IS.

The proof of Theorem 1 is *different* from the standard constructions that reduce the decision version of 3SAT to the decision version of maximum independent set [18], [21, Theorem 9.4]. This is due to the fact that these constructions do not preserve the number of solutions. Furthermore, the arguments for counting problems that have been applied in #P-hardness proofs also fail in our context, as they increase the instance size by more than a constant factor[2] and thus do not preserve subexponential time.

Theorem 1 is proved in the full version of this paper [14] using a, with hindsight, simple reduction from #3-SAT. But for the reasons given in the last paragraph, it is important to work this out precisely. In this way, we close a non-trivial gap to a result that is very important as it concerns a fundamental problem.

The main result of our paper is based on Theorem 1:

Theorem 2. *Let $x \in \mathbb{Q}$, $x \neq 0$. On input $G = (V, E)$, $n = |V|$, evaluating the independent set polynomial at x, i.e. computing*

$$\sum_{\substack{A \subseteq V \\ A \text{ independent set}}} x^{|A|},$$

requires time $\exp(\Omega(n/\log^3 n))$ unless #ETH fails.

This shows that we can not expect that the partition function of the lattice gas with hard-core self-repulsion and hard-core pair interaction can be computed much faster than in exponential time.

Let us state an important consequence of Theorem 2, the case $x = 1$.

Corollary 3. *Every algorithm that, given a graph G with n vertices, counts the independent sets of G has worst-case running time $\exp(\Omega(n/\log^3 n))$ unless #ETH fails.*

Referring to the discussion after Theorem 1, this gives a conditional lower bound for our second example (i.e. counting all independent sets of a graph). The bound of Corollary 3 is not as strong as the bound of Theorem 1 but holds for *every* algorithm, not only for algorithms that can be modified to solve #$\frac{1}{3}$-IS.

Techniques and Relation to Previous Work

Theorem 1 is proved in two steps: First, we reduce from #3-SAT to #X3SAT. #X3SAT is a version of SAT that counts the assignments that satisfy *exactly* one literal per clause. From #X3SAT we can reduce to independent sets using a modified version of a standard reduction from SAT to independent sets [21, Theorem 9.4]. We also use the fact that the exponential time hypothesis with number of variables as a parameter is equivalent to the hypothesis with number of clauses as parameter. Impagliazzo, Paturi, and Zane proved this for the decision version [15]. We use the following version for counting problems:

---

[2] For instance, Valiant's step from perfect matchings to prime implicants [28] includes transforming a $\Theta(n)$ vertex graph into a $\Theta(n^2)$ vertex graph.

Theorem 4 ([7, Theorem 1]). *For all $d \geq 3$, #ETH holds if and only if #d-SAT requires time $exp(\Omega(m))$.*

Our main result (Theorem 2) is inspired by recent work of Dell, Husfeldt, and Wahlén on the Tutte polynomial [7, Theorem 3(ii)]. These authors use Sokal's formula for the Tutte polynomial of generalized Theta graphs [26]. In Section 2, we devise and analyze S-clones, a new graph transformation that can be used with the independent set polynomial in a similar way as generalized Theta graphs with the Tutte polynomial. S-clones are a combination of vertex cloning (used under this name for the interlace polynomial [2], but generally used in different situations for a long time [28, Theorem 1, Reduction 3.], [16], [23, Lemma A.3]) and addition of paths. Having introduced S-clones, we are able to transfer the construction of Dell et al. quite directly to the independent set polynomial. The technical details are more involved than in the previous work on the Tutte polynomial, but the general idea is the same: Use the graph transformation (S-clones) to evaluate the graph polynomial (independent set polynomial) at different points, and use the result for interpolation. An important property of the construction is that the graph transformation increases the size of the graph only polylogarithmically. More details on this can be found at the beginning of Section 3.

Before we start with the detailed exposition, let us mention that the reductions we devise for the independent set polynomial can be used with the interlace polynomial [1, 5] as well [13].

2 S-Clones and the Independent Set Polynomial

In this section, we analyze the effect of the following graph transformation on the independent set polynomial.

Definition 5. *Let S be a finite multiset of nonnegative integers and $G = (V, E)$ be a graph. We define the S-clone $G_S = (V_S, E_S)$ of G as follows:*

- *For every vertex $a \in V$, there are $|S|$ vertices $a(|S|) := \{a_1, \ldots, a_{|S|}\}$ in $V_{\mathcal{E}}$*
- *For every edge $uv \in E$, there are edges in E_S that connect every edge in $u(|S|)$ to every edge in $v(|S|)$.*
- *Let $S = \{s_1, \ldots, s_\ell\}$. For every vertex $a \in V$, we add a path of length s_i to a_i, the ith clone of a. Formally: For every i, $1 \leq i \leq |S|$, and every $a \in V$ there are vertices $a_{i,1}, \ldots, a_{i,s_i}$ in V_S and edges $a_i a_{i,1}, a_{i,1} a_{i,2}, \ldots, a_{i,s_i-1} a_{i,s_i}$ in E_S.*
- *There are no other vertices and no other edges in G_S but the ones defined by the preceding conditions.*

The effect of S-cloning on a single vertex is illustrated in Figure 1. The purpose of S-clones is that $I(G_S; x)$ can be expressed in terms of $I(G; x(S))$, where $x(S)$ is some number derived from x and S. For technical reasons, we restrict ourselves to x that fulfill the following condition:

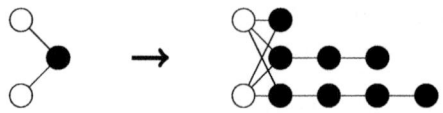

Fig. 1. Effect of a $\{0, 2, 3\}$-clone on a single vertex

Definition 6. *Let $x \in \mathbb{R}$. We say that x is* nondegenerate for path reduction *if $x > -\frac{1}{4}$ and $x \neq 0$. Otherwise, we say that x is* degenerate for path reduction.

Definition 7. *Let $x \in \mathbb{R}$ be nondegenerate for path reduction. Then we define λ_1 and λ_2 to be the two roots of $\lambda^2 - \lambda - x$, i.e.*

$$\lambda_{1,2} = \frac{1}{2} \pm \sqrt{\frac{1}{4} + x}. \tag{2}$$

The following condition ensures that (3) is well-defined (cf. (5)).

Definition 8. *Let $x \in \mathbb{R}$ be nondegenerate for path reduction. We say that a set S of nonnegative integers is* compatible with x *if $\lambda_1^{s+2} \neq \lambda_2^{s+2}$ for all $s \in S$.*

Now we can state the effect of S-cloning on the independent set polynomial:

Theorem 9. *Let $G = (V, E)$ be a graph, x be nondegenerate for path reduction, and S be a finite multiset of nonnegative integers that is compatible with x. Then we have*

$$I(G_S; x) = (\prod_{s \in S} C_s)^{|V|} I(G; x(S)),$$

where

$$x(S) + 1 = \prod_{s \in S} \left(1 + \frac{B_s}{C_s}\right) \quad \text{with} \tag{3}$$

$$B_k = \frac{x}{\lambda_2 - \lambda_1} \cdot \left(-\lambda_1^{k+1} + \lambda_2^{k+1}\right), \tag{4}$$

$$C_k = \frac{1}{\lambda_2 - \lambda_1} \cdot \left(-\lambda_1^{k+2} + \lambda_2^{k+2}\right), \tag{5}$$

and λ_1, λ_2 as in Definition 7.

The rest of this section is devoted to the proof of Theorem 9.

2.1   Notation

We use a *multivariate* version of the independent set polynomial. This means that every vertex has its own variable x. Formally, we define a *vertex-indexed variable* $\mathbf{x}$ to be a set of of independent variables x_a such that, if $G = (V, E)$ is a graph, $\mathbf{x}$ contains $\{x_a \mid a \in V\}$. If $\mathbf{x}$ is a vertex-indexed variable and A is a subset of the vertices of G, we define $x_A := \prod_{a \in A} x_a$. The *multivariate independent set polynomial* [24] is defined as

$$I(G; \mathbf{x}) = \sum_{\substack{A \subseteq V \\ A \text{ independent}}} x_A. \tag{6}$$

We have $I(G; x) = I(G; \mathbf{x})[x_a := x \mid a \in V]$, i.e. the single-variable independent set polynomial is obtained from the multivariate version by substituting every vertex-indexed variable x_a by one and the same ordinary variable x.

We will use the following operation on graphs: Given a graph G and a vertex b of G, $G - b$ denotes the graph that is obtained from G by removing b and all edges incident to b.

2.2   Proof of Theorem 9

Let us first analyze the effect of a single leaf on the independent set polynomial.

Lemma 10. *Let $G = (V, E)$ be a graph and $a \neq b$ be two vertices such that a is the only neighbor of b. Then, as a polynomial equation, we have*

$$I(G, x_a, x_b) = (1 + x_b) I(G - b, x_a / (1 + x_b)), \tag{7}$$

where $I(G, y, z)$ denotes $I(G; \mathbf{x})$ with $x_a = y$ and $x_b = z$ and $I(G - b, z)$ denotes $I(G - b; \mathbf{y})$ with $y_a = z$ and $y_v = x_v$ for all $v \in V \setminus \{a, b\}$.

Proof. Let $V' = V \setminus \{a, b\}$ and $i(A) = 1$ if $A \subseteq V$ is a independent set in G and $i(A) = 0$ otherwise. We have

$$I(G, x_a, x_b) = \sum_{A \subseteq V'} x_A \big(i(A) + x_a i(A \cup \{a\}) + x_b i(A \cup \{b\})\big)$$

$$= \sum_{A \subseteq V'} x_A \big(i(A) + x_a i(A \cup \{a\}) + x_b i(A)\big)$$

$$= \sum_{A \subseteq V'} x_A \big(i(A)(1 + x_b) + x_a i(A \cup \{a\})\big),$$

from which the claim follows.   □

In other words, Lemma 10 states that a single leaf b and its neighbor a can be "contracted" by incorporating the weight of b into a. In a very similar way, two vertices with the same set of neighbors can be contracted:

Lemma 11. *Let $G = (V, E)$ be a graph and $a, b \in V$ two vertices that have the same set of neighbors. Then*

$$I(G; \mathbf{x}) = I(G - b; \mathbf{y}), \tag{8}$$

where $y_v = x_v$ for all $v \in V \setminus \{a, b\}$ and $y_a + 1 = (1 + x_a)(1 + x_b)$.

Proof. Similar to the proof of Lemma 10.   □

Let us now use Lemma 10 to derive a formula that describes how a path, attached to one vertex, influences the independent set polynomial. Basically, we derive an explicit formula from recursive application of (7) (cf. the proof of the formula for the interlace polynomial of a path by Arratia, Bollobás, and Sorkin [1, Proposition 14]).

Theorem 12. *Let $G = (V, E)$ be a graph and $a_0 \in V$ a vertex. For a positive integer k, let $\tau_k G$ denote the graph G with a path of length k added at a_0, i.e. $\tau_k G = (V \cup \{a_1, \ldots, a_k\}, E \cup \{a_0 a_1, a_1 a_2, \ldots, a_{k-1} a_k\})$ with $a_1, \ldots, a_k$ being new vertices. Let $\mathbf{x}$ be a vertex labeling of $\tau_k G$ with variables. Then the following polynomial equation holds:*

$$I(\tau_k G; \mathbf{x}) = C_k I(G; \mathbf{y}), \tag{9}$$

where $y_v = x_v$ for all $v \in V \setminus \{a_0, \ldots, a_k\}$, $y_{a_0} = B_k / C_k$, and $B_0 = x_k$, $C_0 = 1$ and, for $0 \leq i < k$,

$$\begin{pmatrix} B_{i+1} \\ C_{i+1} \end{pmatrix} = M(x_{k-i-1}) \begin{pmatrix} B_i \\ C_i \end{pmatrix}, \tag{10}$$

$$M(x) = \begin{pmatrix} 0 & x \\ 1 & 1 \end{pmatrix}. \tag{11}$$

Let now $x \in \mathbb{R}$ be nondegenerate for path reduction, λ_1, λ_2 be as in Definition 7, $x_a = x$ for all $a \in \{a_0, \ldots, a_k\}$ and $\lambda_1^{k+2} \neq \lambda_2^{k+2}$. Then (9) holds with B_k and C_k defined as in (4) and (5).

Proof. Let us write $I(\tau_k G; x_0, \ldots, x_k)$ for $I(\tau_k G; \mathbf{x})$ where $x_{a_j} = x_j$, $0 \leq j \leq k$. Let us argue that we defined the B_i, C_i such that, for $0 \leq i \leq k$,

$$I(\tau_k G; x_0, \ldots, x_k) = C_i I\left(\tau_{k-i} G; x_0, \ldots, x_{k-i-1}, \frac{B_i}{C_i}\right). \tag{12}$$

This is trivial for $i = 0$. As we have

$$1 + \frac{B_i}{C_i}\left(1 + x_{k-i-1} u^2\right) = \frac{C_i + B_i\left(1 + x_{k-i-1} u^2\right)}{C_i} = \frac{C_{i+1}}{C_i},$$

we see that (12) holds for all $1 \leq i \leq k$: Use Lemma 10 in the following inductive step:

$$\begin{aligned} I(\tau_k G; x_0, \ldots, x_k) &= C_i I(\tau_{k-i} G; x_0, \ldots, x_{k-i-1}, \frac{B_i}{C_i}) \\ &= C_i \frac{C_{i+1}}{C_i} I\left(\tau_{k-i-1} G; x_0, \ldots, x_{k-i-2}, \frac{x_{k-i-1}}{\frac{C_{i+1}}{C_i}}\right) \\ &= C_{i+1} I\left(\tau_{k-i-1} G; x_0, \ldots, x_{k-i-2}, \frac{B_{i+1}}{C_{i+1}}\right). \end{aligned}$$

Thus, (9) holds as a polynomial equality.

Let us now consider x as a real number, $x > -1/4$. Matrix $M(x)$ in (11) can be diagonalized as $M(x) = SDS^{-1}$ with

$$S = \begin{pmatrix} x & x \\ \lambda_1 & \lambda_2 \end{pmatrix}, \quad D = \begin{pmatrix} \lambda_1 & 0 \\ 0 & \lambda_2 \end{pmatrix}, \quad S^{-1} = \frac{1}{x(\lambda_2 - \lambda_1)} \begin{pmatrix} \lambda_2 & -x \\ -\lambda_1 & x \end{pmatrix},$$

$\lambda_{1,2}$ as in (2). Now we substitute variable x_v by real number x for all $v \in \{a_0, \ldots, a_k\}$ and $M(x)$ by SDS^{-1}. This yields the statement of the theorem.

$\square$

Now we see that Theorem 9 can be proved by repeated application of Lemma 11 and Theorem 12.

3   Interpolation via S-Clones

In this section, we give a reduction from evaluation of the independent set polynomial at a fixed point $x \in \mathbb{Q} \setminus \{0\}$ to computation of the coefficients of the independent set polynomial. Thus, given a graph G with n vertices, we would like to interpolate $I(G; X)$, where X is a variable. The degree of this polynomial is at most n, thus it is sufficient to know $I(G; x_i)$ for $n + 1$ different values x_i. Our approach is to modify G in $n + 1$ different ways to obtain $n + 1$ different graphs $G_0, \ldots, G_n$. Then we evaluate $I(G_0; x)$, $I(G_1; x)$, $\ldots$, $I(G_n; x)$. We will prove that $I(G_i; x) = p_i I(G; x_i)$ for $n + 1$ easy to compute x_i and p_i, where $x_i \neq x_j$ for all $i \neq j$. This will enable us to interpolate $I(G; X)$.

If the modified graphs G_i are c times larger than G, we lose a factor of c in the reduction, i.e. a 2^n running time lower bound for evaluating the graph polynomial at x implies only a $2^{n/c}$ lower bound for evaluation at the interpolated points. Thus, we can not afford simple cloning (i.e. S-cloning with S being $\{0,0\}$, $\{0,0,0\}$, $\{0,0,0,0\}$, $\ldots$): To get enough points for interpolation, we would have to evaluate the graph polynomials on graphs of sizes $2n, 3n, \ldots, n^2$. To overcome this problem, we transfer a technique of Dell, Husfeldt, and Wahlén [7], which they developed for the Tutte polynomial to establish a similar reduction: We clone every vertex $O(\log n)$ times and use n different ways to add paths of different (but at most $O(\log^2 n)$) length at the different clones. Eventually, this will lead to the following result:

Theorem 13. *Let $x_0 \in \mathbb{Q}$ such that x_0 is nondegenerate for path reduction and the independent set polynomial I of every n-vertex graph G can be evaluated at x_0 in time $2^{o(n/\log^3 n)}$.*

Then, for every n-vertex graph G, the X-coefficients of the independent set polynomial $I(G; X)$ can be computed in time $2^{o(n)}$. In particular, the independent set polynomial $I(G; x_1)$ can be evaluated in this time for every $x_1 \in \mathbb{Q}$.

Using this theorem, our main result follows with Theorem 1. For $x > -1/4$, this is immediate; for other x, we refer to the full paper [14].

The rest of this section is devoted to the proof of Theorem 13, which is quite technical. The general idea is similar to Dell et al. [7, Lemma 4, Theorem 3(ii)].

Definition 14. *Let S be a set of numbers. Then we define $\|S\| = \sum_{s \in S} s$.*

Remark 15. *The S-clone G_S of a graph $G = (V, E)$ has $|V|(\|S\| + |S|)$ vertices.*

Lemma 16. *Assume that $x \in \mathbb{R}$ is nondegenerate for path reduction. Then there are sets $S_0, S_1, \ldots, S_n$ of positive integers, constructible in time $\mathsf{poly}(n)$, such that*

1. *$x(S_i) \neq x(S_j)$ for all $i \neq j$ and*
2. *$\|S_i\| \in O(\log^3 n)$ and $|S_i| \in O(\log n)$ for all i, $0 \leq i \leq n$.*

Proof. We use the notation from Theorem 12 and assume $\lambda_1 > \lambda_2$.

As $\left|\frac{\lambda_1}{\lambda_2}\right|^k \to \infty$ for $k \to \infty$, there is a positive integer s_0 such that

$$\left(\frac{\lambda_1}{\lambda_2}\right)^s \notin \left\{ \left(\frac{\lambda_2}{\lambda_1}\right)^2, \frac{\lambda_2(x + \lambda_2)}{\lambda_1(x + \lambda_1)} \right\} \quad \forall s \geq s_0.$$

Thus, for every i, $0 \leq i \leq n$, the following set fulfills the precondition on S and T in Lemma 17:

$$S_i = \{s_0 + \Delta(2j + b_j^{(i)}) \mid 0 \leq j \leq \lfloor \log n \rfloor\},$$

where Δ is a positive integer defined later, $\Delta \in \Theta(\log n)$, and $[b_{\lfloor \log n \rfloor}^{(i)}, \ldots, b_1^{(i)}, b_0^{(i)}]$ is the binary representation of i. Note that this construction is very similar to Dell et al. [7, Lemma 4]. It is important that the elements in these sets have distance at least Δ from each other. The sets are $\mathsf{poly}(n)$ time constructible as s_0 does not depend on n. We have $\|S_i\| \leq (1 + \log n)(s_0 + (1 + 2\log n)\Delta)$ and obviously $|S_i| \in O(\log n)$ for all i. Thus, the second statement of the lemma holds.

To prove the first statement, we use Lemma 17. Let $1 \leq i < j \leq n$ and $S = S_i \backslash S_j$, $T = S_j \backslash S_i$. Let s_1 be the smallest number in $S \cup T$ and $A_1 = (S \cup T) \backslash \{s_1\}$. For f as in Lemma 17, let us prove that $|f(A_1)| > \sum_{\substack{A \subseteq S \cup T \\ A \neq A_1}} |f(A)|$, which yields the statement of Lemma 16.

Assume without loss of generality that $s_1 \in S$. As x is nondegenerate, $C_1 := \min\{1, |\lambda_1|, |\lambda_2|, |x + \lambda_1|, |x|, |\lambda_1 - \lambda_2|\}$ is a nonzero constant. As $|S| = |T|$,

$$\begin{aligned}
D(S, T, A_1) &= \lambda_1^{|T|}(x + \lambda_1)^{|S|-1}(x + \lambda_2) - \lambda_1^{|S|-1}\lambda_2(x + \lambda_1)^{|T|} \\
&= \lambda_1^{|S|-1}(x + \lambda_1)^{|S|-1}\big(\lambda_1(x + \lambda_2) - \lambda_2(x + \lambda_1)\big) \\
&= \lambda_1^{|S|-1}(x + \lambda_1)^{|S|-1}x(\lambda_1 - \lambda_2),
\end{aligned}$$

and we have

$$|f(A_1)| \geq |\lambda_1|^{\|S \cup T\| - s_1}|\lambda_2|^{s_1}C_1^{7|S|}. \tag{13}$$

If $A = \emptyset$ or $A = S \cup T$, we have $D(A) = 0$, which implies $f(A) = 0$. For every $A \subseteq S \cup T$, $A \neq \emptyset$, $A \neq S \cup T$, $A \neq A_1$, we have $\|A\| \leq \|S \cup T\| - s_1 - \Delta$. Thus,

$$|f(A)| \leq |\lambda_1|^{\|S \cup T\| - s_1 - \Delta}|\lambda_2|^{s_1 + \Delta}C_2^{7|S|}, \tag{14}$$

where $C_2 = 2\max\{1, |\lambda_1|, |\lambda_2|, |x+\lambda_1|, |x+\lambda_2|\}$. There are less than $2^{\lfloor \log n \rfloor + 1} \leq 2n^2$ such A. Combining this with (13) and (14), it follows that we have proved the lemma if we ensure

$$\left|\frac{\lambda_1}{\lambda_2}\right|^\Delta > \left(\frac{C_2}{C_1}\right)^{7|S|} 2n^2.$$

This holds if

$$\Delta > 7\left((\log n + 1)\log\frac{C_2}{C_1} + 2\log n + 1\right)/\log\frac{\lambda_1}{|\lambda_2|}.$$

As C_1, C_2, λ_1, λ_2 do not depend on n, we can choose $\Delta \in \Theta(\log n)$.   □

Lemma 17. *Let S and T be two sets of positive integers. Let also $x \in \mathbb{R}$ be nondegenerate for path reduction and, for all $s \in S \cup T$,*

$$\left(\frac{\lambda_1}{\lambda_2}\right)^{s+2} \neq 1 \quad and \tag{15}$$

$$\left(\frac{\lambda_1}{\lambda_2}\right)^{s+1} \neq \frac{x+\lambda_2}{x+\lambda_1}, \tag{16}$$

where λ_1, λ_2 are defined as in Theorem 12. Then we have $x(S) = x(T)$ iff

$$\sum_{A \subseteq S \triangle T} f(A) = 0,$$

where

$$f(A) = \lambda_1^{\|A\|}\lambda_2^{\|(S\triangle T)\setminus A\|}(-\lambda_1)^{|A|}\lambda_2^{|(S\triangle T)\setminus A|} \cdot D(S\setminus T, T\setminus S, A),$$
$$D(S,T,A) = c(S,T,A\cap S, A\cap T) - c(T,S,A\cap T,A\cap S),$$
$$c(S,T,S_0,T_0) = \lambda_1^{|T_0|}\lambda_2^{|T\setminus T_0|}(x+\lambda_1)^{|S_0|}(x+\lambda_2)^{|S\setminus S_0|}.$$

The proof of Lemma 17 can be found in the full version of this work [14].

Proof (of Theorem 13). On input $G = (V,E)$ with $|V| = n$, do the following. Construct G_{S_0}, G_{S_1}, ..., G_{S_n} with S_i from Lemma 16. Every G_{S_i} can be constructed in time polynomial in $|G_{S_i}|$, which is poly(n) by Remark 15 and condition 2. of Lemma 16. Thus, the whole construction can be performed in time poly(n).

Again by condition 2. of Lemma 16, there is some $c' > 1$ such that all G_{S_i} have $\leq c'n\log^3 n$ vertices. Evaluate $I(G_{S_0}; x)$, $I(G_{S_1}; x)$, ..., $I(G_{S_n}; x)$. By the assumption of the theorem, one such evaluation can be performed in time

$$2^{c\frac{c'n\log^3 n}{(\log(c'n\log^3 n))^3}} = 2^{\frac{cc'n\log^3 n}{(\log c' + \log n + 3\log\log n)^3}} \leq 2^{\frac{cc'n\log^3 n}{(\log n)^3}} = 2^{cc'n}$$

for every $c > 0$.

Using Theorem 9, we can compute $I(G; x(S_0))$, $I(G; x(S_1))$, ..., $I(G; x(S_n))$ from the already computed $I(G_{S_i}; x)$ in time poly(n).

By condition 1. of Lemma 16, the $n+1$ values $x(S_i)$ are pairwise distinct. As $I(G; X)$ is a polynomial of degree at most n in X, this enables us to interpolate $I(G; X)$. The overall time needed is poly$(n)2^{cc'n} \leq 2^{(cc'+\varepsilon)n}$ for every $\varepsilon > 0$.   □

4   Open Problems

The most important open problem is to find a reduction that does not lose the factor $\Theta(\log^3 n)$ in the exponent of the running time.

Another interesting direction for further research are restricted classes of graphs, for example graphs of bounded maximum degree or regular graphs.

The independent set polynomial is a special case of the two-variable interlace polynomial [1]. It would be interesting to have an exponential time hardness result for this polynomial as well. In this context, the following question arises: Is the upper bound $\exp(O(\sqrt{n}))$ [25] for evaluation of the Tutte polynomial on *planar* graphs sharp?

Acknowledgments

I would like to thank Raghavendra Rao and the anonymous referees for helpful comments.

References

[1] Arratia, R., Bollobás, B., Sorkin, G.B.: A two-variable interlace polynomial. Combinatorica 24(4), 567–584 (2004)

[2] Bläser, M., Hoffmann, C.: On the complexity of the interlace polynomial. In: Albers, S., Weil, P. (eds.) 25th International Symposium on Theoretical Aspects of Computer Science (STACS). Dagstuhl Seminar Proceedings, vol. 08001, pp. 97–108. Internationales Begegnungs- und Forschungszentrum für Informatik (IBFI), Schloss Dagstuhl, Germany (2008); Updated full version: arXiv:cs.CC/0707.4565v3

[3] Bourgeois, N., Escoffier, B., Paschos, V.T., van Rooij, J.M.M.: A bottom-up method and fast algorithms for max independent set. In: Kaplan, H. (ed.) Algorithm Theory - SWAT 2010. LNCS, vol. 6139, pp. 62–73. Springer, Heidelberg (2010)

[4] Cormen, T.H., Leiserson, C.E., Rivest, R.L., Stein, C.: Introduction to algorithms, 2nd edn. MIT Press, Cambridge (2001)

[5] Courcelle, B.: A multivariate interlace polynomial and its computation for graphs of bounded clique-width. The Electronic Journal of Combinatorics 15(1) (2008)

[6] Dahllöf, V., Jonsson, P.: An algorithm for counting maximum weighted independent sets and its applications. In: SODA, pp. 292–298 (2002)

[7] Dell, H., Husfeldt, T., Wahlén, M.: Exponential time complexity of the permanent and the Tutte polynomial. In: Abramsky, S., Gavoille, C., Kirchner, C., auf der Heide, F.M., Spirakis, P.G. (eds.) ICALP 2010, Part I. LNCS, vol. 6198, pp. 426–437. Springer, Heidelberg (2010); Full paper: Electronic Colloquium on Computational Complexity TR10-078

[8] Dyer, M.E., Greenhill, C.S.: On Markov chains for independent sets. J. Algorithms 35(1), 17–49 (2000)

[9] Fomin, F.V., Grandoni, F., Kratsch, D.: A measure & conquer approach for the analysis of exact algorithms. J. ACM 56(5) (2009)

[10] Garey, M.R., Johnson, D.S.: Computers and Intractability – A Guide to the Theory of NP-Completeness. Freeman, New York (1979)
[11] Gutman, I., Harary, F.: Generalizations of the matching polynomial. Utilitas Math. 24, 97–106 (1983)
[12] Hoede, C., Li, X.: Clique polynomials and independent set polynomials of graphs. Discrete Mathematics 125(1-3), 219–228 (1994)
[13] Hoffmann, C.: Computational Complexity of Graph Polynomials. PhD thesis, Saarland University, Department of Computer Science (2010)
[14] Hoffmann, C.: Exponential time complexity of weighted counting of independent sets (2010); arXiv:cs.CC/1007.1146
[15] Impagliazzo, R., Paturi, R., Zane, F.: Which problems have strongly exponential complexity? J. Comput. Syst. Sci. 63(4), 512–530 (2001)
[16] Jerrum, M., Valiant, L.G., Vazirani, V.V.: Random generation of combinatorial structures from a uniform distribution. Theor. Comp. Sc. 43, 169–188 (1986)
[17] Jian, T.: An $O(2^{0.304n})$ algorithm for solving maximum independent set problem. IEEE Trans. Computers 35(9), 847–851 (1986)
[18] Karp, R.M.: Reducibility among combinatorial problems. In: Miller, R.E., Thatcher, J.W. (eds.) Complexity of Computer Computations, pp. 85–103. Plenum Press, New York (1972)
[19] Kneis, J., Langer, A., Rossmanith, P.: A fine-grained analysis of a simple independent set algorithm. In: Kannan, R., Narayan Kumar, K. (eds.) IARCS Annual Conference on Foundations of Software Technology and Theoretical Computer Science (FSTTCS 2009), Dagstuhl, Germany. Leibniz International Proceedings in Informatics (LIPIcs), vol. 4, pp. 287–298 (2009); Schloss Dagstuhl–Leibniz-Zentrum fuer Informatik
[20] Luby, M., Vigoda, E.: Approximately counting up to four (extended abstract). In: STOC, pp. 682–687 (1997)
[21] Papadimitriou, C.H.: Computational Complexity. Addison Wesley Longman, Amsterdam (1994)
[22] Robson, J.M.: Algorithms for maximum independent sets. J. Algorithms 7(3), 425–440 (1986)
[23] Roth, D.: On the hardness of approximate reasoning. Artif. Intell. 82(1-2), 273–302 (1996)
[24] Scott, A.D., Sokal, A.D.: The repulsive lattice gas, the independent-set polynomial, and the Lovász local lemma. J. Stat. Phys. 118, 1151 (2005)
[25] Sekine, K., Imai, H., Tani, S.: Computing the Tutte polynomial of a graph of moderate size. In: Staples, J., Katoh, N., Eades, P., Moffat, A. (eds.) ISAAC 1995. LNCS, vol. 1004, pp. 224–233. Springer, Heidelberg (1995)
[26] Sokal, A.D.: Chromatic roots are dense in the whole complex plane. Combinatorics, Probability & Computing 13(2), 221–261 (2004)
[27] Tarjan, R.E., Trojanowski, A.E.: Finding a maximum independent set. SIAM J. Comput. 6(3), 537–546 (1977)
[28] Valiant, L.G.: The complexity of enumeration and reliability problems. SIAM Journal on Computing 8(3), 410–421 (1979)
[29] Vigoda, E.: A note on the glauber dynamics for sampling independent sets. Electr. J. Comb. 8(1) (2001)
[30] Weitz, D.: Counting independent sets up to the tree threshold. In: Kleinberg, J.M. (ed.) STOC, pp. 140–149. ACM, New York (2006)

The Exponential Time Complexity of Computing the Probability That a Graph Is Connected*

Thore Husfeldt[1,2] and Nina Taslaman[1]

[1] IT University of Copenhagen, Denmark
[2] Lund University, Sweden

Abstract. We show that for every probability p with $0 < p < 1$, computation of all-terminal graph reliability with edge failure probability p requires time exponential in $\Omega(m/\log^2 m)$ for simple graphs of m edges under the Exponential Time Hypothesis.

1   Introduction

Graph reliability is a simple mathematical model of connectedness in networks that are subject to random failure of its communication channels. This type of stochastic networks arise naturally in, e.g., communication or traffic control; see [1] for an extensive survey of application areas.

For a connected graph $G = (V, E)$ and probability p, the *all-terminal reliability* $R(G; p)$ is the probability that there is a path of operational edges between every pair of nodes, given that every edge of the graph fails independently with probability p. For example, with $p = \frac{1}{2}$, the all-terminal reliability of the graph ⬠ is $\frac{3}{16}$, and the all-terminal reliability of the graph ⋈ is $\frac{2}{16}$.

In general, for a connected, undirected graph $G = (V, E)$ with edge-failure probability p, the all-terminal reliability can be given as

$$R(G; p) = \sum_{\substack{A \subseteq E \\ \text{spanning} \\ \text{connected}}} p^{|E \setminus A|}(1-p)^{|A|}. \tag{1}$$

For example $R(⬠; \frac{1}{3}) = (\frac{2}{3})^5 + 5 \cdot \frac{1}{3} \cdot (\frac{2}{3})^4 = \frac{112}{243}$. Computing $R(G; p)$ directly from (1) can take up to $O(2^m)$ operations, where m is the number of edges in G. An algorithm of Buzacott [4] solves the problem in time $3^n n^{O(1)}$ for graphs of n vertices. Further improvements exist [2], but remain exponential in n; this is explained by the $\exp(\Omega(n))$ lower bound of [5]. On the other hand, subexponential time algorithms have been found for some restricted classes of graphs. For example, the problem can be solved in time $\exp(O(\sqrt{n}))$ for planar graphs [14]. A natural question is then whether the complexity can be reduced for for further classes of graphs. Especially, from an applications point of view, the case of simple graphs is interesting.

---

* Partially supported by Swedish Research Council grant VR 2008–2010.

V. Raman and S. Saurabh (Eds.): IPEC 2010, LNCS 6478, pp. 192–203, 2010.

It is clear that $R(G; 0) = 1$ and $R(G; 1) = 0$ for all connected graphs, so for some values of p the problem is trivial. One may ask what the situation is for values close to these extremes. Moreover, for $p = \frac{1}{2}$ it is easily seen that (1) equals the number of connected, spanning subgraphs of G, divided by 2^m. This is an interesting enumeration problem in itself, and one could be tempted to hope for a better algorithm than $\exp(O(n))$, because a related enumeration problem, the number of spanning *trees* of G, can be solved in polynomial time by Kirchhoff's matrix–tree theorem [11].

Result. We give a lower bound on the problem of computing all-terminal graph reliability for the class of simple graphs for all nontrivial p, in the framework recently proposed by Dell *et al.* [5]. In particular, we work under the counting exponential time hypothesis:

(#ETH) There is a constant $c > 0$ such that no deterministic algorithm can
compute #3-Sat in time $\exp(cn)$.

This is a relaxation of the exponential time hypothesis (ETH) of Impagliazzo *et al.* [9], so our results hold under ETH as well. The best current bound for #3-Sat is $O(1.6423^n)$ [12].

Theorem 1. *For any fixed probability p with $0 < p < 1$, computing the all-terminal reliability $R(G; p)$ of a given simple graph G of m edges requires time exponential in $\Omega(m/\log^2 m)$ under #ETH.*

In particular, the bound holds for $p = \frac{1}{2}$, i.e., counting the number of connected spanning subgraphs of a given graph.

We have expressed the lower bound in terms of the parameter m, the number of edges of the input graph. Since $n \leq m$ for connected graphs, the result implies the lower bound $\exp(\Omega(n/\log^2 n))$ in terms of the parameter n, the number of vertices of the input graph. Moreover, the $\Omega(m/\log^2 m)$ lower bound together with the $\exp(O(n))$ algorithm from [4,2] shows that the hard instances have roughly linear density, ruling out a better algorithm than $\exp(O(n/\log^2 n))$ also for the restricted case of sparse graphs.

Our bound does not quite match the best known upper bound $\exp(O(n))$ of [4,2]. This situation is similar to the bounds reported in [5] for related problems on simple graphs, which also fall a few logarithmic factors (in the exponent) short of the best known algorithms. The bound does, however, suffice to separate the complexity of reliability computation from the $\exp(O(\sqrt{n}))$ bound for the planar case [14].

Graph Polynomials. Expression (1), viewed as a function $p \mapsto R(G; p)$ for fixed G, is known as the *reliability polynomial* of G, an object studied in algebraic graph theory [6, Sec. 15.8]. For example, $R(\mathord{\hbox{⌂}}; p) = (1-p)^5 + 5p(1-p)^4$.

Arguably, the most important graph polynomial is the bivariate *Tutte polynomial* $T(G; x, y)$, which encodes numerous combinatorial parameters of the input graph G, and whose restriction to certain lines and curves in the xy-plane specialize to other well known graph polynomials. The reliability polynomial is essentially a restriction of this polynomial to the ray $\{(1, y) : y > 1\}$. The complexity of computing $T(G; x, y)$ at various points (x, y) with respect to an input graph G is very well-studied in various models of computation, and the present paper thus establishes lower bounds for simple graphs along the mentioned ray, which was left open in a recent study [5] to completely map the exponential time complexity of the "Tutte plane".

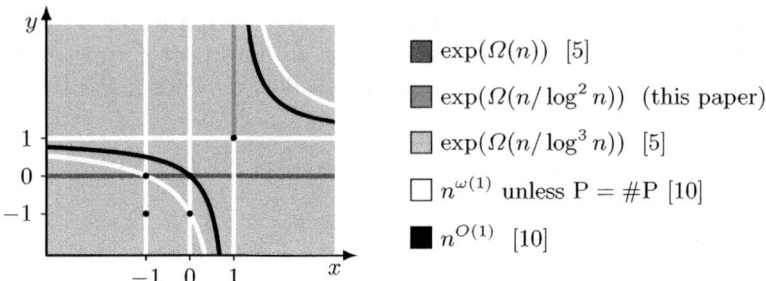

Fig. 1. Exponential time complexity under #ETH of the Tutte plane for simple graphs

Related Work. The structural complexity of all-terminal graph reliability was studied by Provan and Ball [13]. For any probability p, with $0 < p < 1$, it is shown that computing $R(G; p)$ for given G is hard for Valiant's counting class #P [16]. The reductions in [13] do not preserve the parameters n and m, so that the running time bounds under #ETH implicitly provided by their techniques are typically exponential in $\Omega(n^{1/k})$ for some k.

The reliability problem under consideration in this paper admits a number of natural extensions.

1. We can consider the computational problem of finding the reliability polynomial itself, instead of its value at a fixed point p. The input to this problem is a graph, and the output is a list of coefficients. For example, on input ⌂ the output should give $R(\text{⌂}; p) = 4p^5 - 15p^4 + 20p^3 - 10p^2 + 1$.
2. We can associate individual probabilities to every edge. For example, the graph ½ ¼ becomes disconnected with probability $\frac{5}{8}$.
3. We can consider multigraphs like ∘—⟨⟩∘, but with the same edge weight p. As indicated by the examples (for $p = \frac{1}{2}$), the multigraph case is a special case of the individually edge-weighted case, a fact that we will use later.

All of these problems are at least as hard as the problem under consideration in the present paper. Lower bounds of size $\exp(\Omega(m))$ are given in [5] or follow relatively easily (see §2.3).

A recent paper of Hoffman [8] studies the complexity of another graph polynomial, the independent set polynomial, in the same framework.

2   Preliminaries

We will only be concerned with undirected graphs. For a graph $G = (V, E)$ let n denote the number of vertices, m the number of edges, and for any subset $A \subseteq E$ let $\kappa(A)$ denote the number of connected components in the subgraph (V, A) (especially, $\kappa(A) = 1$ means that the edge subset A is spanning and connected). Also, for graph polynomials P and Q, we write $P(G; \mathbf{x}) \sim Q(G'; \mathbf{x}')$ if the two expressions are equal up to an easily computable factor.

2.1   Weighted Reliability

The reliability polynomial can be formulated as a restriction of the *Tutte polynomial*, which for an undirected graph G is given by

$$T(G; x, y) = \sum_{A \subseteq E} (x - 1)^{\kappa(A) - \kappa(E)} (y - 1)^{\kappa(A) + |A| - |V|} .$$

Note that $\kappa(E) = 1$ in our case. We find the reliability polynomial along the ray $\{ (1, y) : y > 1 \}$ in the so called *Tutte plane*, as $R(G; p) \sim T(G; 1, 1/p)$; in full detail:

$$R(G; p) = p^{m-n+1} (1 - p)^{n-1} T(G; 1, 1/p) , \qquad (0 < p < 1) . \tag{2}$$

For complexity analysis of the Tutte polynomial, it has proved a considerable technical simplification to consider Sokal's *multivariate Tutte polynomial* [15]. Here the graph is equipped with some weight function $\mathbf{w} \colon E \to \mathbb{R}$, and the polynomial is given by

$$Z(G; q, \mathbf{w}) = \sum_{A \subseteq E} \mathbf{w}(A) q^{\kappa(A)} ,$$

where $\mathbf{w}(A) = \prod_{e \in A} \mathbf{w}(e)$ is the edge-weight product of the subset A. For constant edge weights $w = y - 1$ we have $Z(G; q, w) \sim T(G; x, y)$ with $q = (x - 1)(y - 1)$. The "reliability line" $x = 1$ in the Tutte plane thus corresponds to $q = 0$ in the weighted setting, where Z vanishes, so instead we will consider the slightly modified polynomial

$$\hat{Z}(G; q, \mathbf{w}) = q^{-1} Z(G; q, \mathbf{w}) .$$

At $q = 0$, this gives a weighted version of the reliability polynomial:

Definition 1 (Weighted reliability polynomial). *For a connected, undirected graph $G = (V, E)$, the weighted reliability polynomial of G is given by*

$$\hat{R}(G; \mathbf{w}) = \hat{Z}(G; q, \mathbf{w})|_{q=0} = \sum_{\substack{A \subseteq E \\ \kappa(A)=1}} \mathbf{w}(A) . \tag{3}$$

For constant edge weight $w > 0$ we have $\hat{R}(G; w) = w^{n-1} T(G; 1, 1 + w)$, so for $0 < p < 1$ we can recover the reliability polynomial through (2) as

$$R(G; p) = p^m \hat{R}(G; 1/p - 1) . \tag{4}$$

2.2   Graph Transformations

A classical technique for investigating the complexity of the Tutte polynomial at a certain point (x', y') of the Tutte plane, is to relate it to some already settled point (x, y) via a graph transformation φ, such that $T(G; x', y') \sim T(\varphi(G); x, y)$. For the weighted setting we have the following rules, which are simple generalizations of [7, Sec. 4.3]. (See Appendix A.1.) For a graph $G = (V, E)$ with edge weights given by $\mathbf{w}$:

Lemma 1. *If $\varphi(G)$ is obtained from G by replacing a single edge $e \in E$ with a simple path of k edges $P = \{e_1, ..., e_k\}$ with $\mathbf{w}(e_i) = w_i$, then*

$$\hat{R}(\varphi(G); \mathbf{w}) = C_P \cdot \hat{R}(G; \mathbf{w}[e \mapsto w']),$$

where

$$\frac{1}{w'} = \frac{1}{w_1} + \cdots + \frac{1}{w_k} \qquad and \qquad C_P = \frac{1}{w'} \prod_{i=1}^{k} w_i.$$

Lemma 2. *If $\varphi(G)$ is obtained from G by replacing a single edge $e \in E$ with a bundle of parallel edges $B = \{e_1, \ldots, e_k\}$ with $\mathbf{w}(e_i) = w_i$, then*

$$\hat{R}(\varphi(G); \mathbf{w}) = \hat{R}(G; \mathbf{w}[e \mapsto w']),$$

where

$$w' = -1 + \prod_{i=1}^{k} (1 + w_i).$$

Corollary 1. *If $\varphi(G)$ is obtained from G by replacing a single edge $e \in E$ with a simple path of k edges of constant weight w, then*

$$\hat{R}(\varphi(G); \mathbf{w}) = kw^{k-1} \cdot \hat{R}(G; \mathbf{w}[e \mapsto w/k]), \tag{5}$$

and if it is obtained from G by replacing $e \in E$ with a bundle of k parallel edges of constant weight w, then

$$\hat{R}(\varphi(G); \mathbf{w}) = \hat{R}(G; \mathbf{w}[e \mapsto (1 + w)^k - 1]). \tag{6}$$

These rules are transitive [7, Lem. 1], and so can be freely combined for more intricate weight shifts. To preserve constant weight functions we need to perform the same transformation to every edge of the graph. This calls for the graph theoretic version of Brylawski's tensor product for matroids [3]. We found the following terminology more intuitive for our setting:

Definition 2 (Graph inflation). *Let H be a 2-terminal undirected graph. For any undirected graph $G = (V, E)$, an H-inflation of G, denoted $G \otimes H$, is obtained by replacing every edge $xy \in E$ by (a fresh copy of) H, identifying x with one of the terminals of H and y with the other.[1]*

---

[1] This can, in general, be done in two different ways, resulting in graphs that need not be isomorphic. However, the Tutte polynomial is blind to this difference. See extensive footnote in [5], Section 5.1.

If H is a simple path of k edges, $G \otimes H$ gives the k-*stretch* of G. Similarly, a bundle of k parallel edges results in a k-*thickening*, of G.

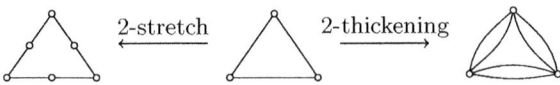

2.3   Hardness of Computing Coefficients

Our pivot for proving Theorem 1 will be the following hardness result, which says that even when restricted to fixed hyperbolas $(x-1)(y-1) = q$, computing the full Tutte polynomial is hard. This is an extension to the case $q = 0$ of Lemma 2 in [5], and the proof is given in Appendix 2.

Lemma 3. *Under #ETH, computing the coefficients of the polynomial $w \mapsto \hat{Z}(G; q, w)$ for given simple graph G and rational number $q \notin \{1, 2\}$ requires time exponential in $\Omega(m)$.*

Since $R(G; p)$ is essentially $\hat{R}(G; \mathbf{w})$ restricted to positive constant weight functions, and since $\hat{R}(G; \mathbf{w}) = \hat{Z}(G; 0, \mathbf{w})$, the following is immediate:

Corollary 2. *Under #ETH, it requires time exponential in $\Omega(m)$ to compute the coefficients of the reliability polynomial.*

3   Bounce Graphs

As a first step towards Theorem 1, we present here a class of graph transformations whose corresponding weight shifts for the reliability polynomial are all distinct. These transformations are mildly inspired by k-byte numbers, in the sense that each has associated to it a sequence of length k, such that the lexicographic order of these sequences determines the numerical order of the corresponding (shifted) weights. Each transformation is a *bounce inflation*:

Definition 3 (Bounce graph). *For positive numbers h (height) and l (length), the (h, l)-bounce is the graph obtained by identifying all the left and all the right endpoints of h simple paths of length l. Given a bounce sequence, $S = \langle s_1, s_2, \dots, s_k \rangle$, of k numbers $s_i > 1$, the corresponding bounce graph, B_S, is the (simple) graph obtained by concatenating k (h, l)-bounces by their endpoints, where the height starts at 1 for the first bounce and then increases by one for each follower, and the length of the ith bounce is s_i.*

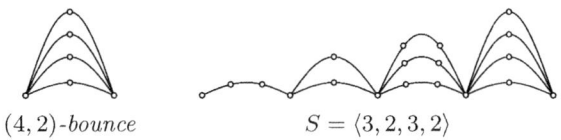

(4, 2)-*bounce* $S = \langle 3, 2, 3, 2 \rangle$

The length *of a bounce graph is the number of bounces in it (or, equivalently, the height of the highest bounce).*

Inflation by a bounce graph has the following weight-shifting effect, from $\hat{R}$'s perspective:

Lemma 4. *For a graph G, bounce sequence $S = \langle s_1, s_2, \ldots, s_k \rangle$ and $w > 0$:*

$$\hat{R}(G \otimes B_S; w) = C_S^m \cdot \hat{R}(G; w_S),$$

where

$$\frac{1}{w_S} = \sum_{i=1}^{k} \frac{1}{(1 + w/s_i)^i - 1} \quad \text{and} \quad C_S = \frac{1}{w_S} \cdot \prod_{i=1}^{k} w^{(s_i - 1)i} \left((w + s_i)^i - s_i^i \right).$$

Proof. Starting out with $G \otimes B_S$, we will look at the effect of replacing one of the m copies of B_S with a single edge e. We show that, with φ denoting this operation:

$$\hat{R}(G \otimes B_S; w) = C_S \cdot \hat{R}(\varphi(G \otimes B_S); \mathbf{w}[e \mapsto w_S]), \tag{7}$$

where w_S has the above form, and $\mathbf{w}$ has the old value w on all unaffected edges. The lemma follows from performing φ for every copy of B_S in $G \otimes B_S$.

The first step towards transforming a bounce graph (say, $\circ\!\!-\!\!\circ\!\!\curvearrowright\!\!\curvearrowright$) into a single edge, is to replace each path of each bounce in it by a single edge. Applying (5) of Corollary 1 to each path of the ith bounce gives a factor $(s_i w^{s_i - 1})^i$ to the polynomial, and each edge in the resulting $(h, 1)$-bounce gets weight w/s_i in the modified graph. Repeating this process for every bounce gives a simplified bounce graph ($\circ\!\!-\!\!\curvearrowright\!\!\curvearrowright$) in a transformed graph $\phi(G \otimes B_S)$ such that

$$\hat{R}(G \otimes B_S; w) = \left(\prod_{i=1}^{k} (s_i w^{s_i - 1})^i \right) \cdot \hat{R}(\phi(G \otimes B_S); \mathbf{w}'),$$

with $\mathbf{w}'$ taking the value w/s_i for every edge in the ith bounce of the simplified bounce graph, and the old value w outside it. Next, we successively replace each of its $(h, 1)$-bounces by a single edge, to get a simple path ($\circ\!\!-\!\!\circ\!\!-\!\!\circ\!\!-\!\!\circ$) of length k (with non-constant edge weights). From (6) of Corollary 1, we know that this does not produce any new factors for the polynomial, but the weight of the ith edge in this path will be given by

$$w_i = (1 + w/s_i)^i - 1.$$

Finally, we compress the path into a single edge e. A single application of Lemma 1 then gives the result in (7). $\qquad\qquad\square$

If Lemma 3 is our pivot for proving Theorem 1, the following result is the lever:

Lemma 5. *For any size m, there exist $m+1$ distinct, simple bounce graphs B_S of size $O(\log^2 m)$, such that for any two associated bounce sequences S and T:*

$$S >_{lex} T \implies w_S < w_T \tag{8}$$

for all $w > 6$.

Proof. The set of bounce sequences $S = \langle s_1, \ldots, s_l \rangle$ of length $l = \log(m + 1)$ and with each $s_i \in \{2, 3\}$, provides $m + 1$ different simple bounce graphs of promised size. To show that any two of them satisfy equation (8) we will look at the difference

$$\Delta_{S,T}(w) = \frac{1}{w_S} - \frac{1}{w_T},$$

and show that $\Delta_{S,T}(w) > 0$ for $w > 6$.

Let k be the first index where the sequences differ, say $s_k = 3$ and $t_k = 2$. We then have

$$\Delta_{S,T}(w) = \frac{1}{(1 + w/3)^k - 1} + \sum_{i=k+1}^{l} \frac{1}{(1 + w/s_i)^i - 1}$$

$$- \frac{1}{(1 + w/2)^k - 1} - \sum_{i=k+1}^{l} \frac{1}{(1 + w/t_k)^i - 1}.$$

This would be minimal if $s_i = 2$ and $t_i = 3$ for all $i > k$, i.e. if

$$\Delta_{S,T}(w) = f(1 + w/3) - f(1 + w/2),$$

where

$$f(x) = \frac{1}{x^k - 1} - \sum_{i=k+1}^{l} \frac{1}{x^i - 1}.$$

If we could show that $f'(x) < 0$ for $x > x_0$ then it would follow (e.g. from the mean value theorem) that $f(x) > f(y)$ for $x < y$ above x_0. In particular, with $x_0 = 3$ this would prove our claim. To see that this is indeed the case, we look at the derivative

$$f'(x) = -\frac{kx^{k-1}}{(x^k - 1)^2} + \sum_{i=k+1}^{l} \frac{ix^{i-1}}{(x^i - 1)^2}.$$

A bit of manipulation shows that the terms of the sum, let us call them T_i, satisfy $T_i > 2T_{i+1}$ for $x > 3$, so

$$f'(x) < \frac{kx^{k-1}}{(x^k - 1)^2} \left(-1 + \sum_{i=k+1}^{l} \frac{1}{2^i} \right) < 0$$

for $x > 3$, and we are done by the above argument. □

4   Evaluating the Reliability Polynomial Is Hard

We are ready to prove Theorem 1. We introduce the following notation for the problem of evaluating a graph polynomial $P(G; \mathbf{x})$ at a given point $\mathbf{x}$

P-VAL($\mathbf{x}$):
   input A simple, connected, undirected graph G
   output The value of $P(G; \mathbf{x})$ (a rational number)

In this notation, the computational problem described in Theorem 1 can be written as R-VAL(p), and the corresponding problem for weighted reliability with constant edge weight w is $\hat{R}$-VAL(w). We will prove Theorem 1 by reducing the problem of computing coefficients of the polynomial $p \mapsto R(G; p)$, to the problem R-VAL(p) for any arbitrary fixed probability p with $0 < p < 1$.

Proof (of Theorem 1). Let G be a simple graph with n vertices and m edges. We prove that $\hat{R}$-VAL(w) requires time exponential in $\Omega(m/\log^2 m)$ for any $w > 0$, which by (4) gives the same bound for R-VAL(p) for any p with $0 < p < 1$.

Suppose we have an algorithm for $\hat{R}$-VAL(w) for some fixed $w > 0$. For simplicity of exposition, first assume $w > 6$. From Lemma 5 we can easily construct $m + 1$ bounce graphs B_S such that each $\hat{R}(G \otimes B_S; w)$ gives us the value of $\hat{R}(G; w_S)$ at some new weight w_S, with all such w_S distinct. Computing $\hat{R}$-VAL(w) for each of these $m + 1$ bounce inflations, we get the value of $\hat{R}(G; w)$ at $m + 1$ distinct w-values. Since the degree of this polynomial is m, that gives us the coefficients by interpolation. By Corollary 2, the whole process must then require time $\exp(\Omega(m))$. By Lemma 5, each inflation $G \otimes B_S$ will have $O(m \log^2 m)$ edges. Thus, for graphs of size $O(m \log^2 m)$, the problem $\hat{R}$-VAL(w) has a lower bound of $\exp(\Omega(m))$. The claimed bound then follows from the fact that

$$\varphi(m) = m \log^2 m \quad \Longrightarrow \quad \varphi^{-1}(m) \in O\left(m/\log^2 m\right).$$

We turn to the case $0 < w \le 6$. Given such a w, choose a number k such that

$$w' := (w/2 + 1)^k - 1 > 6.$$

From the above, we know that for any simple graph G it takes time exponential in $\Omega(m/\log^2 m)$ to evaluate $\hat{R}(G; w')$. Now consider the k-thickening-2-stretch of G, let us call this G'. This will be a simple graph of $O(km)$ edges, and from Corollary 1 it follows that the value of $\hat{R}(G'; w)$ would give us the value of $\hat{R}(G; w')$. Thus, computing the former must also require time exponential in $\Omega(m/\log^2 m)$. Since G' has $O(km)$ edges, this gives a lower bound for $\hat{R}$-VAL(w) exponential in $\Omega\left(m/(k \log^2 m)\right)$, which is $\Omega(m/\log^2 m)$ as a function of m. $\quad\square$

5   Remarks

For the multivariate Tutte polynomial $Z(G; q, w)$, the current lower bound for simple graphs is $\exp(\Omega(m/\log^3 m))$ [5]. One might ask whether the bounce graph construction could be used to improve this. The weight shift corresponding to a bounce inflation is in this case given by (for $w \ne 0$ and $q \notin \{0, -2w\}$)

$$\frac{q}{w_S} = \prod_{i=1}^{k}\left(\frac{q}{\{(q/r_i) + 1\}^i - 1} - 1\right), \quad \text{where } r_i = \left(1 + \frac{q}{w}\right)^{s_i} - 1$$

Unlike the expression in Lemma 4 this is not a sum of powers, so the 'k-byte number' analogy is remote, and there is now also a dependency on q which seems to make it difficult proving something like Lemma 5 under the same constraints.

References

1. Ball, M.O., Colbourn, C.J., Provan, J.S.: Network Reliability. Handbooks in operations research and management science, ch. 11, vol. 7. Elsevier Science, Amsterdam (1995)
2. Björklund, A., Husfeldt, T., Kaski, P., Koivisto, M.: Computing the Tutte polynomial in vertex-exponential time. In: 49th Annual IEEE Symposium on Foundations of Computer Science (FOCS 2008), Philadelphia, Pennsylvania, USA, October 25-28, pp. 677–686. IEEE Computer Society, Los Alamitos (2008)
3. Brylawski, T.H.: The Tutte polynomial. In: Matroid Theory and Its Applications, pp. 125–275. C.I.M.E., Ed. Liguori, Napoli & Birkhäuser (1980)
4. Buzacott, J.A.: A recursive algorithm for finding reliability measures related to the connection of nodes in a graph. Networks 10(4), 311–327 (1980)
5. Dell, H., Husfeldt, T., Wahlén, M.: Exponential time complexity of the permanent and the Tutte polynomial. In: Gavoille, C. (ed.) ICALP 2010, Part I. LNCS, vol. 6198, pp. 426–437. Springer, Heidelberg (2010); Full paper in Electronic Colloquium on Computational Complexity, Report No. 78 (2010)
6. Godsil, C., Royle, G.: Algebraic Graph Theory. Graduate Texts in Mathematics. Springer, Heidelberg (2001)
7. Goldberg, L.A., Jerrum, M.: Inapproximability of the Tutte polynomial. Inform. Comput. 206(7), 908–929 (2008)
8. Hoffmann, C.: Exponential time complexity of weighted counting of independent sets. In: Raman, V., Saurabh, S. (eds.) IPEC 2010. LNCS, vol. 6478, pp. 180–191. Springer, Heidelberg (2010)
9. Impagliazzo, R., Paturi, R., Zane, F.: Which problems have strongly exponential complexity? J. Comput. Syst. Sci. 63(4), 512–530 (2001)
10. Jaeger, F., Vertigan, D.L., Welsh, D.J.: On the computational complexity of the Jones and Tutte polynomials. Math. Proc. Cambridge 108(1), 35–53 (1990)
11. Kirchhoff, G.: Über die Auflösung der Gleichungen, auf welche man bei der Untersuchung der linearen Verteilung galvanischer Ströme geführt wird. Ann. Phys. Chem. 72, 497–508 (1847)
12. Kutzkov, K.: New upper bound for the #3-SAT problem. Inf. Process. Lett. 105(1), 1–5 (2007)
13. Provan, J.S., Ball, M.O.: The complexity of counting cuts and of computing the probability that a graph is connected. SIAM J. Comput. 12(4), 777–788 (1983)
14. Sekine, K., Imai, H., Tani, S.: Computing the Tutte polynomial of a graph of moderate size. In: Staples, J., Katoh, N., Eades, P., Moffat, A. (eds.) ISAAC 1995. LNCS, vol. 1004, pp. 224–233. Springer, Heidelberg (1995)
15. Sokal, A.D.: The multivariate Tutte polynomial (alias Potts model) for graphs and matroids. In: Surveys in Combinatorics, pp. 173–226. Cambridge University Press, Cambridge (2005)
16. Valiant, L.G.: The complexity of enumeration and reliability problems. SIAM J. Comput. 8(3), 410–421 (1979)

A   Supplementary Proofs

A.1   Lemma 1 and 2

Proof (of Lemma 1 and 2). We repeat the arguments from the proof in [10, Sec.4.3]: Let S be the set of subsets $A \subseteq E \setminus \{e\}$ that already span the whole graph G, i.e.

$$S = \{A \subseteq E \setminus \{e\} : \kappa(A) = 1\},$$

and let T be the set of subsets that need the edge e to span the graph:

$$T = \{A \subseteq E \setminus \{e\} : \kappa(A) = 2 \text{ and } \kappa(A \cup \{e\}) = 1\}.$$

With w' denoting the weight of the edge e in the original graph G, (3) gives, for both lemmas,

$$\hat{R}(G; \mathbf{w}[e \mapsto w']) = \sum_{A \in S} \mathbf{w}(A)(1 + w') + \sum_{A \in T} \mathbf{w}(A)w'. \tag{9}$$

We will compare the partial sums here to the corresponding ones obtained when we alter the graph. When φ is the operation described in Lemma 1, we have (with P the set of edges in the path)

$$\hat{R}(\varphi(G); \mathbf{w}) = \sum_{A \in S} \mathbf{w}(A) \left(\mathbf{w}(P) + \sum_{i=1}^{k} \mathbf{w}(P \setminus e_i) \right) + \sum_{A \in T} \mathbf{w}(A)\mathbf{w}(P) =$$

$$\sum_{A \in S} \mathbf{w}(A) \left(\prod_{i=1}^{k} w_i + \sum_{j=1}^{k} \prod_{i \neq j} w_i \right) + \sum_{A \in T} \mathbf{w}(A) \prod_{i=1}^{k} w_i.$$

Comparing corresponding sums to (9), it is easy to check that the expressions for w' and C_p in Lemma 1 indeed make $\hat{R}(\varphi(G); \mathbf{w}) = C_P \cdot \hat{R}(G; \mathbf{w}[e \mapsto w'])$.

When φ is the operation described in Lemma 2, we have (with B the set of edges in the bundle)

$$\hat{R}(\varphi(G); \mathbf{w}) = \sum_{A \in S} \mathbf{w}(A) \left(1 + \sum_{\substack{A' \subseteq B \\ A' \neq \emptyset}} \mathbf{w}(A') \right) + \sum_{A \in T} \mathbf{w}(A) \left(\sum_{\substack{A' \subseteq B \\ A' \neq \emptyset}} \mathbf{w}(A') \right),$$

and Lemma 2 follows since

$$\sum_{\substack{A' \subseteq B \\ A' \neq \emptyset}} \mathbf{w}(A') = \prod_{i=1}^{k} (w_i + 1) - 1.$$

□

A.2   Deletion/Contraction and Lemma 3

For an edge $e = xy \in E$, let $G \setminus e$ be the graph obtained by deleting e, and let G/e be the (multi)graph obtained by *contracting* e, i.e. by identifying the end vertices x and y before removing e . With w_e denoting the weight of edge e, we have the following *deletion/contraction reduction* for the weighted Tutte polynomial (see [15, Sec. 4.3])

$$Z(G; q, \mathbf{w}) = Z(G \setminus e; q, \mathbf{w}) + w_e \cdot Z(G/e; q, \mathbf{w}) \,.$$

Note that if $e \in E$ is a bridge, then $G \setminus e$ has one more component than G, while in any other case both $G \setminus e$ and G/e have the same number of connected components as G. Using the above identity and the fact that $\hat{Z}(G; q, \mathbf{w}) = q^{-1} Z(G; q, \mathbf{w})$, this gives:

$$\hat{Z}(G; q, \mathbf{w}) = \begin{cases} q \cdot \hat{Z}(G \setminus e; q, \mathbf{w}) + w_e \cdot \hat{Z}(G/e; q, \mathbf{w}) & \text{if } e \text{ is a bridge,} \\ \hat{Z}(G \setminus e; q, \mathbf{w}) + w_e \cdot \hat{Z}(G/e; q, \mathbf{w}) & \text{otherwise.} \end{cases} \tag{10}$$

Proof (of Lemma 3). The case $q \neq 0$ is treated in Lemma 2 of [5]. For $q = 0$, we give a reduction to Lemma 1 from the same paper. Under our assumptions it says that $w \mapsto \hat{Z}(G; 0, \mathbf{w})$ cannot be computed faster than $\exp(\Omega(m))$, where $\mathbf{w}$ is given by (for some set T of three edges):

$$\mathbf{w}(e) = \begin{cases} -1, & \text{if } e \in T, \\ w, & \text{otherwise.} \end{cases} \tag{11}$$

The proof of this actually uses the restriction that $G' = (V, E \setminus T)$ is connected, so we can assume that this is the case. Thus, no edge in T is a bridge. Three applications of (10), to delete/contract these edges, gives

$$\hat{Z}(G; 0, \mathbf{w}) = \sum_{C \subseteq \{1,2,3\}} (-1)^{|C|} \hat{Z}(G_C; 0, w) \,,$$

for some graphs G_C of constant edge-weight w. These G_C's may contain loops and multiple edges from the contractions. To address this we look at the 2-stretch $G_C \otimes P_3$ of each G_C. This will give simple graphs of constant weight functions, and m applications of (5) from Corollary 1 gives

$$\hat{Z}(G_C \otimes P_3; 0, w) = (2w)^m \hat{Z}(G_C; 0, w/2) \,.$$

If an algorithm could compute the coefficients of $w \mapsto \hat{Z}(G; 0, w)$ faster than $\exp(\Omega(m))$ for any simple graph G, it could be used to compute eight polynomials $w \mapsto \hat{Z}(G_C \otimes P_3; 0, w)$ (one for each subset C). This would give us first $w \mapsto \hat{Z}(G_C; 0, w)$ and then $w \mapsto \hat{Z}(G; 0, \mathbf{w})$, faster than $\exp(\Omega(m))$, which is impossible according to Lemma 1 from [5].   □

Inclusion/Exclusion Branching for Partial Dominating Set and Set Splitting

Jesper Nederlof[1] and Johan M.M. van Rooij[2]

[1] Department of Informatics, University of Bergen,
N-5020 Bergen, Norway
Jesper.Nederlof@ii.uib.no

[2] Department of Information and Computing Sciences, Utrecht University,
P.O. Box 80.089, 3508 TB Utrecht, The Netherlands
jmmrooij@cs.uu.nl

Abstract. Inclusion/exclusion branching is a way to branch on requirements imposed on problems, in contrast to the classical branching on parts of the solution. The technique turned out to be useful for finding and counting (minimum) dominating sets (van Rooij et al., ESA 2009). In this paper, we extend the technique to the setting where one is given a set of properties and seeks (or wants to count) solutions that have at least a given number of these properties. Using this extension, we obtain new algorithms for Partial Dominating Set and the parameterised problem k-Set Splitting. In particular, we apply the new idea to the fastest polynomial space algorithm for counting dominating sets, and directly obtain a polynomial space algorithm for Partial Dominating Set with the same running time (up to a linear factor). Combining the new idea with some previous work, we also give a polynomial space algorithm for k-Set Splitting that improves the fastest known result significantly.

1 Introduction

Exact exponential time algorithms aim to minimise the worst-case running time of algorithms for $\mathcal{NP}$-hard problems. Since $\mathcal{P}$ is not believed to be equal to $\mathcal{NP}$ it is indeed expected that these worst-case running times are super-polynomial. Even stronger, for some problems it is not even known whether "trivial" brute-force algorithms can be improved significantly. In particular, algorithms with running times of the type $\mathcal{O}^*(2^n)^1$, where n is some quantitive property of the instance, seem often hard to beat for many problems (see for example [13,23]).

In the field of exact exponential time algorithms, the *inclusion/exclusion formula* recently found many new applications (see for example [3,6,19]). The formula operates on a given family of n sets and consists of 2^n summands. Therefore, a typical running time of algorithms using this method is $\mathcal{O}^*(2^n)^1$, where n is some quantitative property of the instance.

---

[1] The $\mathcal{O}^*$ notation suppresses factors polynomial in the input size.

V. Raman and S. Saurabh (Eds.): IPEC 2010, LNCS 6478, pp. 204–215, 2010.

One method useful for improving such algorithms is *trimming*, where one predicts which summands in the inclusion/exclusion formula are non-zero in order to be able to skip the others. This method found applications in [4,5]. An as least as powerful technique as trimming is *inclusion/exclusion branching* (IE-branching), initiated by Bax [2]: the inclusion/exclusion formula has a summand for every subset of a n-sized ground set, and these subsets are enumerated by a branching algorithm. Now, trimming is equivalent to IE-branching in combination with a halting rule that returns zero when it can be predicted that all summands enumerated from the current branch are zero.

Branch and reduce is one of the oldest techniques in the field of exact exponential time algorithms. A branch and reduce algorithm consists of a set of *branching*, *reduction* and *halting rules*. A rule is a (usually) polynomial time algorithm that solves an instance given the solution of b "easier" instances. If $b \geq 2$, we speak of a branching rule; if $b = 1$, we speak of a reduction rule; and if $b = 0$, we speak of a halting rule. The worst-case running time of branch and reduce algorithms is often hard to determine. Recently, a nice tool for giving upper bounds on this running time was developed, named *Measure & Conquer* [11]: here, one assigns a clever potential function to each instance of the problem that intuitively resembles how "difficult" the instance is. Now, an upper bound on the worst-case running time can be obtained by finding lower bounds on the reduction in the potential function for the subproblems generated by the branching rules.

In [22], the current authors proposed (among others) an algorithm that combines IE-branching with Measure & Conquer for the DOMINATING SET problem. The IE-branching here is used to count objects that meet a given set of requirements. In this paper, we extend this approach to the setting where we have to find objects that meet *at least* (or exactly) t requirements from a given set.

Our first application is the PARTIAL DOMINATING SET (PDS) problem. PDS is a natural extension of the well-known DOMINATING SET problem where we are given a graph G and integer t, and are asked to compute a vertex set of minimum size that dominates at least t vertices in the graph (see Section 3.2).

In [14], it is shown that this problem is Fixed Parameter Tractable (FPT) parameterised by t, and in [15] even an $\mathcal{O}^*(2^t)$ algorithm is provided. [1] shows that the problem is FPT when restricted to planar graphs and parameterised by the solution size. In [12], the latter has been improved to a sub-exponential time algorithm. The previously fastest exact exponential time algorithm runs in $\mathcal{O}(1.6183^n)$ time and polynomial space [16]. As an application of our new branching rule, we give algorithms for solving PDS in $\mathcal{O}(1.5012^n)$ and $\mathcal{O}(1.5673^n)$ time using, respectively, exponential space and polynomial space. The polynomial space algorithm is directly obtained by extending the algorithm in [20] for counting dominating sets with the new branching rule.

Our second application is a faster algorithm for parameterised k-SET SPLITTING. Given a set family $\mathcal{S}$ over a universe $\mathcal{U}$, k-SET SPLITTING asks whether it is possible to partition $\mathcal{U}$ in two colour classes such that at least k sets contain elements of both colour classes. The SET SPLITTING problem has been studied

Table 1. List of know results on parameterised SET SPLITTING

Authors		Time	Space
F. K. H. A. Dehne, M. R. Fellows, F. A. Rosamond	[8]	$\mathcal{O}^*(72^k)$	poly
F. K. H. A. Dehne, M. R. Fellows, F. A. Rosamond, P. Shaw	[9]	$\mathcal{O}^*(8^k)$	poly
D. Lokshtanov and C. Sloper	[18]	$\mathcal{O}^*(2.6499^k)$	poly
J. Chen and S. Lu (randomized algorithm)	[7]	$\mathcal{O}^*(2^k)$	poly
D. Lokshtanov and S. Saurabh	[17]	$\mathcal{O}^*(2^k)$	poly
D. Lokshtanov and S. Saurabh	[17]	$\mathcal{O}^*(1.9630^k)$	exp
J. Nederlof and J. M. M. van Rooij		$\mathcal{O}^*(1.8213^k)$	poly

extensively both combinatorially and algorithmically; see [17] for a long list of pointers to the literature. For the parameterised version of the problem, there has been a long sequence of improvements; see Table 1. Recently an efficient kernel for this problem has been found [17]. Furthermore, the kernel has been used to obtain an $\mathcal{O}(1.9630^k)$ algorithm. We will use the same kernel and then apply an algorithm based on our new branching rule to obtain an $\mathcal{O}(1.8213^k)$ algorithm.

Notations and Definitions. For a set U, denote 2^U for the *power set* of U, i.e., the set consisting of all subsets of U.

If $G = (V, E)$ is graph and $v \in V$, we denote $N(v) = \{w \in V : (v, w) \in E\}$, $N[v] = N(v) \cup \{v\}$ and $d(v) = |N(v)|$. For a set $X \subseteq V$, denote $N_X(v) = N(v) \cap X$, $d_X(v) = |N_X(v)|$, $N(X) = (\bigcup_{v \in X} N(v))$ and $N[X] = N(X) \cup X$. Also, we will use $N^2(v) = N(N(v)) \setminus \{v\}$ (i.e. all vertices at distance exactly 2 from v). A pair of vertices $u, v \in V$ is a pair of *false twins* if $N(u) = N(v)$.

For a matrix A, we let $\texttt{shiftdown}(A)$ denote the matrix obtained by inserting an empty row vector on top and shifting all rows one row down. Also, we let $\texttt{shiftxtimestotheleft}(A, x)$ and $\texttt{shifttotheleft}(A)$ be the matrix obtained by removing the first x columns and the first column from A, respectively.

2   Extended IE-Branching

Let U be a set and let $S_1, S_2 \ldots, S_n \subseteq U$. In the context of this paper, the sets $S_1, S_2, \ldots, S_n$ can be thought of as "properties". Following Bax [2], we define for a partition (R, O, F) of $\{1, 2, \ldots, n\}$:

$$a(R, O, F) = \left| \left(\bigcap_{i \in R} S_i \right) \setminus \left(\bigcup_{i \in F} S_i \right) \right|$$

That is, $a(R, O, F)$ can be thought of as the number of $e \in U$ that have all the properties in R (the *required properties*), and none of the properties in F (the *forbidden properties*). Since (R, O, F) partitions $\{1, 2, \ldots, n\}$, all constraints that are neither required nor forbidden are in O, which naturally can be thought of as the *optional properties*. Now, it is easy to see that the following holds:

$$a(R, O \cup \{i\}, F) = a(R \cup \{i\}, O, F) + a(R, O, F \cup \{i\}) \tag{1}$$

It is not surprising that this equation is used in many branching algorithms by branching on an optional property of a solution (for example, does it contain vertex v) by making it required in one branch and forbidden in another branch. However, the same equation can also be used to branch on required properties imposed on a solution to be found by rearranging the formula:

$$a(R \cup \{i\}, O, F) = a(R, O \cup \{i\}, F) - a(R, O, F \cup \{i\}) \qquad (2)$$

The reader familiar with inclusion/exclusion may notice that the inclusion/exclusion formula can be obtained by expanding the recurrence (see [2,22]).

Let us extend this idea to the setting of this paper. Here, we are interested in all elements that have *exactly* t of the properties in R. Similarly, we define:

$$a_t(R, O, F) = \left| \{ e \in U : |R[e]| = t \wedge F[e] = 0 \} \right| \qquad (3)$$

where we denote $T[e] = \{ i \in T : e \in S_i \}$ for any subset $T \subseteq \{1, 2, \ldots, n\}$. Informally, we will say that $e \in U$ is counted by $a_t(R, O, F)$ if it $|R[e]| = t$ and $|F[e]| = 0$.

Now, we state the branching rule that we will use in this paper:

$$
\begin{aligned}
a_t(R \cup \{i\}, O, F) = a_t(R, O, F \cup \{i\}) + \\
(a_{t-1}(R, O \cup \{i\}, F) - a_{t-1}(R, O, F \cup \{i\})) \qquad (4)
\end{aligned}
$$

To see why the equation holds, first note that $a_t(R \cup \{i\}, O, F)$ is the number of elements of U that are in exactly t of the sets S_j with $j \in R \cup \{i\}$ and in no S_j with $j \in F$. Let $1 \leq i \leq n$ and consider a set S_i. Then, $a_t(R, O, F \cup \{i\})$ counts exactly the $e \in U$ also counted in $a_t(R \cup \{i\}, O, F)$ such that $e \notin S_i$. Moreover, $a_{t-1}(R, O \cup \{i\}, F) - a_{t-1}(R, O, F \cup \{i\})$ counts exactly all the $e \in U$ counted in $a_t(R \cup \{i\}, O, F)$ such that $e \in S_i$, analogous to Equation 2.

At first, the above branching rule does not look particularly efficient, since the branching algorithm has to recurse on 3 cases. However, the first and last branch only differ in the subscript parameter, which we can exploited by computing a_s for all $0 \leq s \leq n$ simultaneously whenever we need to compute a_t. This will slow down the algorithm by only a factor n, since we just have to consider the cases $a_t(R, O, F \cup \{i\})$ and $a_{t-1}(R, O \cup \{i\}, F)$ as $a_{t-1}(R, O, F \cup \{i\})$ will be computed when the algorithm computes $a_t(R, O, F \cup \{i\})$.

If one expands the recurrence of Equation 4, the following natural variation on the inclusion/exclusion formula could be obtained for computing $a_i(R, O, F)$:

$$a_t(R, O, F) = \sum_{X \subseteq R} (-1)^{|X|-t} \binom{|X|}{t} a_0(\emptyset, O \cup X, F \cup (R \setminus X)) \qquad (5)$$

This can be useful for covering problems where one relaxes covering constraints by penalising solutions that do not satisfy this constraint (e.g. one could think of partial GRAPH COLOURING, TSP, STEINER TREE). Note that this observation is, for example, also implicit in [6].

3   Partial Dominating Set

In this section, we will give a small modification of the current fastest polynomial space algorithm for counting dominating sets to obtain a polynomial space algorithm for PARTIAL DOMINATING SET. The curious fact is that the new algorithm solves a more general and before separately studied problem in the same running time as used by the old algorithm for counting dominating sets, up to a linear factor. In the full paper, we will also give an improved exponential space algorithm for PARTIAL DOMINATING SET. Here, we will only state this result and omit the details due to space restrictions.

To obtain the result of this section, we first apply our new branching rule to the following problem:

PARTIAL RED BLUE DOMINATING SET (PRBDS)
Input A bipartite graph $(V = R \cup B, E)$; k, t such that $0 \le k \le |R|$ and $0 \le t \le |B|$.
Question Is there a subset $X \subseteq R$ such that $|N(X)| \ge t$ and $|X| \le k$?

Note that for $t = |B|$, we obtain the ordinary RED BLUE DOMINATING SET problem (find a subset $X \subseteq R$ such that $|X| \le k$ and $N(X) = B$).

Define for any $0 \le s \le |R|$ and $0 \le j \le |B|$:

$$b_{s,j}(R, B) = |\{X \subseteq R : |X| = s \wedge |N[X] \cap B| = j\}|$$

The answer to PRBDS is yes if and only if there exist $s \le k$ and $j \ge t$ such that $b_{s,j}(R, B) > 0$. Now, we set up a recurrence similar to Equation 4: Let $U = 2^R$ and $B = \{v_1, v_2, \ldots, v_{|B|}\}$. For every $0 \le i \le |B|$, let $S_i \subseteq U$ contain all the subsets $X \subseteq R$ such that $v_i \in N[X]$, i.e., $S_i = \{X \subseteq R \mid v_i \in N(X)\}$. Hence, S_i can be thought of as the property "dominating v_i" of subsets of R that can be thought of as the elements of U. Then, we can restate Equation 4 as:

$$b_{s,j}(R, B) = b_{s,j}(R \setminus N[v], B \setminus \{v\}) +$$
$$(b_{s,j-1}(R, B \setminus \{v\}) - b_{s,j-1}(R \setminus N[v], B \setminus \{v\})) \qquad (6)$$

3.1   The Algorithm

Now, we are ready to give our algorithm for PRBDS: Algorithm 1. We want to stress that this algorithm is exactly the algorithm described in [20] except that we extended the inclusion/exclusion branching rules as described in Section 2.

Let $G = (R \cup B, E)$ be a bipartite graph. The algorithm accepts a set of red vertices R, a set of blue vertices B, and a set of *annotated vertices* $A \subseteq (R \cup B)$ as input. The algorithm returns a matrix containing for each $0 \le k \le |R|$ and $0 \le t \le |B|$ the number of subsets $X \subseteq R$ such that $|N(X)| = t$ and $|X| = k$.

The annotated vertices have the following role (see also [20]): whenever the algorithm makes any decision through branching, annotated vertices are treated like the other vertices. However, when selecting a vertex to branch on, annotated vertices are ignored: both as candidate to branch on and as a neighbour

---

Algorithm 1. $\mathcal{T}(R, B, A))$

Input: Vertex sets R, B and $A \subseteq (R \cup B)$.
Output: The matrix with values $b_{s,j}(R, B)$ for every $0 \le s \le |R|$, $0 \le j \le |B|$.
1: Let $K = (R \cup B) \setminus A$
2: **if** $\exists v \in K$ such that $d_K(v) = 1$ **then**
3:     **return** $\mathcal{T}(R, B, A \cup \{v\})$
4: **else if** $\exists$false twins $v_1, v_2 \in K : d_K(v_1) \le 2$ and $d_K(v_2) \le 2$ **then**
5:     **return** $\mathcal{T}(R, B, A \cup \{v_1\})$
6: **else**
7:     Let $s \in R \setminus A$ be a vertex maximising $\delta_0 = d_K(s)$
8:     Let $e \in B \setminus A$ be a vertex maximising $\delta_1 = d_K(e)$
9:     **if** $\delta_0 \le 2$ and $\delta_1 \le 2$ **then**
10:       **return** Polytime-DP(R, B, A)
11:     **else if** $\delta_0 > \delta_1$ **then**
12:       Let $L_{\text{take}} = \mathcal{T}(R \setminus \{s\}, B \setminus N(s), A \setminus N(s))$
13:       Let $L_{\text{discard}} = \mathcal{T}(R \setminus \{s\}, B, A)$
14:       **return** shiftdown(shiftxtimestotheleft($|N(s)|, L_{\text{take}}$)) + L_{discard}
15:     **else**
16:       Let $L_{\text{optional}} = \mathcal{T}(R, B \setminus \{e\}, A)$
17:       Let $L_{\text{forbidden}} = \mathcal{T}(R \setminus N(e), B \setminus \{e\}, A \setminus N(e))$
18:       **return** $L_{\text{forbidden}}$ + shifttotheleft($L_{\text{optional}} - L_{\text{forbidden}}$)
19:     **end if**
20: **end if**

---

of another vertex effectively changing the degrees of their neighbours. From Lemma 1 in [20], it follows that when the procedure Polytime-DP is provoked, the treewidth of the graph $G[R \cup B \cup A]$ is at most 2. Using this, it is not hard to see that Polytime-DP can be implemented by a polynomial time dynamic programming algorithm.

Notice that the first branching rule starting on Line 12 uses Equation 1: if v is discarded, simply remove it; if v is taken, its neighbours are dominated, and so we remove its neighbourhood. Moreover, we update the counting machinery by shifting the entries in the computed matrix such that they correspond to taking one red vertex and dominating the blue vertices. The second branching rule (Line 16-18) is correct as it equals Equation 6, slightly reformulated.

3.2 Using PRBDS to Solve PARTIAL DOMINATING SET

It remains to show how to solve PARTIAL DOMINATING SET using Algorithm 1. Let us first state PARTIAL DOMINATING SET formally:

PARTIAL DOMINATING SET (PDS)
Input A graph (V, E); integers k, t such that $0 \le k, t \le |V|$.
Question Is there a subset $X \subseteq V$ such that $|N(X)| \ge t$ and $|X| \le k$.

We will use a fairly standard reduction (as in [10,11,22]) from PARTIAL DOMINATING SET to PRBDS: for each vertex v in G, introduce a red vertex and a blue vertex, each connected to all vertices with the opposite colour that represent the neighbours of v. A blue vertex represents that v must be dominated, and a red vertex represents that v can dominate others. Now, each t-red blue dominating set corresponds to an equivalent t-dominating set and vice versa.

Now, PDS can be solved by modelling it as PRBDS and running Algorithm 1. Clearly, the running time of this algorithm is the same as the one written in [20], up to a polynomial factor. One can improve the running time of this algorithm at the cost of exponential space by switching to a dynamic programming approach when the maximum degree in the graph is small. This combination was introduced by Fomin et al. [10] (also used in [22]). The details this exponential space algorithm will be given in the full paper: here, we only state the result[2].

Resuming, we proved the first part of the the following theorem while the proof of the second part is omitted due to space restrictions.

Theorem 1. *The* PARTIAL DOMINATING SET *problem can be solved by an algorithm using* $\mathcal{O}(1.5673^n)$ *time and polynomial space or* $\mathcal{O}(1.5014^n)$ *time and exponential space.*

3.3   Symmetry in the PRBDS Problem

The reduction from (PARTIAL) DOMINATING SET to the more general (PARTIAL) RED BLUE DOMINATING SET problem seems somewhat redundant, and it is a natural question how one can exploit the fact that the more specific problem has to be solved. An approach could be to exploit *symmetry* of the incidence graphs defined by subproblems solved by the branching algorithm.

In this sense a natural question also is whether the "flipped instance", obtained by changing roles of the red and blue vertices, is computationally different from the original. Indeed, the presented algorithm is completely symmetric (besides Line 11), so the two instances don't differ much for this algorithm concerning it's running time. The relation between the two appears to be even more intimate:

Lemma 2.

$$b_{s,j}(R, B) = \sum_{l=0}^{|R|} \sum_{m=0}^{|B|} (-1)^{m-j} \binom{m}{j} \binom{|R| - l}{s} b_{m,l}(B, R)$$

As a consequence, we have that when we are given the solution for the flipped instance, the solution of the original instance can be computed in polynomial time. This can be proved by using Equation 5 and grouping summands. The full proof is omitted due to space restrictions.

---

[2] Although this gives even a faster algorithm for DOMINATING SET than claimed in [22], we do not claim the fastest algorithm for this problem as recent submissions give better results. The recently submitted journal version of [21] gives a faster algorithm for DOMINATING SET, and even the journal version in preparation of [22] gives a slightly faster algorithm for counting dominating sets.

4 The k-SET SPLITTING Problem

In this section, we consider the parameterised SET SPLITTING problem. This problem is also known as parameterised MAX HYPERGRAPH 2-COLOURING.

k-SET SPLITTING
Input A family of sets S over a universe U.
Parameter Integer $k > 0$.
Question Can we partition the elements of U into two sets (Red/Green) such that at least k sets from S contain at least one element from both partitions.

The current fastest algorithms for this problem run in $\mathcal{O}^*(2^k)$ time using polynomial space or $\mathcal{O}^*(1.9630^k)$ time and space [17]. In this section, we will improve these results by giving an $\mathcal{O}^*(1.8213^k)$ time and polynomial space algorithm.

Our algorithm uses the following result from [17]:

Theorem 3 ([17]). k-SET SPLITTING *admits a kernel with at most $2k$ sets and at most k elements.*

This result allows us to perform the following preprocessing in linear time: either an input instance is a YES-instance, or it can be transformed into an equivalent instance that has at most $2k$ sets and k elements.

After this first preprocessing step, we apply the following test that directly decides that some instances are YES-instances.

Lemma 4 ([7]). *Let s_i be the number of sets of cardinality i in a given k-SET SPLITTING instance. The instance is a YES-instance if:*

$$k \le \frac{1}{2}s_2 + \frac{3}{4}s_3 + \frac{7}{8}s_4 + \frac{15}{16}s_5 + \frac{31}{32}s_6 + \frac{63}{64}s_7 + \cdots + \left(1 - \frac{1}{2^{i-1}}\right)s_i + \cdots$$

The above lemma can be proved by considering a random colouring and arguing that there must exist a colouring that splits at least as many sets as the expected number of sets split by a random colouring, which exactly is the right-hand side. See also Lemma 2.4 in [7] (note that we slightly modified the statement for our purposes).

After preprocessing, we apply a branching algorithm: Algorithm 2. This algorithm computes a list of numbers containing for each $0 \le l \le 2k$ the number of ways to colour the elements of the remaining instance such that exactly l sets are split. During the execution of our algorithm a partitioning of the remaining sets in three different types of sets is maintained: a set is in U (*uncoloured*) if none of its element have been coloured thus far, it is in R (*coloured red*) or in G (*coloured green*) if the thus far coloured elements all have the corresponding colour. If a set has elements of two different colours, then the set is split: it is removed from the instance and the entries in the computed list are shifted such that the numbers correspond to splitting one more set. Elements occur in only one set E (*uncoloured elements*): they are removed after a colour has been assigned and the partitioning of the sets has been updated.

Algorithm 2. $\text{SetSpl}(E, U, R, G, A)$

Input: The incidence graph of a SET SPLITTING instance with $V = E \cup U \cup R \cup G$
  where E contains the element vertices, U the uncoloured set vertices, and R and G
  the coloured set vertices with colour red and green, respectively, and a set $A \subset V$
  of annotated vertices.

Output: A list containing for each k the number of colourings splitting k sets.

1: Let $K = (E \cup U \cup R \cup G) \setminus A$
2: **if** $\exists v \in K$ such that $d_K(v) = 1$ **then**
3:     **return** $\text{SetSpl}(E, U, R, G, A \cup \{v\})$
4: **else if** $\exists$false twins $v_1, v_2 \in K : d_K(v_1) \leq 2$ and $d_K(v_2) \leq 2$ **then**
5:     **return** $\text{SetSpl}(E, U, R, G, A \cup \{v_1\})$
6: **else**
7:     Let $e \in E \setminus A$ be a vertex maximising $\delta_e = d_K(e)$.
8:     Let $u \in U \setminus A$ be a vertex maximising $\delta_u = d_K(u)$.
9:     Let $c \in (R \cup G) \setminus A$ be a vertex maximising $\delta_c = d_K(c)$.
10:    **if** $\max(\delta_e, \delta_u, \delta_c) \leq 2$ **then**
11:       **return** $\text{Polytime-DP}(E, U, R, G, A)$
12:    **else if** $\delta_e = \max(\delta_e, \delta_u, \delta_c)$ **then**
13:       Let $L_{\text{green}} = \text{SetSpl}(E \setminus \{e\}, U \setminus N(e), R \setminus N(e), G \cup (N(e) \cap U), A \setminus (N(e) \cap R))$
14:       Let $L_{\text{red}} = \text{SetSpl}(E \setminus \{e\}, U \setminus N(e), R \cup (N(e) \cap U), G \setminus N(e), A \setminus (N(e) \cap G))$
15:       **return** $\text{shiftxtimestotheleft}(L_{\text{green}}, |R \cap N(e)|)$
            $+ \text{shiftxtimestotheleft}(L_{\text{red}}, |G \cap N(e)|)$
16:    **else if** $\delta_u = \max(\delta_e, \delta_u, \delta_c)$ **then** $//\delta_u > \delta_e$
17:       $L_{\text{optional}} = \text{SetSpl}(E, U \setminus \{u\}, R, G, A)$
18:       Create a new element vertex x that is adjacent to all vertices in $N^2(u)$.
19:       $L_{\text{forbidden}} = \text{SetSpl}(E \cup \backslash N(u) \cup \{x\}, U \setminus \{u\}, R, G, A \setminus N[u])$
20:       **return** $L_{\text{forbidden}} + \text{shifttotheleft}(L_{\text{optional}} - L_{\text{forbidden}})$
21:    **else** $//\delta_c > \delta_e, \delta_u$
22:       $L_{\text{optional}} = \text{SetSpl}(E, U, R \setminus \{c\}, G \setminus \{c\}, A)$
23:       **if** $c \in R$ **then**
24:          $L_{\text{forbidden}} =$
               $\text{SetSpl}(E \setminus N(c), U \setminus N^2(c), (R \setminus \{c\}) \cup (N^2(c) \cap U), G \setminus N^2(c), A \setminus N[u])$
25:       **else** $//c \in G$
26:          $L_{\text{forbidden}} =$
               $\text{SetSpl}(E \setminus N(c), U \setminus N^2(c), R \setminus N^2(c), (G \setminus \{c\}) \cup (N^2(c) \cap U), A \setminus N[u])$
27:       **end if**
28:       **return** $L_{\text{forbidden}} + \text{shifttotheleft}(L_{\text{optional}} - L_{\text{forbidden}})$
29:    **end if**
30: **end if**

It should be noted that we have abused notation slightly since the incidence graph is also given as input and even modified (on Line 19).

Algorithm 2 looks a lot like Algorithm 1. It uses the same annotation procedure that marks low degree vertices that represent sets and elements such that they will be ignored by the procedure that selects a vertex to branch on. Hence, the procedure Polytime-DP(E, U, R, G, A) can be implemented in polynomial time by Lemma 1 from [20] using dynamic programming.

Our algorithm uses three different branching rules, two of which are based on the branching rule as defined in Section 2:

1. **Branching on an element.** (lines 13-15) Recursively solve two subproblems, one in which the element is coloured red, and one where it is coloured green. After this, shift the computed entries to update the number of split sets and join the results by summing the two lists component wise.
2. **Branching on an uncoloured set.** (lines 17-20) We apply Equation 4. Because we are computing lists of numbers with for each l the number of solutions splitting exactly l sets, we only need to compute two of these lists: one where it is *optional* to split the set, and one where it is *forbidden* to split the set. In the optional branch, we remove the set. In the forbidden branch, we remove the set and merge all its elements to a single element: a solution to this instance corresponds to a solution of the original instance in which all elements in the set have the same colour, i.e., in which the set is not split.
3. **Branching on a coloured set.** (lines 21-26) Similar to branching on uncoloured sets, we apply Equation 4 and generate only two subproblems. In the optional branch, we remove the set. In the forbidden branch, we generate an instance equivalent to making sure that the set will not be split. Since the set is already coloured, either red or green, this means that we colour all elements in the set with the colour that is already present in this set.

For the running time, we use *Measure & Conquer* [11] with weight functions $v, w, x : \mathbb{N} \to \mathbb{R}_+$ and the following measure μ on subproblems (E, U, R, G, A):

$$\mu = \sum_{e \in E, e \notin A} v(d_K(e)) + \sum_{u \in U, u \notin A} w(d_K(u)) + \sum_{c \in (R \cup G), c \notin A} x(d_K(c))$$

where $K = (E \cup U \cup R \cup G) \setminus A$ and with the following values for the weights:

i	≤ 1	2	3	4	5	6	> 6
$v(i)$	0.000000	0.726515	1.000000	1.000000	1.000000	1.000000	1.000000
$w(i)$	0.000000	0.525781	0.903938	1.054594	1.084859	1.084885	1.084885
$x(i)$	0.000000	0.387401	0.617482	0.758071	0.792826	0.792852	0.792852

Lemma 5. *Algorithm 2 runs in time $\mathcal{O}(1.31242^{\mu})$ on an input of measure μ.*

Proof (Sketch). Standard measure and conquer analysis. Let $N(\mu)$ be the number of subproblem generated on an input of measure μ. We generate a large set of recurrence relations of the form $N(\mu) \leq N(\mu - \Delta\mu_1) + N(\mu - \Delta\mu_1)$ representing all possible situations in which Algorithm 2 can branch. The solution of this set of recurrence relations satisfies $N(\mu) \leq 1.31242^{\mu}$ proving the lemma.   □

Theorem 6. *There exists an $\mathcal{O}^*(1.8213^k)$ time and polynomial space algorithm for k-*SET SPLITTING.

Proof. First, apply the preprocessing of Theorem 3 and Lemma 4. Then, apply Algorithm 2 to the reduced instance. From the resulting list containing for each $0 \leq l \leq 2k$ the number of colourings of the remaining elements that split exactly l sets, we can see if there exist a colouring splitting at least k sets.

The running time is dominated by the exponential time used by the call to Algorithm 2 since the preprocessing is done in polynomial time. We use Lemma 5 to compute this running time. By Theorem 3 an instance can have at most k elements that each have measure at most 1. By Lemma 4 a reduced instance must satisfy $k > \sum_{i=2}^{\infty} \left(1 - \frac{1}{2^{i-1}}\right) s_i$ where s_i is the number of sets of cardinality i. Hence, the maximum measure z of a reduced instance is the solution of:

$$z = k + \max \sum_{i=2}^{\infty} w(i)s_i \quad \text{with the restriction} \quad \sum_{i=2}^{\infty} \left(1 - \frac{1}{2^{i-1}}\right) s_i < k$$

We obtain $z = 2.205251k$. This completes the proof as $1.31242^z < 1.8213^k$. □

It should be noted that an algorithm for the more general k-NOT ALL EQUAL SAT problem with the same running time can be obtained by extending the algorithm. The kernel of [17] also extends to this problem (as noted in [17]).

Acknowledgements. We thank Daniel Lokshtanov for suggesting Lemma 4 and useful discussions. We also thank Thomas C. van Dijk for useful discussions.

References

1. Amini, O., Fomin, F.V., Saurabh, S.: Implicit branching and parameterized partial cover problems. In: IARCS Annual Conference on Foundations of Software Technology and Theoretical Computer Science, FSTTCS 2008. LIPIcs, vol. 2, pp. 1–12 (2008)
2. Bax, E.: Recurrence-based reductions for inclusion and exclusion algorithms applied to #$\mathcal{P}$ problems. Technical report, California Institute of Technology (1996)
3. Björklund, A., Husfeldt, T., Kaski, P., Koivisto, M.: Fourier meets möbius: fast subset convolution. In: 39th Annual ACM Symposium on Theory of Computing, STOC 2007, pp. 67–74 (2007)
4. Björklund, A., Husfeldt, T., Kaski, P., Koivisto, M.: The travelling salesman problem in bounded degree graphs. In: Aceto, L., Damgård, I., Goldberg, L.A., Halldórsson, M.M., Ingólfsdóttir, A., Walukiewicz, I. (eds.) ICALP 2008, Part I. LNCS, vol. 5125, pp. 198–209. Springer, Heidelberg (2008)
5. Björklund, A., Husfeldt, T., Kaski, P., Koivisto, M.: Trimmed moebius inversion and graphs of bounded degree. In: 25th Symposium on Theoretical Aspects of Computer Science, STACS 2008. LIPIcs, vol. 1, pp. 85–96 (2008)
6. Björklund, A., Husfeldt, T., Koivisto, M.: Set partitioning via inclusion-exclusion. SIAM Journal on Computing 39(2), 546–563 (2009)
7. Chen, J., Lu, S.: Improved parameterized set splitting algorithms: A probabilistic approach. Algorithmica 54(4), 472–489 (2009)

8. Dehne, F.K.H.A., Fellows, M.R., Rosamond, F.A.: An FPT algorithm for set splitting. In: Bodlaender, H.L. (ed.) WG 2003. LNCS, vol. 2880, pp. 180–191. Springer, Heidelberg (2003)

9. Dehne, F.K.H.A., Fellows, M.R., Rosamond, F.A., Shaw, P.: Greedy localization, iterative compression, modeled crown reductions: New FPT techniques, an improved algorithm for set splitting, and a novel $2k$ kernelization for vertex cover. In: Downey, R.G., Fellows, M.R., Dehne, F. (eds.) IWPEC 2004. LNCS, vol. 3162, pp. 271–280. Springer, Heidelberg (2004)

10. Fomin, F.V., Gaspers, S., Saurabh, S., Stepanov, A.A.: On two techniques of combining branching and treewidth. Algorithmica 54(2), 181–207 (2009)

11. Fomin, F.V., Grandoni, F., Kratsch, D.: A measure & conquer approach for the analysis of exact algorithms. Journal of the ACM 56(5) (2009)

12. Fomin, F.V., Lokshtanov, D., Raman, V., Saurabh, S.: Subexponential algorithms for partial cover problems. In: IARCS Annual Conference on Foundations of Software Technology and Theoretical Computer Science, FSTTCS 2009. LIPIcs, vol. 4, pp. 193–201 (2009)

13. Impagliazzo, R., Paturi, R.: On the complexity of k-sat. Journal of Computer and System Sciences 62(2), 367–375 (2001)

14. Kneis, J., Mölle, D., Rossmanith, P.: Partial vs. complete domination: t-dominating set. In: van Leeuwen, J., Italiano, G.F., van der Hoek, W., Meinel, C., Sack, H., Plášil, F. (eds.) SOFSEM 2007. LNCS, vol. 4362, pp. 367–376. Springer, Heidelberg (2007)

15. Koutis, I., Williams, R.: Limits and applications of group algebras for parameterized problems. In: 36th International Colloquium on Automata, Languages and Programming, ICALP 2009. LNCS, vol. 5555, pp. 653–664. Springer, Heidelberg (2009)

16. Liedloff, M.: Algorithmes exacts et exponentiels pour les problèmes NP-difficiles: domination, variantes et généralisation, PhD thesis (2007)

17. Lokshtanov, D., Saurabh, S.: Even faster algorithm for set splitting! In: 4th International Workshop on Parameterized and Exact Computation, IWPEC 2009. LNCS, vol. 5917, pp. 288–299. Springer, Heidelberg (2009)

18. Lokshtanov, D., Sloper, C.: Fixed parameter set splitting, linear kernel and improved running time. In: 1th Algorithms and Complexity in Durham Workshop, ACiD 2005. Texts in Algorithmics, vol. 4, pp. 105–113 (2005)

19. Nederlof, J.: Fast polynomial-space algorithms using möbius inversion: Improving on steiner tree and related problems. In: 36th International Colloquium on Automata, Languages and Programming, ICALP 2009. LNCS, vol. 5555, pp. 713–725. Springer, Heidelberg (2009)

20. van Rooij, J.M.M.: Polynomial space algorithms for counting dominating sets and the domatic number. In: Calamoneri, T., Diaz, J. (eds.) Algorithms and Complexity. LNCS, vol. 6078, pp. 73–84. Springer, Heidelberg (2010)

21. van Rooij, J.M.M., Bodlaender, H.L.: Design by measure and conquer, a faster exact algorithm for dominating set. In: 25th Annual Symposium on Theoretical Aspects of Computer Science, STACS 2008. LIPIcs, vol. 1, pp. 657–668 (2008)

22. van Rooij, J.M.M., Nederlof, J., van Dijk, T.C.: Inclusion/exclusion meets measure and conquer. In: Fiat, A., Sanders, P. (eds.) ESA 2009. LNCS, vol. 5757, pp. 554–565. Springer, Heidelberg (2009)

23. Woeginger, G.J.: Exact algorithms for np-hard problems: a survey. In: Combinatorial optimization - Eureka, you shrink!, pp. 185–207 (2003)

Small Vertex Cover Makes Petri Net Coverability and Boundedness Easier

M. Praveen

The Institute of Mathematical Sciences, Chennai, India

Abstract. The coverability and boundedness problems for Petri nets are known to be EXPSPACE-complete. Given a Petri net, we associate a graph with it. With the vertex cover number k of this graph and the maximum arc weight W as parameters, we show that coverability and boundedness are in PARAPSPACE. This means that these problems can be solved in space $\mathcal{O}\left(ef(k, W)poly(n)\right)$, where $ef(k, W)$ is some exponential function and $poly(n)$ is some polynomial in the size of the input. We then extend the PARAPSPACE result to model checking a logic that can express some generalizations of coverability and boundedness.

1 Introduction

Petri nets, introduced by C. A. Petri [18], are popularly used for modelling concurrent infinite state systems. Using Petri nets to verify various properties of concurrent systems is an ongoing area of research, with abstract theoretical results like [2] and actually constructing tools for C programs like [13]. Reachability, coverability and boundedness are some of the most fundamental questions about Petri nets. All three of them are EXPSPACE-hard [16]. Coverability and boundedness are in EXPSPACE [20]. Reachability is known to be decidable [17, 14] but no upper bound is known.

In this paper, we study the parameterized complexity of coverability and boundedness problems. The parameters we consider are vertex cover number k of the underlying graph of the given Petri net and the maximum arc weight W. We show that both problems can be solved in space exponential in the parameters and polynomial in the size of the input. Such algorithms are called PARAPSPACE algorithms. Fundamental complexity theory of such parameterized complexity classes have been studied [9], but parameterized PTIME (popularly known as Fixed Parameter Tractable, FPT) is the most widely studied class. Usage of other parameterized classes such as PARAPSPACE is rare in the literature.

As mentioned before, one of the uses of Petri nets is modelling software. It is desirable to have better complexity bounds for certain classes of Petri nets that may have some simple underlying structure due to human designed systems that the nets model. For example, it is known that well structured programs have small treewidth [23]. Unfortunately, the Petri net used by Lipton in the reduction in [16] (showing EXPSPACE-hardness) has a constant treewidth. Hence, we cannot hope to get better bounds for coverability and boundedness with treewidth as parameter. Same is the case with many other parameters like pathwidth, cycle

V. Raman and S. Saurabh (Eds.): IPEC 2010, LNCS 6478, pp. 216–227, 2010.

rank, dagwidth etc. Hence, we are forced to look for stronger parameters. In [19], we studied the effect of a newly introduced parameter called benefit depth. In this paper, we study the effect of using vertex cover as parameter, using different techniques. The class of Petri nets with bounded benefit depth is incomparable with the class of Petri nets with bounded vertex cover.

Feedback vertex set of a graph is a set of vertices whose removal leaves the graph without any cycles. The smallest feedback vertex set of the Petri net used in the lower bound proof of [16] is large (as opposed to treewidth, pathwidth, cycle rank etc., which are small). In the context of modelling software, smallest feedback vertex set can be thought of as control points covering all loop structures. In fact, the Petri net in the lower bound proof of [16] models a program that uses a large number of loops to manipulate counters that can hold doubly exponential values. Removal of a feedback vertex set leaves a Petri net without any cycles. It would be interesting to explore the complexity of coverability and boundedness problems with the size of the smallest feedback vertex set as parameter. We have not been able to extend our results to the case of feedback vertex set yet, but hope that these results will serve as a theoretically interesting intermediate step.

In a tutorial article [6], Esparza argues that for most interesting questions about Petri nets, the rule of thumb is that they are all EXPSPACE-hard. Despite this, the introduction of the same article contains an excellent set of reasons for studying finer complexity classification of such problems. We will not reproduce them here but note some relevant points — many experimental tools have been built that solve EXPSPACE-complete problems that can currently handle small instances. Also, a knowledge of complexity of problems helps in answering other questions. In such a scenario, having an "extended dialog" with the problem is beneficial, and parameterized complexity is very good at doing this [4].

Related work. In [22], Rosier and Yen study the complexity of coverability and boundedness problems with respect to different parameters of the input instance, such as number of places, transitions, arc weight etc. In particular, they show that the space required for boundedness is exponential in the number of unbounded places and polynomial in the number of bounded places. If for a Petri net, the smallest vertex cover is the set of all places, our results coincide with those found in [22]. Hence, our results refine those of Rosier and Yen. In [12], Habermehl shows that the problem of model checking linear time μ-calculus formulas on Petri nets is PSPACE-complete in the size of the formula and EXPSPACE-complete in the size of the net. However, the μ-calculus considered in [12] cannot express coverability and boundedness. In [24], Yen extends the induction strategy used by Rackoff in [20] to give EXPSPACE upper bound for deciding many other properties. Another work closely related to Yen's above work is [1].

One-counter automata are closely related to Petri nets. Precise complexity of reachability and many other problems of this model have been recently obtained in [11, 10]. We have adapted some of the techniques used in [11, 10], in particular the use of [15, Lemma 42].

The effect of treewidth and other parameters on the complexity of some pebbling problems on digraphs have been considered in [5, Section 5]. These problems relate to the reachability problem in a class of Petri nets (called *Elementary Net Systems*) with semantics that are different from the ones used in this paper (see [21] for details of different Petri Net semantics).

2   Preliminaries

Let $\mathbb{Z}$ be the set of integers and $\mathbb{N}$ the set of natural numbers. A Petri net is a 4-tuple $\mathcal{N} = (P, T, Pre, Post)$, where P is a set of places, T is a set of transitions and Pre and $Post$ are the incidence functions: $Pre : P \times T \to [0 \ldots W]$ (arcs going from places to transitions) and $Post : P \times T \to [0 \ldots W]$ (arcs going from transitions to places), where $W \geq 1$. In diagrams, places will be represented by circles and transitions by thick bars. Arcs are represented by weighted directed edges between places and transitions.

A function $M : P \to \mathbb{N}$ is called a *marking*. A marking can be thought of as a configuration of the Petri net, with every place p having $M(p)$ tokens. Given a Petri net $\mathcal{N}$ with a marking M and a transition t such that for every place p, $M(p) \geq Pre(p, t)$, the transition t is said to be *enabled* at M and can be *fired*. After firing, the new marking M' (denoted as $M \overset{t}{\Rightarrow} M'$) is given by $M'(p) = M(p) - Pre(p, t) + Post(p, t)$ for every place p. A place p is an *input* (*output*) place of a transition t if $Pre(p, t) \geq 1$ ($Post(p, t) \geq 1$) respectively. We can think of firing a transition t resulting in $Pre(p, t)$ tokens being deducted from every input place p and $Post(p', t)$ tokens being added to every output place p'. A sequence of transitions $\sigma = t_1 t_2 \cdots t_r$ (called *firing sequence*) is said to be enabled at a marking M if there are markings $M_1, \ldots, M_r$ such that $M \overset{t_1}{\Rightarrow} M_1 \overset{t_2}{\Rightarrow} \cdots \overset{t_r}{\Rightarrow} M_r$. $M, M_1, \ldots, M_r$ are called *intermediate markings*. The fact that firing σ at M results in M_r is denoted by $M \overset{\sigma}{\Rightarrow} M_r$.

We assume that a Petri net is presented as two matrices for Pre and $Post$. In the rest of this paper, we will assume that a Petri net $\mathcal{N}$ has m places, n transitions and that W is the maximum of the range of Pre and $Post$. We define the size of the Petri net to be $|\mathcal{N}| = 2mn \log W + m \log |M_0|$ bits, where $|M_0|$ is the maximum of the range of the initial marking M_0.

Definition 2.1 (Coverability and Boundedness). *Given a Petri net with an initial marking M_0 and a target marking M_{cov}, the Coverability problem is to determine if there is a firing sequence σ such that $M_0 \overset{\sigma}{\Rightarrow} M'$ and for every place p, $M'(p) \geq M_{cov}(p)$ (this is denoted as $M' \geq M_{cov}$). The boundedness problem is to determine if there is a number $c \in \mathbb{N}$ such that for every firing sequence σ enabled at M_0 with $M_0 \overset{\sigma}{\Rightarrow} M$, $M(p) \leq c$ for every place p.*

In the Petri net shown in Fig. 1, the initial marking M_0 is given by $M_0(p_1) = 1$ and $M_0(p_2) = M_0(p_3) = 0$. If M_{cov} is defined as $M_{cov}(p_1) = M_{cov}(p_2) = 1$ and $M_{cov}(p_3) = 0$, then M_{cov} is not coverable since p_1 and p_2 cannot have tokens simultaneously. Since for any $c \in \mathbb{N}$, the Petri net in Fig. 1 can reach a marking where p_3 has more than c tokens (by firing the sequence $t_1 t_2$ repeatedly), this

Petri net is not bounded. Lipton proved both coverability and boundedness problems to be EXPSPACE-hard [16, 6]. Rackoff provided EXPSPACE upper bounds for both problems [20]. In the definition of the coverability problem, if we replace $M' \geq M_{cov}$ by $M' = M_{cov}$, we get the *reachability* problem. Lipton's EXPSPACE lower bound applies to the reachability problem too, and this is the best known lower bound. Though the reachability problem is known to be decidable [17, 14], no upper bound is known. Many of the problems that are decidable for bounded Petri nets are undecidable for unbounded Petri nets. Model checking some logics extending the one defined in section 6 fall into this category. Esparza and Nielsen survey such results in [7]. Reachability, coverability and boundedness are few problems that remain decidable for unbounded Petri nets.

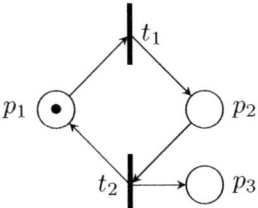

Fig. 1. An example of a Petri net

Proofs marked with (*) are skipped due to lack of space. A full version of this paper with the same title is available in arXiv, which contains all the proofs.

3   Vertex Cover for Petri Nets

In this section, we introduce the notion of vertex cover for Petri nets and intuitively explain how small vertex covers help in getting better algorithms. We will also state and prove the key technical lemma used in the next two sections.

For a normal graph $G = (V, E)$ with set of vertices V and set of edges E, a vertex cover $VC \subseteq V$ is a subset of vertices such that every edge has at least one of its vertices in VC. Given a Petri net $\mathcal{N}$, we associate with it an undirected graph $G(\mathcal{N})$ whose set of vertices is the set of places P. Two vertices are connected by an edge if there is a transition connecting the places corresponding to the two vertices. To be more precise, if two vertices represent two places p_1 and p_2, then there is an edge between the vertices in $G(\mathcal{N})$ iff in $\mathcal{N}$, there is some transition t such that $Pre(p_1, t) + Post(p_1, t) \geq 1$ and $Pre(p_2, t) + Post(p_2, t) \geq 1$. If a place p is both an input and an output place of some transition, the vertex corresponding to p has a self loop in $G(\mathcal{N})$. Any vertex cover of $G(\mathcal{N})$ should include all vertices that have self loops.

Suppose VC is a vertex cover for some graph G. If $v_1, v_2 \notin VC$ are two vertices not in VC that have the same set of neighbours (neighbours of a vertex v are vertices that have an edge connecting them to v), v_1 and v_2 have similar properties. This fact is used to obtain FPT algorithms for many hard problems, e.g., see [8]. The same phenomenon leads to PARAPSPACE algorithms for Petri net coverability and boundedness. In the rest of this section, we will define the formalisms needed to prove these results.

Let the places of a Petri net $\mathcal{N}$ be $p_1, p_2, \ldots, p_m$. Suppose there is a vertex cover VC consisting of places $p_1, \ldots, p_k$. We say that two transitions t_1 and t_2 are of the same type if $Pre(p_i, t_1) = Pre(p_i, t_2)$ and $Post(p_i, t_1) = Post(p_i, t_2)$ for all i between 1 and k. In Fig. 2, transitions t_1 and t_5 are of the same type. Intuitively, two transitions of the same type behave similarly as far as places in the vertex cover are concerned. Since there can be $2k$ arcs between a transition and places in VC and each arc can have weight between 0 and W, there can be at most $(W + 1)^{2k}$ different types of transitions.

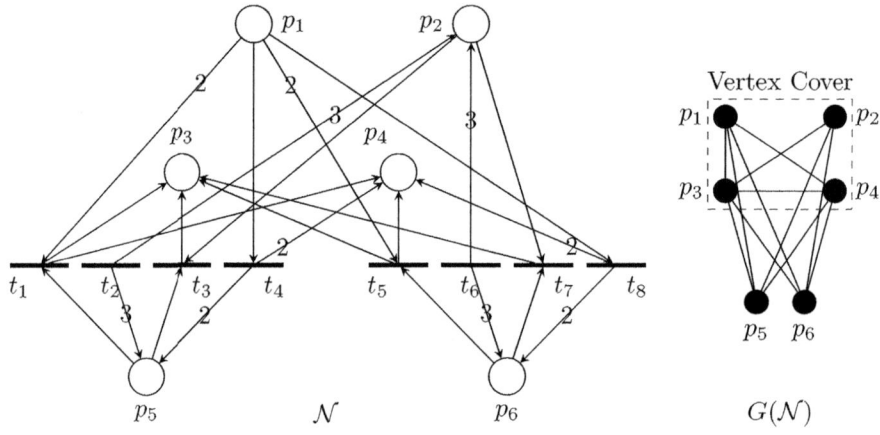

Fig. 2. A Petri net with vertex cover $\{p_1, \ldots, p_4\}$

Let p be a place not in the vertex cover VC. Suppose there are $l \leq (W+1)^{2k}$ types of transitions. Place p can have one incoming arc from or one outgoing arc to each transition of the net (it cannot have both an incoming and an outgoing arc since in that case, p would have a self loop and would be in VC). If p' is another place not in VC, then no transition can have arcs to both p and p', since otherwise, there would haven been an edge between p and p' in $G(\mathcal{N})$ and one of the places p and p' would have been in VC. Hence, places not in VC cannot interact with each other directly. Places not in VC can only interact with places in VC through transitions and there are at most l types of transitions. Suppose p and p' have the following property: for every transition t that has an arc to/from p with weight w, there is another transition t' of the same type as t that has an arc to/from p' with weight w. Then, p and p' interact with VC in the same way in the following sense: whenever a transition involving p fires, an "equivalent" transition can be fired that involves p' instead of p, provided there are enough tokens in p'. In Fig. 2, places p_5 and p_6 satisfy the property stated above. Transition t_5 can be fired instead of t_1, t_6 can be fired instead of t_2 etc.

Definition 3.1. *Suppose $\mathcal{N}$ is a Petri net with vertex cover VC and l types of transitions. Let $p \notin VC$ be a place not in the vertex cover. The variety $var[p]$ of p is defined as the function[1] $var[p] : \{1, \ldots l, \} \to 2^{\{-W, \ldots, W\} \setminus \{0\}}$, where for every j between 1 and l and every $w \neq 0$ between $-W$ and W, there is a transition t_j of type j such that $w = -Pre(p, t_j) + Post(p, t_j)$ iff $w \in var[p]$. We denote varieties of places by v, v' etc.*

In the above definition, since $p \notin VC$, at most one among $Pre(p, t_j)$ and $Post(p, t_j)$ will be non-zero.

The fact that transitions can be exchanged between two places of the same variety can be used to obtain better bounds on the length of firing sequences. For example, suppose a firing sequence σ is fired in the Petri net of Fig. 2, with an initial marking that has no tokens in p_5 and p_6. Let c be the maximum number of tokens in any place in any intermediate marking during the firing of σ. Since there are 6 places and each intermediate marking has at most c tokens in every place, the number of possible distinct intermediate markings is $(c+1)^6$. This is also an upper bound on the length of σ (if two intermediate markings are equal, then the subsequence between those two markings can be removed without affecting the final marking reached). Now, suppose that in the final marking reached, p_5 and p_6 do not have any tokens and we replace all occurrences of t_5, t_6, t_7 and t_8 in σ by t_1, t_2, t_3 and t_4 respectively. After this replacement, the final marking reached will be same as the one reached after firing σ. Number of tokens in p_5 will be at most $2c$ in any intermediate marking and there will be no tokens at all in p_6. Variation in the number of tokens in p_1, p_2, p_3 and p_4 do not change (since as far as these places are concerned, transitions t_5, t_6, t_7 and t_8 behave in the same way as do t_1, t_2, t_3 and t_4 respectively). Hence, in any intermediate marking, each of the places p_1, p_2, p_3 and p_4 will still have at most c tokens. When we exchange the transitions as mentioned above, there might be some intermediate markings that are same, so that we can get a shorter firing sequence achieving the same effect as the original one. These duplicate markings signify the "redundancy" that was present in the original firing sequence σ, but was not apparent to us due to the distribution of tokens among places. After removing such redundancies, the new upper bound on the length of the firing sequence is $(2c + 1).(c + 1)^4$, which is asymptotically smaller than the previous bound $(c + 1)^6$. A careful observation of the effect of this phenomenon on Rackoff's induction strategy in [20] leads us to the main results of this paper.

Definition 3.2. *Let p_1 and p_2 be two places of the same variety. Let σ be a firing sequence. A sequence of transitions $\sigma' = t_1 \ldots t_r$ is said to be a sub-word of σ if there are positions $i_1 < \cdots < i_r$ in σ such that for each j between 1 and r, $i_j{}^{th}$ transition of σ is t_j. Suppose σ' is a sub-word of σ made up of transitions that have an arc to/from p_1. **Transferring** σ' **from** p_1 **to** p_2 means replacing every transition t of σ' (which has an arc to/from p_1 with some weight w) with another transition t' of the same type as t which has an arc to/from p_2 with*

---

[1] The author acknowledges an anonymous IPEC referee for pointing out an error here in the submitted version.

weight w. The sub-word σ' is said to be **safe for transfer** from p_1 if for every prefix σ'' of σ', the effect of σ'' on p_1 (i.e., the change in the number of tokens in p_1 as a result of firing all transitions in σ'') is greater than or equal to 0.

Intuitively, if some sub-word σ' is safe for transfer from p_1, it never removes more tokens from p_1 than it has already added to p_1. So if we transfer σ' from p_1 to p_2, the new transitions will always add tokens to p_2 before removing them from p_2, so there is no chance of number of tokens in p_2 becoming negative due to the transfer. However, the number of tokens in p_1 may become negative due to some old transitions remaining back in the "untransferred" portion of the original firing sequence σ. The following lemma says that if some intermediate marking has very high number of tokens in some place, then a suitable sub-word can be safely transfered without affecting the final marking reached or introducing negative number of tokens in any place, but reducing the maximum number of tokens accumulated in any intermediate marking. The proof is a simple consequence of [15, Lemma 42], which is about one-counter automata. A full proof of the following lemma can be found in the Appendix of the full version.

Lemma 3.3 (Truncation lemma, [15]). *Let p_1 and p_2 be places of the same variety. Let $e \in \mathbb{N}$ be any number and σ be a firing sequence. Suppose during the firing of σ, there are intermediate markings M_1 and M_3 such that $M_1(p_1) = e$ and $M_3(p_1) \leq e$. Suppose M_2 is an intermediate marking between M_1 and M_3 such that $M_2(p_1) \geq e + W^2 + W^3$ is the maximum number of tokens in p_1 at any intermediate marking between M_1 and M_3. Then, there is a sub-word σ' of σ that is safe for transfer from p_1 to p_2 such that*

1. *The total effect of σ' on p_1 is 0.*
2. *After transferring σ' to p_2, the number of tokens in p_1 at M_2 is strictly less than the number of tokens in p_1 at M_2 before the transfer.*
3. *No intermediate marking will have negative number of tokens in p_1 after the transfer.*

There can be at most $(2^{2W})^l \leq 2^{2W(W+1)^{2k}}$ varieties of places that are not in the vertex cover VC, if the number of places in the vertex cover is k. For each variety v, we designate one of the places having v as its variety as special, and use p_v to denote it. We will call $S = VC \cup \{p_v \mid v$ is the variety of a place not in $VC\}$ the set of special places. We will denote the set $P \setminus S$ using I and call the places in I independent places. We will use k' for the cardinality of S and note that $k' \leq k + 2^{2W(W+1)^{2k}}$. If k and W are parameters, then k' is a function of the parameters only. Hence, in the rest of the paper, we will treat k' as the parameter.

4   ParaPspace Algorithm for the Coverability Problem

In this section, we will show that for a Petri net $\mathcal{N}$ with a vertex cover of size k and maximum arc weight W, the coverability problem can be solved in space $\mathcal{O}(ef(k,W)poly(|\mathcal{N}| + \log |M_{cov}|))$. Here, ef is some computable function exponential in k and W while $poly(|\mathcal{N}|+\log|M_{cov}|)$ is some polynomial in the size of the net and the marking to be covered. We will need the following definition, which is Definition 3.1 from [20] adapted to our notation.

Definition 4.1. *Let* $Q \subseteq P$ *be some subset of places such that* $I \subseteq Q$. *For a transition* t *and functions* $M, M' : P \to \mathbb{Z}$, *we write* $M \xrightarrow[Q]{t} M'$ *if* $M'(p) = M(p) - Pre(p,t) + Post(p,t)$ *for all* $p \in P$ *and* $M(q), M'(q) \geq 0$ *for all* $q \in Q$. *Let* M_{cov} *be some marking to be covered. For a function* $M_0 : P \to \mathbb{Z}$, *a firing sequence* $\sigma = t_1 t_2 \cdots t_r$ *is said to be* Q-**covering** *from* M_0 *if there are intermediate functions* $M_1, M_2, \ldots, M_r$ *such that* $M_0 \xrightarrow[Q]{t_1} M_1 \xrightarrow[Q]{t_2} \cdots \xrightarrow[Q]{t_r} M_r$ *and* $M_r(q) \geq M_{cov}(q)$ *for all* $q \in Q$. *The firing sequence* σ *is further said to be* Q, e-*covering if for all* i *between* 0 *and* $r - 1$, *the functions* M_i *above satisfy* $M_i(q) \leq e$ *for all* $q \in Q$. *For a function* $M : P \to \mathbb{Z}$, *let* $lencov(Q, M, M_{cov})$ *be the length of the shortest firing sequence that is* Q-*covering from* M. *Define* $lencov(Q, M, M_{cov})$ *to be* 0 *if there is no such sequence. Define* $\ell(i) = \max\{lencov(Q, M, M_{cov}) \mid I \subseteq Q \subseteq P, |Q \setminus I| = i, M : P \to \mathbb{Z}\}$.

Intuitively, a Q-covering sequence does not care about places that are not in Q, even if some intermediate markings have "negative number of tokens". The number $\ell(i)$ is an upper bound on the length of covering sequences that only care about independent places and i special places. Obviously, we are only interested in $\ell(k')$, but other values help in obtaining it. With slight abuse of terminology, we will call functions $M : P \to \mathbb{Z}$ also as markings. It will be clear from context what is meant.

Let R be the maximum of the range of M_{cov}, the marking to be covered. We will denote $R + W + W^2 + W^3$ by R'. Recall that m is the number of places in the given Petri net. The following lemmas give an upper bound on $\ell(k')$.

Lemma 4.2. $\ell(0) \leq mR$.

Proof. $\ell(0)$ is the length of the shortest I-covering sequence. Recall that all places in I are independent of each other, so if a transition has an arc to one of the places in I, it does not have arcs to any other place in I. Since an I-covering sequence does not care about places in S, it only has to worry about adding tokens to places in I. If a transition adds a token to some place p in I, it does not remove tokens from any other place in I. Hence, this transition can be repeated R times to add at least R tokens to the place p, which is all that is needed for p. Arguing similarly for other places in I, a total of mR transitions are enough to add all required tokens to all places in I, since there are less than m places in I. $\qquad\square$

Lemma 4.3 (*). $\ell(i + 1) \leq R'^m(W\ell(i) + R)^{i+1} + \ell(i)$.

The following lemma gives an upper bound on $\ell(i)$ using the recurrence relation obtained above.

Lemma 4.4 (*). $\ell(i) \leq (2mWRR')^{m(i+1)!}$.

Theorem 4.5. *With the vertex cover number k and maximum arc weight W as parameters, the Petri net coverability problem can be solved in* PARAPSPACE.

Proof. From the Lemma 4.4, we get $\ell(k') \leq (2mWRR')^{m(k'+1)!}$. To guess and verify a covering sequence of length at most $\ell(k')$, a non-deterministic Turing machine needs to maintain a counter and intermediate markings, which can be done using memory size $\mathcal{O}(m(k'+1)!(m \log |M_0| + \log m + \log W + \log R + \log R'))$. An application of Savitch's theorem then gives us the PARAPSPACE algorithm. □

5   The Boundedness Problem

In this section, we will show that with vertex cover number and maximum arc weight as parameters, the Petri net boundedness problem can be solved in PARA-PSPACE. If there is a firing sequence σ such that $M_0 \xrightarrow{\sigma} M_1$ and an intermediate marking M such that $M < M_1$ (i.e., $M \leq M_1$ and $M \neq M_1$), then σ is called a *self-covering sequence*. It is well known that a Petri net is unbounded iff the initial marking enables a self-covering sequence. Similar to the recurrence relation for the length of covering sequences, Rackoff gave a recurrence relation for the length of self-covering sequences also in [20]. We will again use truncation lemma to prove that this recurrence relation grows slowly for Petri nets with small vertex cover. The following lemma formalizes the way truncation lemma is used in boundedness.

Definition 5.1. *Let $Q \subseteq P$ be a subset of places with $I \subseteq Q$. Let $M_0 : P \to \mathbb{Z}$ be some function. A firing sequence $\sigma = t_1 t_2 \cdots t_r$ is said to be a Q-enabled self-covering sequence if there are intermediate functions $M_1, M_2, \ldots, M_{r'}, \ldots, M_r$ with $r' < r$ such that $M_0 \xrightarrow[Q]{t_1} M_1 \xrightarrow[Q]{t_2} \cdots \xrightarrow[Q]{t_{r'}} M_{r'} \longrightarrow \cdots \xrightarrow[Q]{t_r} M_r$ and $M_{r'} < M_r$. We call the subsequence between $M_{r'}$ and M_r as the pumping portion of the self-covering sequence.*

Lemma 5.2 (*). *Suppose $Q \subseteq P$ is a subset of places with $I \subseteq Q$. Let U be the maximum of the range of the initial marking. If there is a Q-enabled self-covering sequence, then there is a Q-enabled self-covering sequence in which none of the places in I will have more than $U + W + W^2 + W^3$ tokens in any intermediate marking.*

Before we can use Lemma 5.2, we need the following technical lemmas. The first one is an adaptation of Lemma 4.5 in Rackoff's paper [20] to our setting.

Lemma 5.3 (*). *Let $Q \subseteq P$ with $I \subseteq Q$ and $U' \in \mathbb{N}$ be such that there is a Q-enabled self-covering sequence from some M_0 in which all intermediate markings have at most U' tokens in any independent place. Also suppose that all intermediate markings have at most e tokens in any place in $Q \setminus I$. Then, there is a Q-enabled self-covering sequence of length at most $8k'(2e)^{c'k'^3}(U'W)^{c'm^4}$ for some constant c'.*

Definition 5.4. *Let $U' \in \mathbb{N}$ be some fixed number (we will later use it to denote $U + W + W^2 + W^3$, as in Lemma 5.2). For $j \in \mathbb{N}$, $Q \subseteq P$ with $I \subseteq Q$ and a function $M : P \to \mathbb{Z}$, let $slencov(Q, j, M)$ be the length of the shortest Q-enabled self-covering sequence from M if there is a Q-enabled self-covering sequence from M in which all intermediate markings have at most $U' + jW$ tokens in any independent place. Let $slencov(Q, j, M)$ be 0 if there is no such sequence. Define $\ell_1(i, j) = \max\{slencov(Q, j, M) \mid I \subseteq Q \subseteq P, |Q \setminus I| = i, M : P \to \mathbb{Z}\}$.*

The following lemma is an immediate consequence of Lemma 4.5 in [20].

Lemma 5.5. *There is a constant d such that $\ell_1(0, j) \leq (U' + jW)^{m^d}$.*

Lemma 5.6 (*). *$\ell_1(i + 1, j) \leq 8k'(2W\ell_1(i, j + 1))^{ck'^3}((U' + jW)W)^{c'm^4}$ for some appropriately chosen constants c and c'.*

Now using Lemma 5.2, we can conclude that if there is a self-covering sequence, there is one of length at most $\ell_1(k', 1)$, setting $U' = U + W^2 + W^3$ in the definition of ℓ_1. The following lemma gives an upper bound on this quantity. We use h to denote $c'k'^3$.

Lemma 5.7 (*). *$\ell_1(i, j) \leq (8k')^{(1+h)^i}(2W)^{poly_1(h^i)}(U' + (j + i)W)^{poly_2(h^i)}$ where $poly_1(h^i)$ and $poly_2(h^i)$ are polynomials in h^i, c', k' and m.*

Theorem 5.8. *With the vertex cover number k and maximum arc weight W as parameters, the Petri net boundedness problem can be solved in* PARAPSPACE.

Proof. A non-deterministic Turing machine can test for unboundedness by guessing and verifying the presence of a self-covering sequence of length at most $\ell_1(k', 1)$. By Lemma 5.7, the memory needed by such a Turing machine is bounded by $\mathcal{O}(m \log |M_0| + m + \log W + (1 + c'k'^3)^{k'} \log k' + poly_1(c'^{k'}k'^{3k'}) \log W + poly_2(c'^{k'}k'^{3k'}) \log(U'k'W))$, or $\mathcal{O}(m \log |M_0| + m + poly(c'^{3k'}k'^{3k'}) \log(U'k'W))$ for some polynomial *poly*. An application of Savitch's theorem now gives us the PARAPSPACE algorithm for boundedness. □

6 A Logic Based on Coverability and Boundedness

Following is a logic (borrowed from [19]) of properties such that its model checking can be reduced to coverability (κ) and boundedness (β) problems, but is designed to avoid expressing reachability. This is a fragment of Computational Tree Logic (CTL).

$$\tau ::= p, \ p \in P \mid \tau_1 + \tau_2 \mid c\tau, \ c \in \mathbb{N}$$
$$\kappa ::= \tau \geq c, \ c \in \mathbb{N} \mid \kappa_1 \wedge \kappa_2 \mid \kappa_1 \vee \kappa_2 \mid \mathbf{EF}\kappa$$
$$\beta ::= \{\tau_1, \ldots, \tau_r\} < \omega \mid \neg\beta \mid \beta_1 \vee \beta_2$$
$$\phi ::= \beta \mid \kappa \mid \phi_1 \wedge \phi_2 \mid \phi_1 \vee \phi_2$$

The semantics of the above logic is explained in the full version with examples.

Theorem 6.1. *Given a Petri net with an initial marking and a formula ϕ, if the vertex cover number k and the maximum arc weight W of the net are treated as parameters and the nesting depth D of **EF** modality in the formula is treated as a constant, then there is a* PARAPSPACE *algorithm that checks if the net satisfies the given formula.*

The details of model checking κ formulas is given in the full version of this paper. While reading [3], we realized that there is a mistake in the reduction from model checking β formulas to checking the presence of self-covering sequences that we gave in [19]. However, it can be corrected using the notion of *disjointness sequences* introduced by Demri in [3]. The full version of this paper contains the details of a PARAPSPACE algorithm for model checking β formulas using ideas borrowed from [3].

7   Conclusion

With the vertex cover number of the underlying graph of a Petri net and maximum arc weight as parameters, we proved that the coverability and boundedness problems can be solved in PARAPSPACE. A fragment of CTL based on these two properties can also be model checked in PARAPSPACE. Since vertex cover is better studied than the parameter benefit depth we introduced in [19], the results here might lead us towards applying other techniques of parameterized complexity to these problems. Whether coverability and boundedness are in PARAPSPACE with the size of the smallest feedback vertex set and maximum arc weight as parameters is an open problem.

Acknowledgements. The author acknowledges Kamal Lodaya and Saket Saurabh for helpful discussions and feedback on the draft.

References

[1] Atig, M.F., Habermehl, P.: On Yen's path logic for Petri nets. In: Bournez, O., Potapov, I. (eds.) RP 2009. LNCS, vol. 5797, pp. 51–63. Springer, Heidelberg (2009)

[2] Atig, M.F., Bouajjani, A., Qadeer, S.: Context-bounded analysis for concurrent programs with dynamic creation of threads. In: TACAS 2009. LNCS, vol. 5505, pp. 107–123. Springer, Heidelberg (2009)

[3] Demri, S.: On selective unboundedness. In: Infinity (to appear, 2010)

[4] Downey, R.: Parameterized complexity for the skeptic. In: CCC 2003, pp. 147–170 (2003)

[5] Downey, R.G., Fellows, M.R., Stege, U.: Parameterized complexity: A framework for systematically confronting computational intractability. In: Contemporary Trends in Discrete Mathematics: From DIMACS and DIMATIA to the Future. DIMACS, vol. 49, pp. 49–100 (1999)

[6] Esparza, J.: Decidability and complexity of Petri net problems — An introduction. In: Reisig, W., Rozenberg, G. (eds.) APN 1998. LNCS, vol. 1491, pp. 374–428. Springer, Heidelberg (1998)

[7] Esparza, J., Nielsen, M.: Decidability issues for Petri nets — a survey. J. Inform. Process. Cybernet. 30(3), 143–160 (1994)

[8] Fellows, M.R., Lokshtanov, D., Misra, N., Rosamond, F.A., Saurabh, S.: Graph layout problems parameterized by vertex cover. In: Hong, S.-H., Nagamochi, H., Fukunaga, T. (eds.) ISAAC 2008. LNCS, vol. 5369, pp. 294–305. Springer, Heidelberg (2008)

[9] Flum, J., Grohe, M.: Describing parameterized complexity classes. Inf. Comput. 187(2), 291–319 (2003)

[10] Göller, S., Haase, C., Ouaknine, J., Worrell, J.: Model checking succinct and parametric one-counter automata. In: Gavoille, C. (ed.) ICALP 2010, Part II. LNCS, vol. 6199, pp. 575–586. Springer, Heidelberg (2010)

[11] Haase, C., Kreutzer, S., Ouaknine, J., Worrell, J.: Reachability in succinct and parametric one-counter automata. In: Bravetti, M., Zavattaro, G. (eds.) CONCUR 2009 - Concurrency Theory. LNCS, vol. 5710, pp. 369–383. Springer, Heidelberg (2009)

[12] Habermehl, P.: On the complexity of the linear-time μ-calculus for Petri-nets. In: Azéma, P., Balbo, G. (eds.) ICATPN 1997. LNCS, vol. 1248, pp. 102–116. Springer, Heidelberg (1997)

[13] Kavi, K.M., Moshtaghi, A., Chen, D.-J.: Modeling multithreaded applications using petri nets. Int. J. Parallel Program. 30(5), 353–371 (2002)

[14] Kosaraju, S.R.: Decidability of reachability in vector addition systems. In: Proc. 14th STOC, pp. 267–281 (1982)

[15] Lafourcade, P., Lugiez, D., Treinen, R.: Intruder deduction for AC-like equational theories with homomorphisms. In: Giesl, J. (ed.) RTA 2005. LNCS, vol. 3467, pp. 308–322. Springer, Heidelberg (2005)

[16] Lipton, R.: The reachability problem requires exponential space. Technical report, Yale university (1975)

[17] Mayr, E.W.: An algorithm for the general Petri net reachability problem. In: Proc. 13th STOC, pp. 238–246 (1981)

[18] Petri, C.A.: Kommunikation mit Automaten. PhD thesis, Inst. Instrumentelle Math. (1962)

[19] Praveen, M., Lodaya, K.: Modelchecking counting properties of 1-safe nets with buffers in parapspace. In: FSTTCS 2009. LIPIcs, vol. 4, pp. 347–358 (2009)

[20] Rackoff, C.: The covering and boundedness problems for vector addition systems. Theoret. Comp. Sci. 6, 223–231 (1978)

[21] Reisig, W., Rozenberg, G.: Informal introduction to Petri nets. In: Reisig, W., Rozenberg, G. (eds.) APN 1998. LNCS, vol. 1491, pp. 1–11. Springer, Heidelberg (1998)

[22] Rosier, L.E., Yen, H.-C.: A multiparameter analysis of the boundedness problem for vector addition systems. J. Comput. Syst. Sci. 32(1), 105–135 (1986)

[23] Thorup, M.: All structured programs have small tree width and good register allocation. Inf. and Comp. 142(2), 159–181 (1998)

[24] Yen, H.-C.: A unified approach for deciding the existence of certain petri net paths. Inf. Comput. 96(1), 119–137 (1992)

Proper Interval Vertex Deletion

Yngve Villanger*

Department of Informatics, University of Bergen, N-5020 Bergen, Norway
yngve.villanger@uib.no

Abstract. Deleting a minimum number of vertices from a graph to obtain a proper interval graph is an *NP*-complete problem. At *WG* 2010 *van Bevern et al.* gave an $O((14k+14)^{k+1}kn^6)$ time algorithm by combining iterative compression, branching, and a greedy algorithm. We show that there exists a simple greedy $O(n + m)$ time algorithm that solves the Proper Interval Vertex Deletion problem on $\{claw, net, tent, C_4, C_5, C_6\}$-free graphs. Combining this with branching on the forbidden structures $claw, net, tent, C_4, C_5$, and C_6 enables us to get an $O(kn^66^k)$ time algorithm for Proper Interval Vertex Deletion, where k is the number of deleted vertices.

1 Introduction

Many problems that are *NP*-hard on general graphs may be polynomial-time solvable on restricted graph classes. Examples of this are maximum induced clique and independent set in chordal graphs [8], and maximum induced forest on *AT*-free graphs [9]. Most famous of these examples is Courcelle's theorem for problems that can be expressed in monadic second order logic on graphs of bounded treewidth [6]. These algorithms can often be generalized to also work for graphs that are somehow close to one of these restricted graph classes.

The closeness of a graph to a graph class can be defined in several ways. One way is to measure the difference in edges, that is, the minimum number of edges to add or remove in order to get a specific graph class. Examples of such algorithms can be found in [17] and [13]. Alternatively the closeness can be defined as a few number of vertices that can be removed, so that a specific graph structure is obtained. A fundamental problem here is to ask if k vertices can be deleted such that a clique or independent set remains. By [11] it is *NP*-complete to check if k vertices can be deleted, such that a proper interval graph remains. Some of these problems are referred to as Fixed Parameter Tractable (FPT), which means that there exists an algorithm that solves the problem in $O(f(k) \cdot n^c)$ time for some function f only depending on k and a constant c. A long range of vertex deletion problems have this kind of algorithms. Some examples are deleting k vertices to get an Independent Set [5], Forest [4], Chordal Graph [13], or Proper Interval Graph [16]. Not all k-vertex deletion problems are expected to have such an algorithm. For instance the problem of deleting k vertices such that a *wheel*-free graph is obtained is $W[2]$-hard [12]. Thus, it is not expected that an FPT algorithm exists for this problem.

---

* This work is supported by the Research Council of Norway.

V. Raman and S. Saurabh (Eds.): IPEC 2010, LNCS 6478, pp. 228–238, 2010.
© Springer-Verlag Berlin Heidelberg 2010

All the graph classes mentioned above can be characterized by a set of forbidden induced subgraphs. Independent set is exactly the class of graphs where the vertices are pairwise nonadjacent, and forests is the class of graphs without cycles. Whereas an edge is an induced subgraph of constant size, the forbidden cycles may be of any length from 3 to n. Cai [3] showed that deleting k vertices such that any hereditary graph class remains is FPT, as long as the number of forbidden induced subgraphs is bounded by some function of k. The result of Cai leaves the complexity open for all hereditary graph classes that have an infinite family of forbidden induced subgraphs. A typical example is the forest, where all induced cycles of length three or more are forbidden.

A *hole* is an induced cycle of length at least four, and chordal graphs are exactly the graphs that are *hole*-free. Despite the simple characterization it was not until 2006 that Marx settled the complexity of deleting k vertices to get a chordal graph to be FPT [13]. Wegner [18] (see also Brandstädt et al. [2]) showed that proper interval graphs are exactly the class of graphs that are $\{claw, net, tent, hole\}$-free. *Claw, net*, and *tent* are graphs containing at most 6 vertices. Proper interval graphs are hereditary[1], so by combining the results of Wegner, Cai, and Marx, it can be shown that the problem of deleting k vertices to get a proper interval graph is FPT. At WG 2010 van Bevern et al. [16] presented a new algorithm for proper interval vertex deletion using the structure of a problem instance that is already $\{claw, net, tent, C_4, C_5, C_6\}$-free. By this approach an algorithm with running time $O((14k + 14)^{k+1}kn^6)$ was obtained. Furthermore they showed that the problem of deleting a minimum number of vertices from a $\{claw, net, tent\}$-free graph to get a proper interval graph is *NP*-hard. Proper interval graphs are also known as *unit interval* graphs and *indifference* graphs.

Our result. We present a simpler and faster algorithm for the proper interval vertex deletion problem, having running time $O(6^k kn^6)$. The main result is a proof showing that a connected component of a $\{claw, net, tent, C_4, C_5, C_6\}$-free graph is a proper circular arc graph. Given the structure of a proper circular arc graph it is not hard to show that the problem can be solved by a greedy approach. The FPT algorithm is obtained by combining this algorithm with a simple branching algorithm that produces a family of $\{claw, net, tent, C_4, C_5, C_6\}$-free problem instances. Like the algorithm in [17] this algorithm is a straightforward branching algorithm, where the leaves contain polynomial-time solvable problem instances. Due to space limitations, some of the proofs are not included in this extended abstract.

2   Preliminaries

All graphs considered in this text are simple and undirected. For a graph $G = (V, E)$, we use n as the number of vertices and m as the number of edges. Two vertices $u, v \in V$ are *adjacent* if $\{u, v\} \in E$. The neighborhood of a vertex u

---

[1] Since all induced subgraphs of a proper interval graph is also $\{claw, net, tent, hole\}$-free.

is denoted $N(u)$ and vertex $v \in N(u)$ if $\{u,v\} \in E$. The *closed* neighborhood for vertex u is denoted $N[u] = N(u) \cup \{u\}$. A *path* is a sequence of vertices $v_1, v_2, \ldots, v_r$ such that $\{v_i, v_{i+1}\} \in E$ for $1 \le i < r$. The path is called *induced* if there is no edge $\{v_i, v_j\} \in E$ such that $i + 1 < j$. A *cycle* is a path $v_1, v_2, \ldots, v_r$ where $\{v_1, v_r\} \in E$, and the cycle is induced if 1 and r are the only numbers for i and j where $i + 1 < j$ and $\{v_i, v_j\} \in E$. An induced cycle containing $4 \le r$ vertices will be denoted C_r, and referred to as a *hole*.

A graph is an *interval* graph if each vertex $v \in V$ can be assigned an interval I_v on the real line, such that two vertices are adjacent if and only if their corresponding intervals intersect. The collection of intervals is referred to as the interval model $\mathcal{I}$. The graph G is called a *proper interval* graph if it can be represented by an interval model $\mathcal{I}$ where no interval is a sub-interval of another interval. Alternatively, a graph is a proper interval graph if and only if it is *claw, net, tent,* and *hole*-free [18]. The forbidden graphs are given in Figure 1.

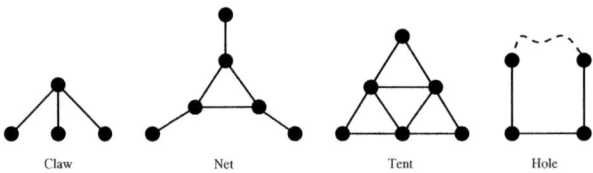

Fig. 1. The hole is any induced cycle of length 4 or more

A *circular arc* graph is a graph G, where each vertex $v \in V$ is assigned an interval (arc) I_v on a circle, such that two vertices are adjacent if and only if their corresponding intervals intersect. Like for interval graphs the model $\mathcal{I}$ consist of the collection of intervals. A circular arc graph is a *proper circular arc* graph if it can be represented by a model $\mathcal{I}$ where no interval of the model is a sub-interval of another interval. Figure 2 gives a proper circular arc model of the *tent* graph.

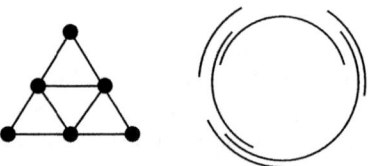

Fig. 2. The figure presents a proper circular arc model of the *tent* graph. The *tent* graph is a proper circular arc graph but not an interval graph or a unit circular arc graph. The three degree-two vertices of the *tent* graph from a so-called asteroidal triple, and thus by [10] it is not an interval graph. Tucker [15] has shown that the *tent* graph is not a unit circular arc graph.

For a proper circular arc model $\mathcal{I}$, *left* direction, is defined as the *counter-clockwise* direction along the circle, and *right* direction as the *clockwise* direction. Each interval I_v in model $\mathcal{I}$, has a left *start* point v^s and a right *end* point v^e.

Notice that this is well defined in a proper circular arc model, since no interval is a subset of another, and thus no interval completes the circle of the model. We will also use the notation $v^s < v^e$ to indicate that v^s is to the left of v^e. A interval I_v is left of interval I_w if $v^s < w^s < v^e < w^e$, and it is possible to construct an interval I_x, such that $w^e < x^s < x^e < v^s$. For short we will say that vertex v is left of vertex w in the model if interval I_v is left of interval I_w.

For two intersecting intervals I_v and I_w where I_v is left of I_w we define the *union* to be an interval I_{vw} having start point v^s and endpoint w^e. Let $X \subset V$ be a vertex set such that $G[X]$ is connected, then there exists a vertex ordering $v_1, v_2, \ldots, v_{|X|}$ of the vertices in X such that $G[v_1, v_2, \ldots, v_i]$ is connected for $1 \leq i \leq |X|$. Let $\mathcal{X}$ be the set of intervals such that $I_v \in \mathcal{X}$ if $v \in X$. The *union* of intervals in $\mathcal{X}$ is defined as the interval $I_{1,|X|}$, where $I_{1,2}$ is the union of intervals I_{v_1} and I_{v_2}, and by induction $I_{1,i}$ is the union of the intervals $I_{1,i-1}$ and I_{v_i}. We say that an interval I_v *covers* the circle of the model if there exists no interval I_w where $w^s < w^e$ such that I_v and I_w do not intersect.

In the proper circular arc model two intervals can not have the same start point since one interval would be a subset of the other. The same goes for the end point. Notice that no point on the circle is both a start and end point since in this case it is not clear if the intervals intersect or not. In the model we can assume that the $2n$ start and end points are evenly distributed over the circle.

For a connected proper circular arc graph G we will define $min(\mathcal{I})$ as a minimum vertex set $X \subset V$ such that $G[V \setminus X]$ is *hole*-free.

3   Almost Proper Interval Graphs

An *almost proper interval* graph will be defined as a $\{claw, net, tent, C_4, C_5, C_6\}$-free graph, containing an *induced cycle* of length at least 7. The main result of this section is a proof showing that every connected component of an almost proper interval graph is a proper circular arc graph. Before giving the proof we need some properties of almost proper interval graphs.

Proposition 1. *Let graph G be a net, where vertices x, y, z are the vertices of the central clique, and x', y', z' are their corresponding degree one neighbors. Then any $\{net, C_4\}$-free graph H obtained by only adding edges incident to z' in G, contains edge $\{z', x\}$ or edge $\{z', y\}$.*

Lemma 1. *Let $\mathcal{I}$ be a proper circular arc model of a graph G, and let r be the length of the longest induced cycle in G, where $4 \leq r$. Then $\lfloor r/2 + 1/2 \rfloor \leq |\mathcal{X}|$ for any set of intervals $\mathcal{X}$ contained in the model $\mathcal{I}$ where the union of the intervals in $\mathcal{X}$ covers entire the circle of the model.*

Corollary 1. *Let $\mathcal{I}$ be a proper circular arc model of a graph G, and let r be the length of the longest induced cycle in G, where $4 \leq r$. Then $\lfloor r/2 + 1/2 \rfloor \leq |C|$ for any induced cycle C in G where $4 \leq |C|$.*

In the claims 1 to 6 following below, we consider the case where G is a connected almost proper interval graph, and u is a vertex of G, such that graph $G' = G[V \setminus \{u\}]$ is a proper circular arc graph and not a proper interval graph. Let $\mathcal{I}$ be a circular arc model of G'. Since G' is not a proper interval graph, it contains an induced cycle $C = w_1, w_2, \ldots, w_r$ where $7 \leq r$. Claims 1, 2, and 3 have previously been proved in the special case where the graph has diameter at least 4 [1].

Claim 1. *Each vertex x of the proper circular arc graph G' is a vertex of cycle C or adjacent to at least two consecutive vertices on C.*

Claim 2. *Vertex u has two consecutive neighbors on the cycle C.*

Claim 3. *All neighbors of u on C are consecutive, and there are at most 4 of them.*

From now on let i, j be numbers such that for each number q where $i \leq q \leq j$ vertex w_q is a neighbor of u.

Claim 4. *There exist two vertices w_t, w_{t+1} on the cycle C such that $w_t, w_{t+1} \in N(u)$ and $N(u) \subseteq (N(w_t) \cup N(w_{t+1}))$.*

Claim 5. *Vertex set $G[N(u)]$ is connected, and the union of the intervals representing vertices in $N(u)$ does not cover the entire circle of the model.*

Claim 6. *Every induced cycle C in G' where the union of the intervals representing vertices of C covers the entire circle of the model contains at least 7 vertices.*

Lemma 2. *A connected almost proper interval graph G is a proper circular arc graph.*

Proof. By definition, an almost proper interval graph is not a proper interval graph, and thus there exists an induced cycle of length at least 7. Let C be this cycle. The proof is by induction, and the *base case* is the cycle C, which clearly is a proper circular arc graph and thus has a proper circular arc model. Let $v_{r+1}, v_{r+2}, \ldots, v_n$ be an ordering of the vertices $V \setminus V(C)$ such that $G[V(C) \cup \{v_{r+1}, v_{r+2} \ldots, v_k\}]$ is connected for $r+1 \leq k \leq n$. Let G_k be the graph $G[V(C) \cup \{v_{r+1}, v_{r+2} \ldots, v_k\}]$. Assume now that G_{k-1} is a proper circular arc graph, and let us argue that G_k is also a proper circular arc graph. This will be done by modifying the proper circular arc model $\mathcal{I}_{k-1}$ of G_{k-1} such that an interval I_{v_k} for vertex v_k can be added without violating the properties of a proper circular arc model.

By Claim 4, $N(v_k)$ contains two consecutive vertices w_i, w_{i+1} of cycle C, where all vertices of $N(v_k)$ are contained in $N(w_i) \cup N(w_{i+1})$. Let z_1 be the vertex in $N(v_k)$ with the leftmost end point z_1^e ($z_1 \in N[w_i]$), and let z_2 be the vertex in $N(v_k)$ with the rightmost start point z_2^s ($z_2 \in N[w_{i+1}]$). Let I_{v_k} be an interval with start point v_k^s and end point v_k^e. Interval I_{v_k} will first be placed on the model $\mathcal{I}_{k-1}$, and then the model will be adapted such that a proper circular arc model $\mathcal{I}_k$ for G_k is obtained. If $\{z_1, z_2\} \notin E$ then v_k^s and v_k^e are placed on model $\mathcal{I}_{k-1}$ such that there exists no start or end point p_1 in model $\mathcal{I}_{k-1}$ where $v_k^s < p_1 < z_1^e$ and there exists no start or end point p_2 where $z_2^s < p_2 < v_k^e$. If $\{z_1, z_2\} \in E$ then start point v_k^s and end point v_k^e are placed such that there exists no start or end point p_1 in model $\mathcal{I}_{k-1}$ where $v_k^s < p_1 < z_2^s$ and there exists no start or end point p_2 where $z_1^e < p_2 < v_k^e$. If v_k is adjacent to a vertex x if and only if intervals I_{v_k} and I_x intersect, and there is no interval I_x such that $v_k^s < x^s < x^e < v_k^e$ or $x^s < v_k^s < v_k^e < x^e$, then adding interval I_{v_k} to model $\mathcal{I}_{k-1}$ makes a proper circular arc model for G_k, and thus G_k is a proper circular arc graph.

Let us list the possible obstructions for getting a proper circular arc model when adding interval I_{v_k}. Consider now a vertex x and the interval I_x starting at x^s and ending at x^e. Among the four points z_1^e, z_2^s, x^s, x^e there are 4! permutations. All permutations that list x^e before x^s are not plausible in the model $\mathcal{I}_{k-1}$. Furthermore the cases where both x^s and x^e are left or right of z_1^e and z_2^s can be ignored, since $x \notin N(v_k)$ by the definition of z_1 and z_2, and I_x does not intersect I_{v_k}. This leaves the eight cases listed in Figure 3.

1. $x^s < z_1^e < z_2^s < x^e$,
2. $x^s < z_1^e < x^e < z_2^s$,
3. $x^s < z_2^s < z_1^e < x^e$,
4. $x^s < z_2^s < x^e < z_1^e$,
5. $z_1^e < x^s < z_2^s < x^e$,
6. $z_1^e < x^s < x^e < z_2^s$,
7. $z_2^s < x^s < z_1^e < x^e$, and
8. $z_2^s < x^s < x^e < z_1^e$.

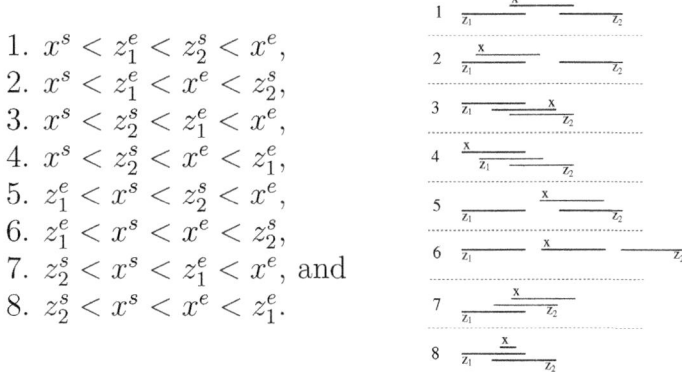

Fig. 3. The figure enumerates the 8 remaining cases

For each of these cases we have two sub-cases depending on the existence of the edge $\{v_k, x\}$. We will argue for the proof case by case. Before starting we can notice that due to Claim 5 the intervals I_{z_1} and I_{z_2} can only intersect if $z_2^s < z_1^e$. Furthermore, due to Claim 6 we need the union of at least 7 intervals in order to cover the entire circle of the model. This means that the model will appear as an proper interval model as long as at most 6 intervals are considered.

Case 1:$(x^s < z_1^e < z_2^s < x^e)$ $\mathbf{x} \notin \mathbf{N(v_k)}$. Vertices z_1 and z_2 are not adjacent, but both of them are adjacent to x. In this case $G[\{z_1, x, z_2, v_k\}]$ induces a C_4, a contradiction. $\mathbf{x} \in \mathbf{N(v_k)}$. If x has neighbors x_1 and x_2, such that $x_1^s < x^s < x_1^e < z_1^e < z_2^s < x_2^s < x^e < x_2^e$, then $x_1, x_2 \notin N(v_k)$ by definition of z_1 and z_2. By definition vertices x_1 and x_2 are not adjacent, since $x_1^e < x_2^s$. Contradiction is now obtained since $G[\{x_1, x, x_2, v_k\}]$ is a *claw*.

Otherwise vertex x_1 or vertex x_2 do not exists, which implies that that z_1 is the leftmost neighbor of x or z_2 is the rightmost. Without loss of generality assume that z_1 is the leftmost neighbor of x. This implies that only start points appear between x^s and v_k^s, since otherwise a vertex x_1 would exists. Let y_1^s be the point directly to the left of v_k^s on the model $\mathcal{I}_{k-1}$ where points v_k^s and v_k^e are added. Interval I_{y_1} is not a sub-interval of I_x, so $x^s < y_1^s < z_1^s < z_2^s < x^e < y_1^e$. Move v_k^s to the left of y_1^s, and obtain a model with one obstruction less. Repeat until no such obstruction exists.

Case 2: $(x^s < z_1^e < x^e < z_2^s)$ $\mathbf{x} \notin \mathbf{N(v_k)}$. If vertices $N_{G_{k-1}}[z_1] = N_{G_{k-1}}[x]$ then we can simply swap the intervals assigned to z_1 and x such that z_1 is assigned interval I_x and x is assigned interval I_{z_1}. After the swap there will be one obstruction less, since $x^s < z_1^s < x^e < v_k^s < z_1^e$ and thus I_x do not intersect I_{v_k}. Note that the swap may result in a vertex z_1' being the left most neighbor of v_k, but any such vertex have a starting and end points before the swap where $z_1^s < z_1'^s < x^s < z_1^e < z_1'^e < x^e$ and thus $N_{G_{k-1}}[z_1] = N_{G_{k-1}}[x] = N_{G_{k-1}}[z_1']$. So $N_{G_{k-1}}[z_1] \neq N_{G_{k-1}}[x]$. There are two cases, either there exists a vertex $z_{1'} \in N(z_1) \setminus N(x)$ or there exists a vertex $x' \in N(x) \setminus N(z_1)$. Vertex $z_{1'}$ has its interval assigned to the left of the interval for z_1, so by the definition of z_1, we get $z_{1'} \notin N(v_k)$, and thus we have a contradiction since $G[\{z_{1'}, z_1, x, v_k\}]$ is a *claw*. In the remaining case x has a neighbor x' to the right not adjacent to z_1. Vertex x' is not adjacent to v_k, since this would make $G[\{v_k, z_1, x, x'\}]$ an induced 4-*cycle*. By the existence of the cycle C, there exists a vertex to the left of z_1 in the model. Let x_ℓ be the rightmost vertex to the left of z_1^s and let w_ℓ be the vertex of C such that $x_\ell^s < w_\ell^s < x_\ell^e < z_1^s < x^s < w_\ell^e$. Interval I_{w_ℓ} can not intersect interval $I_{x'}$ to the right since this makes I_{z_1} a sub-interval of I_{w_ℓ}, and by Claim 6, it can not intersect to the left. Interval I_{x_ℓ} can only intersect I_{w_ℓ} to the right, and by Claim 6, I_{x_ℓ} does not intersect I_x or $I_{x'}$ to the left. Now we have a contradiction since $G[\{x_\ell, w_\ell, z_1, x, x', v_k\}]$ is a *net*. $\mathbf{x} \in \mathbf{N(v_k)}$. We only have to argue that I_x is not a sub-interval of I_{v_k}, since I_{v_k} contains point z_2^s and I_x does not contain this point. Initially v_k^s is just left of z_1^e and v_k^e is just to the right of z_2^s, so in this case I_x is not a sub-interval of I_{v_k}. Since we only move v_k^s to the left or v_k^e to the right of a point y^s or y^e in the case where I_{v_k} is a sub-interval of I_y, this will never occur for I_x and vertex x.

Case 3: $(x^s < z_2^s < z_1^e < x^e)$ $\mathbf{x} \notin \mathbf{N(v_k)}$. Let v_a be the rightmost vertex such that $v_a^e < z_1^s$, let v_b be the leftmost vertex such that $z_2^e < v_b^s$, let w_a and w_b be vertices of C such that $w_a^s < v_a^e < z_1^s < w_a^e$ and $w_b^s < z_2^e < v_b^s < w_b^e$. Vertices w_a, w_b, v_a, v_b exists due to the fact that the intervals of C cover the entire circle of the model, and v_a, v_b are the closest vertices on either side. Vertex w_b is adjacent to z_2 but also to x, since otherwise $G[\{x, z_2, v_k, w_b\}]$ is a *claw*. Vertex w_a is adjacent to z_1 but also to x, since otherwise $G[\{w_a, z_1, x, v_k\}]$ is a *claw*. One of the edges $\{w_a, z_2\}, \{w_a, w_b\}, \{z_1, w_b\}$ are present since otherwise $G[\{w_a, x, w_b, z_1, z_2, v_k\}]$ is a *tent*. If edge $\{w_a, z_2\}$ is present, then $G[\{w_a, z_2, w_b, v_k\}]$ is a *claw* unless edge $\{w_a, w_b\}$ is present. If edge $\{w_b, z_1\}$ is present, then $G[\{w_a, z_1, w_b, v_k\}]$ is a *claw* unless edge $\{w_a, w_b\}$ is present. Thus, we can conclude that edge $\{w_a, w_b\}$

is present. Edge $\{w_a, z_2\}$ is present to prevent $G[\{w_a, z_2, w_b, v_b\}]$ from inducing a *claw*, and edge $\{w_b, z_1\}$ is present to prevent $G[\{w_a, z_1, w_b, v_a\}]$ from inducing a *claw*. Now we have a contradiction since $G[\{v_a, w_a, w_b, v_b, z_1, v_k\}]$ induces a *net*. $\mathbf{x} \in \mathbf{N(v_k)}$. We have to argue that I_{v_k} is not a sub-interval of I_x. There are no two vertices x_1 and x_2 such that $x_1^s < x^s < x_1^e < v_k^s$ and $v_k^e < x_2^s < x^e < x_2^e$, since $G[\{x_1, x, x_2, v_k\}]$ is a *claw*. Let us without loss of generality assume that all points between x^s and v_k^s are start points. Let y^s be the point to the left of v_k^s in the model, then $x^s < y^s < v_k^s < v_k^e < x^e < y^e$. Vertex $y \notin N(v_k)$ is covered by first part of this case. Swap points y^s and v_k^s and reduce the number of conflicting vertices by one. Repeat until x^s and v_k^s are swapped.

Case 4: $(x^s < z_2^s < x^e < z_1^e)$ $\mathbf{x} \notin \mathbf{N(v_k)}$. Let v_a be the rightmost vertex such that $v_a^e < x^s$, let v_b be the leftmost vertex such that $z_2^e < v_b^s$, let w_a and w_b be vertices of C such that $w_a^s < v_a^e < x^s < w_a^e$ and $w_b^s < z_2^e < v_b^s < w_b^e$. Vertices w_a and w_b exist due to the fact that the intervals of C cover the entire circle of the model, and v_a, v_b are the closest vertices on either side. Vertex w_b is adjacent to z_2 and thus also to x, since otherwise $G[\{x, z_2, v_k, w_b\}]$ is a *claw*. Vertex w_a is adjacent to x. In order to prevent $G[\{w_a, x, w_b, v_b, z_2, v_k\}]$ from inducing a *net*, edge $\{w_a, w_b\}$ or edge $\{w_a, z_2\}$ is present since v_b is only adjacent to w_b, and v_k is only adjacent to z_2. Edge $\{w_a, w_b\}$ makes $G[\{w_a, w_b, v_b, z_2\}]$ a *claw* and forces edge $\{w_a, z_2\}$, and edge $\{w_a, z_2\}$ makes $G[\{w_a, w_b, z_2, v_k\}]$ a *claw* and forces edge $\{w_a, w_b\}$. In order to prevent $G[\{v_a, w_a, x, w_b, v_b, z_1\}]$ from inducing a *net*, one of the edge $\{w_a, z_1\}$ or edge $\{z_1, w_b\}$ is present since v_b is only adjacent to w_b, v_a is only adjacent to w_a. Edge $\{w_a, z_1\}$ makes $G[\{w_a, w_b, v_a, z_1\}]$ a *claw* and forces edge $\{z_1, w_b\}$, and edge $\{z_1, w_b\}$ makes $G[\{w_a, w_b, z_1, v_b\}]$ a *claw* and forces edge $\{w_a, z_1\}$. Now we have a contradiction since $G[\{v_a, w_a, w_b, v_b, z_1, v_k\}]$ induces a *net*. $\mathbf{x} \in \mathbf{N(v_k)}$. This is a contradiction to the definition of z_1 as the leftmost neighbor of v_k.

Case 5: $(z_1^e < x^s < z_2^s < x^e)$ This case is symmetric to Case 2.

Case 6: $(z_1^e < x^s < x^e < z_2^s)$ $\mathbf{x} \notin \mathbf{N(v_k)}$. Vertices z_1 and z_2 are not adjacent to each other or to vertex x. By Claim 4, there exists one or two vertices of $C \cap N(v_k)$ that make a path from z_1 to z_2. No vertex in $N(v_k)$ has an interval starting before z_1^e and ending after z_2^s since this vertex would have a super-interval of I_x in the model $\mathcal{I}_{k-1}$. Thus, there is an induced path z_1, w_t, w_{t+1}, z_2 in $N(v_k)$ such that w_t, w_{t+1} are vertices of C. Interval I_x is not a sub-interval of I_{w_t} or $I_{w_{t+1}}$, so x is adjacent to both w_t and w_{t+1}. Now we have a contradiction since $G[\{z_1, z_2, w_t, w_{t+1}, x, v_k\}]$ is a *tent*. $\mathbf{x} \in \mathbf{N(v_k)}$. Vertices z_1 and z_2 are not adjacent to each other or to vertex x. This is a contradiction since $G[\{z_1, x, z_2, v_k\}]$ is a *claw*.

Case 7: $(z_2^s < x^s < z_1^e < x^e)$ This case is symmetric to Case 4.

Case 8: $(z_2^s < x^s < x^e < z_1^e)$ This case is not plausible since I_x is a sub-interval of both I_{z_1} and I_{z_2}, which is a contradiction to $\mathcal{I}_{k-1}$ being a proper circular arc model of G_{k-1}. □

4   FPT Algorithm for Proper Interval Graphs

First we give a polynomial-time algorithm that finds a vertex set of minimum cardinality in an almost proper interval graph, such that the removal of this set makes the graph *hole*-free. Figure 4 gives a full description of the algorithm for general graphs, including the branching to get the set of $\{claw, net, tent, C_4, C_5, C_6\}$-free problem instances. The problem addressed is defined as follows:

PROPER INTERVAL k-VERTEX DELETION PROBLEM
Problem instance: A graph G, and parameter k
Question: Does there exist a vertex set $X \subset V$ such that $|X| \leq k$ and $G[V \setminus X]$ is a proper interval graph?

For a connected almost proper interval graph $G = (V, E)$, a vertex set X such that $G[V \setminus X]$ is a proper interval graph is called a *hole cut*. A *minimal* hole cut is a hole cut X, such that $G[V \setminus X']$ contains a hole for every proper subset X' of X.

Lemma 3. *A connected almost proper interval graph G contains at most n minimal hole cuts.*

Lemma 4. *For a connected almost proper interval graph G, there exists an algorithm that finds a minimum hole cut X in $O(n + m)$ time.*

Theorem 1. *There exists an algorithm that solves the PROPER INTERVAL k-VERTEX DELETION PROBLEM in $O(kn^6 \cdot 6^k)$ time.*

$MaxProperIntervalSubgraph(G, k)$
Input: A graph G, and parameter k
Output: A vertex set X of size at most k
             such that $G[V \setminus X]$ is a proper interval graph or answer "No"

```
if G[U] is a claw, net, tent, C₄, C₅ or C₆ for U ⊂ V then
    if 0 < k then
        for each vertex v of U
            X = MaxProperIntervalSubgraph(G[V \ {v}], k − 1)
            if X ≠ "No" then return X ∪ {v}
    else
        return "No"
else
    X = ∅
    for each connected component C of G
        if G[C] is not a proper interval graph then
            Let I be a proper circular arc model of G[C]
            X = X ∪ min(I)
    if |X| ≤ k then return X
    else return "No"
```

Fig. 4. The algorithm checks if there exists a vertex set X of size at most k, such that $G[V \setminus X]$ is a proper interval graph

5   Conclusion

By recognizing that $\{claw, net, tent, C_4, C_5, C_6\}$-free graphs are indeed the disjoint union of proper circular arc graphs, we have obtained a simple $O(6^k kn^6)$ time algorithm for proper interval vertex deletion.

Finally we can mention some related open problems. Is the problem of deleting k vertices from a graph G to get an interval graph or proper circular arc graph fixed parameter tractable? Deciding the complexity for interval graphs is probably the more interesting of the two.

Acknowledgement. The author would like to thank two anonymous referees for excellent reports, which contributed to improve the presentation of the paper.

References

1. Brandstädt, A., Dragan, F.F.: On linear and circular structure of (claw, net)-free graphs. Discrete Applied Mathematics 129, 285–303 (2003)
2. Brandstädt, A., Le, V.B., Spinrad, J.P.: Graph classes: a survey. In: Society for Industrial and Applied Mathematics, Philadelphia, PA, USA (1999)
3. Cai, L.: Fixed-parameter tractability of graph modification problems for hereditary properties. Inf. Process. Lett. 58, 171–176 (1996)
4. Cao, Y., Chen, J., Liu, Y.: On feedback vertex set new measure and new structures. In: Kaplan, H. (ed.) Algorithm Theory - SWAT 2010. LNCS, vol. 6139, pp. 93–104. Springer, Heidelberg (2010)
5. Chen, J., Kanj, I.A., Xia, G.: Improved parameterized upper bounds for vertex cover. In: Královič, R., Urzyczyn, P. (eds.) MFCS 2006. LNCS, vol. 4162, pp. 238–249. Springer, Heidelberg (2006)
6. Courcelle, B.: Graph rewriting: an algebraic and logic approach, pp. 193–242 (1990)
7. Deng, X., Hell, P., Huang, J.: Linear-time representation algorithms for proper circular-arc graphs and proper interval graphs. SIAM J. Comput. 25, 390–403 (1996)
8. Gavril, F.: The intersection graphs of subtrees in trees are exactly the chordal graphs. J. Comb. Theory, Ser. B 16, 47–56 (1974)
9. Kratsch, D., Müller, H., Todinca, I.: Feedback vertex set on AT-free graphs. Discrete Applied Mathematics 156, 1936–1947 (2008)
10. Lekkerkerker, C., Boland, J.: Representation of a finite graph by a set of intervals on the real line. Fundamentals of Math. 51, 45–64 (1962)
11. Lewis, J.M., Yannakakis, M.: The node-deletion problem for hereditary properties is NP-complete. J. Comput. Syst. Sci. 20, 219–230 (1980)
12. Lokshtanov, D.: Wheel-free deletion is W[2]-hard. In: Grohe, M., Niedermeier, R. (eds.) IWPEC 2008. LNCS, vol. 5018, pp. 141–147. Springer, Heidelberg (2008)
13. Marx, D.: Chordal deletion is fixed-parameter tractable. Algorithmica 57, 747–768 (2010)
14. Rose, D.J., Tarjan, R., Lueker, G.S.: Algorithmic aspects of vertex elimination on graphs. SIAM J. Comput. 5, 266–283 (1976)
15. Tucker, A.: Structure theorems for some circular-arc graphs. Discrete Mathematics 7, 167–195 (1974)

16. van Bevern, R., Komusiewicz, C., Moser, H., Niedermeier, R.: Measuring indifference: Unit interval vertex deletion. In: Thilikos, D.M. (ed.) WG 2010. LNCS, vol. 6410, pp. 232–243. Springer, Heidelberg (2010)
17. Villanger, Y., Heggernes, P., Paul, C., Telle, J.A.: Interval completion is fixed parameter tractable. SIAM J. Comput. 38, 2007–2020 (2009)
18. G. Wegner, Eigenschaften der Nerven homologisch-einfacher Familien im R^n, PhD thesis, Dissertation Gttingen (1967)

Author Index